高等职业教育装备制造大类专业系列教材

现代电气控制技术

赵佳萌　彭定文　主　编

中国建筑工业出版社

图书在版编目（CIP）数据

现代电气控制技术：汉英对照/赵佳萌，彭定文主编. —北京：中国建筑工业出版社，2020.2（2025.1重印）
高等职业教育装备制造大类专业系列教材
ISBN 978-7-112-24623-6

Ⅰ.①现… Ⅱ.①赵…②彭… Ⅲ.①电气控制-高等职业教材-教材-汉、英 Ⅳ.①TM921.5

中国版本图书馆 CIP 数据核字（2020）第 011079 号

　　本教材项目一对如何用 PLC 与触摸屏构建一个现代电气控制系统进行了介绍。项目二对几种典型的机床控制电路及如何对其现代电气改造以及如何排除机床电路故障的方法进行了介绍。项目三介绍了用 PLC 和触摸屏对三相电机和直流电机进行调速控制的方法。项目四介绍了用 PLC 和触摸屏对步进电机和伺服电机的运动控制的方法。项目五介绍了数控机床电气故障维修技术。

　　本教材可作为高职高专院校电气类及自动化类专业的实训教材及指导书，也可以供从事工业自动化控制或其他相关行业的国内外技术人员参考。

责任编辑：司　汉
责任校对：王　瑞

高等职业教育装备制造大类专业系列教材
现代电气控制技术
赵佳萌　彭定文　主编
＊
中国建筑工业出版社出版、发行（北京海淀三里河路 9 号）
各地新华书店、建筑书店经销
北京科地亚盟图文设计有限公司制版
建工社（河北）印刷有限公司印刷
＊
开本：880×1230 毫米　1/16　印张：16¾　字数：523 千字
2020 年 5 月第一版　2025 年 1 月第五次印刷
定价：**52.00** 元
ISBN 978-7-112-24623-6
（35296）

本教材编委会

主　编：赵佳萌　彭定文

副主编：梁正智　廖金团　陈翰雯

编　委：韩立星　梁增提　蒙万才　谢帮灵

　　　　郭　伟　覃继兵　黄丝雪

前 言
Preface

随着"一带一路"倡议的提出，越来越多的中国企业走出国门投资建厂，企业对机电、电气类国际化高技能人才的需求也快速增长。为了建设推动与中国企业和产品"走出去"相配套的职业教育发展模式，加强国际交流与合作，满足国际学生学习现代电气控制技术的需求，编写本教材。

With the proposal of "The Belt and Road" initiative, more and more Chinese enterprises are going abroad to invest and set up factories, and the demand for international high-skilled talents in mechanical, electrical and other fields is also growing rapidly. In order to build a vocational education mode that is compatible with "going out" of Chinese enterprises and products, strengthen international exchanges and cooperation, and meet the needs of international students in learning modern electrical control technology, this textbook is prepared.

本教材是国际、国内现代学徒制教学模式改革的建设成果之一，在全国首批现代学徒制试点项目建设中，学校与上汽通用五菱汽车股份有限公司联合开展"上汽通用五菱印尼基地国际化人才培养项目"，在广西首批现代学徒制试点项目建设中，学校与广西汽车集团开展校企双主体育人的"广西汽车集团现代学徒制工匠英才班"，培养具备国际竞争力的高素质复合型新工科人才，是在实践教学过程中校企共同开发的专业教材。

The textbook is one of the construction achievements in the reform of the modern apprenticeship teaching mode at home and abroad. In the construction of the first batch of modern apprenticeship pilot projects in the country，the school and SAIC-GM-Wuling Automotive Co.，Ltd. （SGMW） jointly carried out the "SGMW-Indonesia Base International Talents Training Project"，and during the construction of that in Guangxi, the school and Guangxi Automobile Group launched the "Guangxi Automobile Modern Apprenticeship Craftsmanship Talent Class" with both the school and the enterprise as the double main bodies to train high-quality inter-disciplinary engineering talents with international competitiveness. And the textbook is a professional teaching material jointly developed by the school and the enterprises in the practical teaching process.

本教材是以职业教育伴随企业"走出去"的国际化项目"上汽通用五菱—印尼"和"中印 SGMW 汽车学院"为载体，根据海内外学生和企业员工培训需求而编制的双语教材。教材以读者够学够用为原则，对《PLC应用技术》《人机界面组态技术》《电机拖动技术》《数控机床电气维修技术》等课程的知识点进行了整合，运用任务驱动的教学模式来进行编写，降低学习的难度。教材中所有中文的内容皆有与之相对应的英文，有利于我国职业教育课程的国际化。

This textbook is a bilingual textbook based on the internationalization project "SGMW-Indonesia" and the project "China-India SGMW Automobile College" with vocational education accompanying the enterprises to "go out", and according to the training needs of students and employees at home and abroad. Based on the principle of sufficiency, this book integrates the knowledge points in the courses such as *PLC Application Technology*, *Human-Machine Interface Configuration Technology*, *Motor Drive Technology* and *Electrical Maintenance Technology of CNC Machine Tools*, and uses the task-driven teaching mode to write, which has reduced the difficulty in learning. All the Chinese contents in this book have corresponding English, which is beneficial to the internationalization of vocational education curriculum in China.

本教材由柳州城市职业学院的赵佳萌、彭定文担任主编，梁正智、廖金团、陈翰雯担任副主编。其中，项目一、二、三、四由赵佳萌负责编写及校对。项目五由彭定文负责编写与校对。全书由赵佳萌进行统稿并校对。

This textbook was edited by Zhao Jiameng and Peng Dingwen, while Liang Zhengzhi Liao Jintuan served as the associate editors. Among them, Item I, II, III and IV are edited and proofread by Zhao Jiameng. Item V is edited and proofread by Peng Dingwen. The whole book is finally compiled and proofread by Zhao Jiameng.

本教材为高等职业教育装备制造大类专业规划推荐教材。本书突出能力培养，注重以实用为原则，将理论知识和实际应用结合起来，是一本实用且全面的教材。

This textbook is a series of teaching material for equipment manufacturing major in higher vocational education. This book focuses on the cultivation of ability, pays attention to the principle of practicality, and combines theoretical knowledge with practical application. It is a practical and comprehensive textbook.

本教材在编写过程中得到了上汽通用五菱汽车股份有限公司、广西汽车集团的大力支持，柳州城市职业学院机电与汽车工程系电气教学团队也为本书编写做了大量工作，在此表示衷心感谢。此外还要感谢柳州城市职业学院机电与汽车工程系的刘宇、韦振钧、黎道兵、曹青方四位同学，他们为本书的编写提供了不少帮助，使得本书的编写工作能够按时完成。

In the preparation of the textbook, vigorous support had been obtained from SGMW and Guangxi Automobile Group. The Electrical Engineering Teaching Team of the Mechatronics and Automotive Engineering Department of Liuzhou City Vocational College had also done lots of work for the preparation of the book, and I would like to express sincere gratitude here. In addition, I would also express my gratitude to Liu Yu, Wei Zhen jun, Li Daobing and Cao Qingfang, four students from the Department of Mechanical and Electrical Engineering and Automotive Engineering of Liuzhou City Vocational College, for their help in the preparation of this book so that this book can be completed on time.

由于编者的经验、水平有限，加之时间仓促，书中难免在内容和文字上存在不足和缺陷，敬请广大读者批评指正。

Due to the limited experience and level of editors, as well as the haste of time, there are inadequacies and defects in contents and texts in the book. We would like to invite the readers to give critical comments.

目　录

Contents

Contents

项目一 PLC 与触摸屏
Item I PLC and Touch Screen

任务一 用触摸屏建立一个三相电机启停的远程控制和运行监控系统

Task I To Build a Remote Start-stop Control and Operation Monitoring System for Three-phase Motor via the Touch Screen

一、任务分析
I. Task Analysis

用人机界面（HMI）、PLC 及相关电气器件组态一个电机启停的监控系统，要求系统具有以下功能：

Assemble the human-machine interface (HMI), PLC and relevant electric devices into the motor start-stop monitoring system, which should have the following functions:

（1）用按钮可以控制电机的启停。

(1) The start-stop operation of motor can be controlled with one button.

（2）能通过人机界面中电机上的指示灯实时监视到电机的运行或停止状态，并可以用人机界面远程操作启停电机，监控画面如图 1.1.1 所示。

(2) The operation or stop state of motor can be monitored in real time, and also the motor can be remotely enabled and disabled with the indicators on HMI, and the monitoring interface is shown in Figure 1.1.1.

（3）若电机过载，热继电器动作，则自动弹出报警画面"电机过载，系统停止"，电机停止运行。热继电器恢复后，报警画面自动消失。

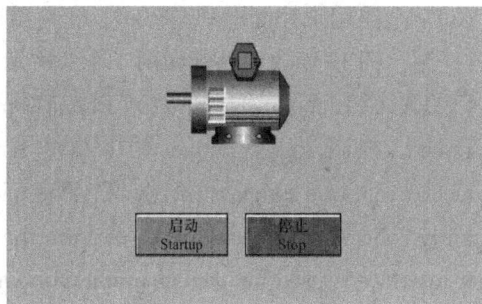

图 1.1.1 触摸屏监控画面

Figure 1.1.1 Monitoring interface of touch screen

(3) When the motor is overloaded, the thermal relay is actuated, the alarm dialog box "Motor Overloaded, System stopped" will automatically appear, and then the motor will be stopped. After the recovery of thermal relay, the alarm dialog box will automatically disappear.

为了完成本任务，首先需要对涉及的设备进行选型，根据现有条件，人机界面选择昆仑通态的 MCGS 触摸屏，PLC 选择西门子 S7-1200 系列 PLC。通过完成本任务，需要学习如何建立一个人机界面工程和用 S7-1200 系列进行简单的编程。

For the purpose of this task, it is firstly necessary to select the required types of devices involved, of which MCGS touch screen, and Siemens S7-1200 PLC will be selected under the current conditions. By completing this task, you can learn how to establish one HMI engineering design and how to use S7-1200 to perform the simple programming.

二、相关知识技能

Ⅱ. Relevant Knowledge & Skills

（一）西门子 S7-1200 系列 PLC 编程入门

（Ⅰ）Basic Knowledge of Siemens S7-1200 PLC Programming

1. S7-1200 系列 PLC 简介

1. Simple introduction of S7-1200 PLC

S7-1200 系列是一款可编程逻辑控制器（PLC，Programmable Logic Controller），可以控制各种自动化应用。S7-1200 系列设计紧凑、成本低廉且具有功能强大的指令集，这些特点使它成为控制各种应用的完美解决方案。S7-1200 系列 PLC 主要由 CPU 模块、信号板、信号模块、通信模块和编程软件组成。

S7-1200 is one Programmable Logic Controller (PLC), which can control various automation applications. With the features of compact design, low cost and powerful instruction set, S7-1200 can be called as the perfect solution to control various applications. S7-1200 PLC mainly consists of CPU module, signal board, signal module, communication module and programming software.

S7-1200 系列的 CPU 模块将微处理器、电源、数字量输入/输出（DI/DQ）电路、模拟量输入/输出（AI/AQ）电路、PROFINET 以太网接口、高速运动控制功能组合到一个设计紧凑的外壳中。微处理器相当于人的大脑和心脏，它不断地采集输入信号，执行用户程序，刷新系统的输出。PROFINET 以太网接口用于与编程计算机、HMI（人机界面）、其他 PLC 或其他设备通信。此外它还通过开放的以太网协议支持与第三方设备的通信。数字量输入（DI）电路用来接收从按钮、选择开关、数字拨码开关、限位开关、接近开关、光电开关、压力继电器等输入信号。模拟量输入（AI）电路用来接收电位器、测速发电机和各种变送器提供的连续变化的模拟量电流、电压信号，或者直接接收热电阻、热电偶提供的温度信号。数字量输出（DQ）电路用来控制接触器、电磁阀、电磁铁、指示灯、数字显示装置和报警装置等输出设备，模拟量输出（AQ）电路用来控制电动调节阀、变频器等执行器。高速运动控制功能用于步进电机或伺服电机的速度和位置控制。CPU 模块的外观如图 1.1.2 所示。

The CPU module of S7-1200 PLC assembles the microprocessor, power source, digital input/output (DI/DQ) circuits, analog input/output (AI/AQ) circuits, PROFINET Ethernet interface and high-speed motion control function into one compact enclosure. The microprocessor is equivalent to the human brain and heart, and can constantly collect and input signals, perform the user program, and refresh the system output. The PROFINET Ethernet interface is used for the communication with the programming computer, HMI and other PLC or devices. What's more, it can support the communication with the third-party devices via the open Ethernet protocol. The digital input (DI) circuit is used for receiving these input signals from buttons, selective switch, digital dial switch, limit switch, approach switch, optoelectronic switch, pressure relay, etc. The analog input (AI) circuit is used for receiving the analog current and voltage signals of continuous changes from the potentiometer, tachometer generator and various transmitters or directly receiving the temperature signal from thermal resistance and thermocouple. The digital output (DQ) circuit is used for controlling the contactor, solenoid valve, electromagnet, indicators, digital display device, warning device and other output devices, and the analog output (AQ) is used for controlling the electric control valve, frequency converter and other actuators. The high-speed motion control function is used for controlling the speed and position of stepping motor or servo motor. The appearance of CPU module is shown in Figure 1. 1. 2.

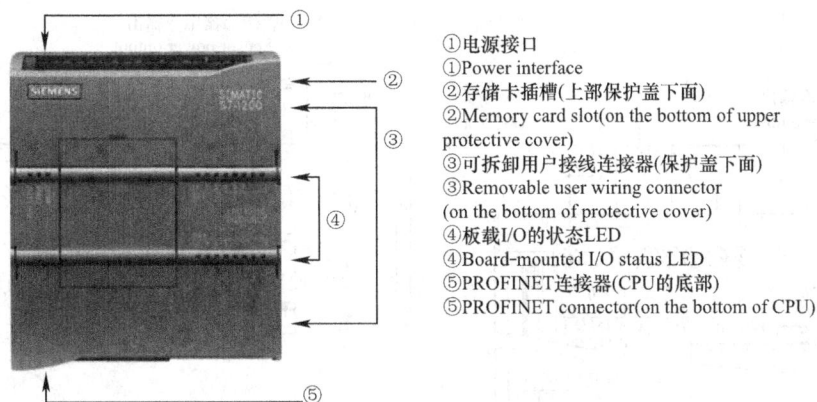

①电源接口
①Power interface
②存储卡插槽(上部保护盖下面)
②Memory card slot(on the bottom of upper protective cover)
③可拆卸用户接线连接器(保护盖下面)
③Removable user wiring connector (on the bottom of protective cover)
④板载I/O的状态LED
④Board-mounted I/O status LED
⑤PROFINET连接器(CPU的底部)
⑤PROFINET connector(on the bottom of CPU)

图 1.1.2　CPU 模块的外观

Figure 1.1.2　Appearance of CPU module

S7-1200 系列的 CPU 模块有 CPU 1211C、CPU 1212C、CPU 1214C、CPU 1215C 和 CPU 1217C 几种类型。每种 CPU 有 3 种具有不同电源电压和输入、输出电压的版本，见表 1.1.1。

S7-1200 CPU module is classified into CPU 1211C，CPU 1212C，CPU 1214C，CPU 1215C and CPU 1217C. Each type of CPU includes three versions with different power supply voltage，input voltage and output voltage，as shown in Table 1.1.1.

S7-1200 CPU 模块的不同版本　　　　　　　　　　表 1.1.1

Different versions of S7-1200 CPU module　　　　Table 1.1.1

版本 Version	电源电压 Power supply voltage	DI 输入电压 DI Input voltage	DQ 输出电压 DQ Output voltage	DQ 输出电流 DQ Output current
DC/DC/DC	DC 24V	DC 24V	DC 24V	0.5A，MOSFET
DC/DC/RLY	DC 24V	DC 24V	DC 5～30V，AC 5～250V	2A，DC 30W/AC 200W
AC/DC/RLY	AC 85～264V	DC 24V	DC 5～30V，AC 5～250V	2A，DC 30W/AC 200W

2. S7-1200 系列 PLC 的外部接线

2. External wiring of S7-1200 PLC

以 CPU 1212C AC/DC/RLY 为例，来说明 AC/DC/RLY 版本的外部接线。接线图如图 1.1.3 所示。输入电路一般使用 CPU 内置的 DC 24V 传感器电源。漏型输入时，需要去除图中的外接 DC 电源，将输入电路的 1M 端子与 DC 24V 传感器电源的 M 端子连接起来，将内置的 24V 电源的 L＋端子接到外接触点的公共端。

The external wiring of AC/DC/RLY is illustrated with an example of CPU 1212C AC/DC/RLY. The wiring is shown in Figure 1.1.3. The input circuit is generally sourced from the built-in DC 24V sensor power supply of CPU. In case of leakage-pattern，it is necessary to remove the external DC power source as shown in the wiring diagram，and connect the 1M terminal of input circuit to the M terminal of DC 24V sensor power supply and the L＋ terminal of built-in 24V power supply to the public terminal of external contact point.

以 CPU 1212C DC/DC/DC 为例，来说明 DC/DC/DC 版本的外部接线。接线图如图 1.1.4 所示，其电源电压、输入电路电压、输出电路电压均为 24V。

The external wiring of DC/DC/DC is illustrated with an example of CPU 1212C DC/DC/DC. As shown in Figure 1.1.4，its power supply voltage，input voltage and output voltage are 24V.

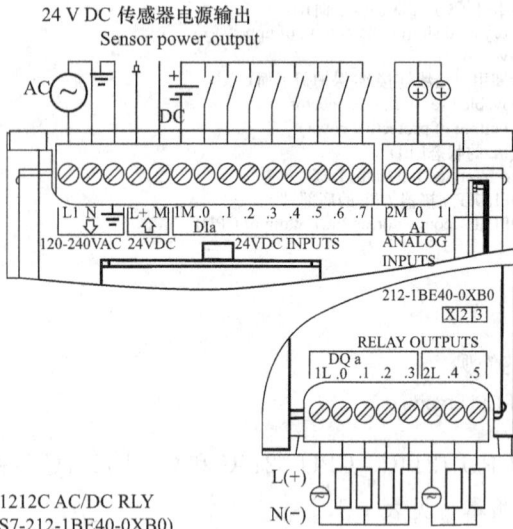

图 1.1.3 CPU 1212C AC/DC/RLY 的外部接线图

Figure 1.1.3 External wiring of CPU 1212C AC/DC/RLY

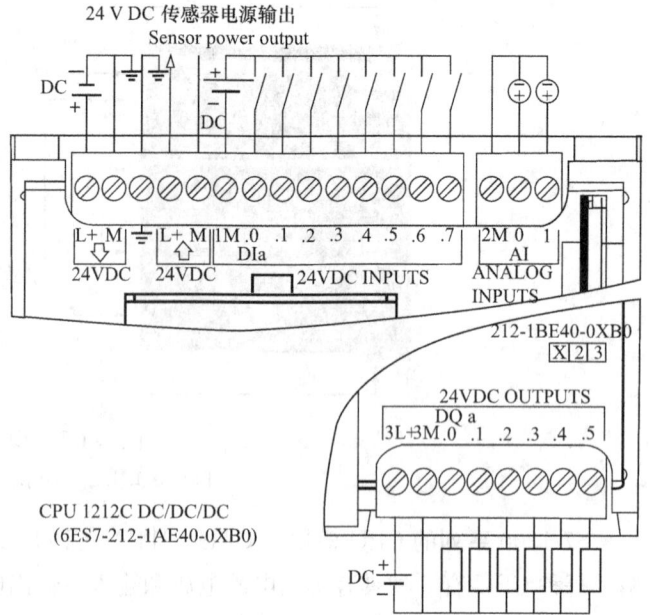

图 1.1.4 CPU 1212C DC/DC/DC 的外部接线图

Figure 1.1.4 External wiring of CPU 1212C DC/DC/DC

3. S7-1200 系列 PLC 的编程环境

3. Programming environment of S7-1200 PLC

TIA 是 Totally Integrated Automation（全集成自动化）的简称，TIA 博途（TIA Portal）是西门子全新的自动化工程设计软件平台。S7-1200 是用 TIA 博途中的 STEP 7 Basic（基本版）或 STEP 7 Profession（专业版）编程。STEP 7 提供了两种不同的视图：门户视图和项目视图，如图 1.1.5 所示。只需通过单击就可以切换门户视图和项目视图。

TIA is the shortened form of Totally Integrated Automation，and TIA Portal is the brand-new automation engineering design software platform developed by Siemens. S7-1200 is programmed with STEP 7 Basic (basic version) or STEP 7 Profession (professional version) of TIA Portal. STEP 7 can provide two different views：Portal view and item view as shown in Figure 1.1.5. You can switch over the portal view and item view only by a single click.

4. S7-1200 系列 PLC 的存储区

4. Storage zone of S7-1200 PLC

CPU 提供了各种专用存储区，其中包括过程映像输入（I）、过程映像输出（Q）和位存储器（M）。用户程序可以利用这些地址访问存储单元中的信息。

CPU provides various special storage zones，including the process image input （I），process image output （Q） and bit memory （M）. The user program can access the information in the storage cell via these addresses.

过程映像输入在用户程序中的标识符为 I，用来存放输入信号的状态，是 PLC 接收外部输入的数字量信号的窗口。在扫描周期开始时，CPU 读取数字量输入点的外部输入电流的状态，并将它们存入过程映像输入区。

The process image input identified as I in the user program is used for storing the input signal status, and it is the window of PLC for receiving the external digital signal. At the beginning of scanning cycle, CPU reads out the external output current status at the digital input point，and stores them into the process image input area.

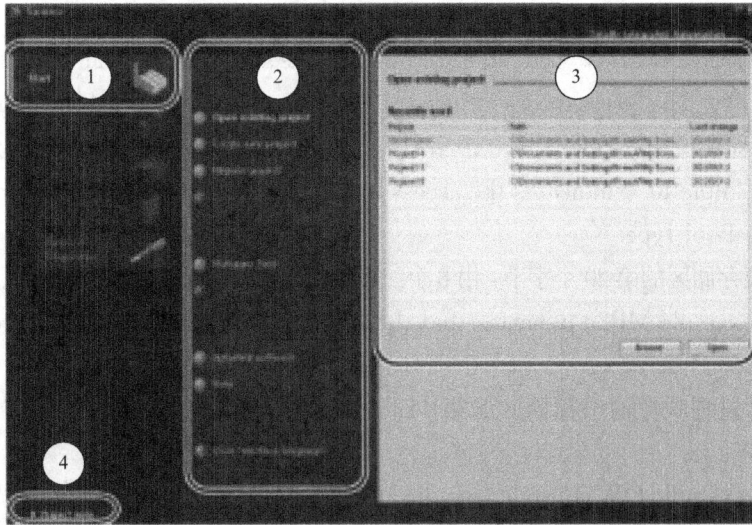

(a) 门户视图
(a) Portal view

门户视图
Portal view
①不同任务的门户
①Portal for different tasks
②所选门户的任务
②Tasks of the selected portal
③所选操作的选择面板
③Option panels of the selected operations
④切换到项目视图
④Switch to the item view

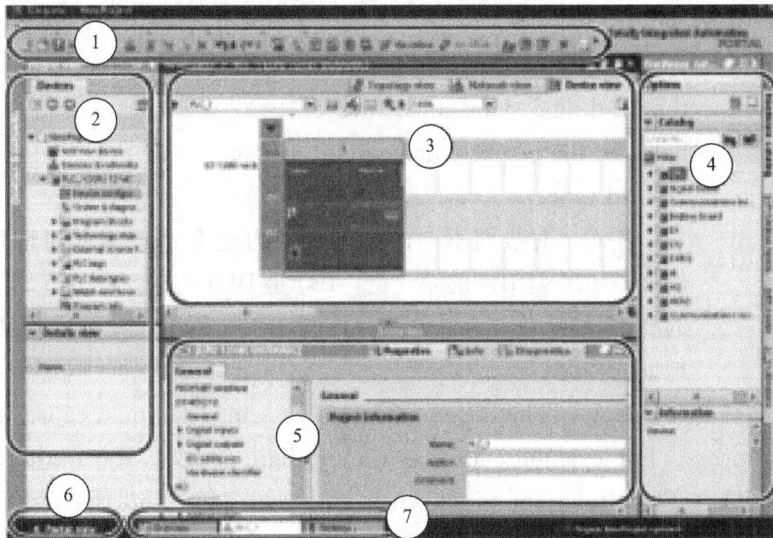

(b) 项目视图
(b) Item view

项目视图
Item view
①菜单和工具栏
①Menus and toolbars
②项目浏览器
②Project browser
③工作区
③Working area
④任务卡
④Task card
⑤巡视窗口
⑤Inspector window
⑥切换到门户视图
⑥Switch to portal view
⑦编辑器栏
⑦Editor bar

图 1.1.5 STEP 7 的不同视图
Figure 1.1.5 Different views of STEP 7

过程映像输出在用户程序中的标识符为 Q，用来存放输出信号的状态。在一个扫描周期中，用户程序计算输出值，并将它们存入过程映像输出区。在下一个扫描周期中，过程映像输出区的内容被写到数字量输出点，再由后者驱动外部负载。

The process image output identified as Q in the user program is used for storing the status of output signal. During one scanning cycle，the user program calculates and stores the output value into the process image output area. During the next scanning cycle，the content in the process image output area is written into the digital output point，and then the later will drive the external load.

位存储器区（M 存储器）用来存储运算的中间操作状态或其他控制信息。

The bit memory（M）is used for storing the intermediate operation status or other control information in the operation process.

S7-1200 的每个存储单元都有唯一的地址，可以在这些存储区标识符后面加上地址标识来对它们进行访

5

问，如 I0.0、QW4、MD10 等。访问的方式可以按位访问、按字节访问、按字访问和按双字访问。

Each storage cell of S7-1200 has a unique address, and can be accessed via the storage area identifier followed by address ID, such as I0.0, QW4 and MD10. The access modes include the access by bit, byte, word and double word.

按位访问，例如 I0.0，表示过程映像输入存储区第 0 个字节的第 0 位。这样的数据类型被记为 Bool 型。

In case of access by bit, for example I0.0 indicates the access to 0 bit of 0 byte in the process image input storage area. Such data is called as Bool type.

按字节访问，例如 MB20，表示位存储区的第 20 个字节，由 8 个二进制位组成。这样的数据类型被记为 Byte 型。

In case of access by byte, for example MB20 indicates that the 20th byte in the bit storage area consists of 8 binary digits. Such data is called as Byte type.

按字访问，例如 QW112，表示过程映像输出存储区的第 112 个字节和第 113 个字节组成的 16 位数据。这样的数据类型被记为 Word 型。

In case of access by word, for example QW112 indicates the 16-digit data consisting of the 112th and 113th bytes in the process image output storage area. Such data is called as Word type.

按双字访问，例如 ID1000，表示过程映像输入存储区的第 1000、1001、1002 和 1003 个字节组成的 32 位数据。这样的数据类型被记为 DWord 型。

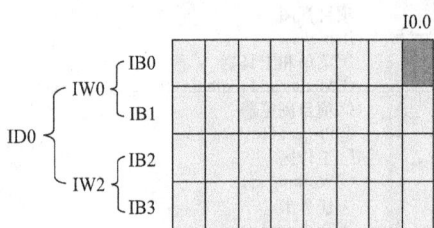

图 1.1.6　存储单元的不同访问方式

Figure 1.1.6　Different access modes of ftorage cell

In case of access by double word, for example ID1000 indicates the 32-ditigal data consisting of 1000th, 1001th, 1002th and 1003th bytes in the process image input storage area. Such data is called as DWord type.

注意：对数据不同的访问方式可能会造成编程时地址的冲突。例如，I0.0 是 IB0 的第 0 位，IB0 是 IW0 的高 8 位字节，IW0 是 ID0 的高 16 位字节。它们的关系如图 1.1.6 所示，在编程时要注意避免地址的冲突。

Note: The different data access modes would cause the address conflict during the programming process. For example, I0.0 is the 0th digit of IB0, IB0 is the 8th byte of IW0, and IW0 is the 16th byte of ID0. Their relations is shown in Figure 1.1.6, and the address conflict should be avoided in the programming process.

5. S7-1200 的用户程序结构

5. User Program Structure of S7-1200

S7-1200 采用模块化编程，它通过调用各种"块"来组织程序，一个"块"可以完成一个较小的子任务，这些"块"主要有组织块（OB）、函数块（FB）、函数（FC）和数据块（DB），其中的组织块、函数块、函数都包含程序，统称为代码块。

S7-1200 is based on the modular programming and calls various "blocks" to organize the program, one "block" can complete a smaller sub-task, and these "blocks" mainly include the organization block (OB), function block (FB), function (FC) and data block (DB), of which the DB, FB and FC include the program and therefore are collectively called as the code block.

组织块是操作系统与用户程序的接口，用于控制扫描循环和中断程序的执行、PLC 的启动和错误处理等。组织块的程序是用户编写的。其中，OB1 是主程序，每次循环扫描都执行一次。此外，当 CPU 的工作模式从 STOP 切换到 RUN 时，还会执行一次启动组织块。如果出现中断事件，例如诊断中断和时间延迟中断等，还会执行中断组织块。

The organization block (OB) is the interface between the operation system and user program, and used for controlling the scanning cycle and termination of program run, PLC startup, error handling, etc. The organization block (OB) is programmed by user. Of which, OB1 is the main program, and executed every scanning cycle. Additionally, the organization block (OB) startup will be executed once when the working mode of CPU is switched from STOP to RUN. In case of interruption events such as the diagnosis interruption and time delay interruption, the organization block will be interrupted.

函数和函数块是用户编写的子程序，其区别是函数块有专用的背景数据块而函数没有。

The function and function block are the subprograms developed by user, their difference is that function block has the special background data block, while the function doesn't have.

数据块是用于存放执行代码块时所需的数据的数据区。数据块没有任何指令。

The data block is the data area for storing the data required for executing the code block. It is free of any instructions.

6. S7-1200 系列 PLC 的工作模式

6. Working mode of S7-1200 PLC

S7-1200 的 CPU 有三种工作模式：停止（STOP）模式、启动（STARTUP）模式和运行（RUN）模式。CPU 前面的状态 LED 指示灯指示当前工作模式。

The CPU of S7-1200 can operate in three working modes: STOP, STARTUP and RUN modes. The status indicator LED on the front of CPU indicates the current working mode.

在 STOP 模式下，CPU 不执行程序，仅处理通信请求和进行自诊断，可用于下载项目。上电后，CPU 进入 STARTUP 模式，执行启动 OB（如果存在）一次。在 RUN 模式，程序循环 OB 重复执行。西门子 S7-1200 的 CPU 没有用于更改工作模式（STOP 或 RUN）的物理开关。请使用编程软件的工具栏按钮"启动 CPU"（Start CPU）// 和"停止 CPU"（Stop CPU）// 更改 CPU 的工作模式。

Under the STOP mode, CPU will not execute the program, only handle the communication request and self-diagnosis, and can download the items. After power up, CPU enters into the STARTUP mode, and execute the startup of OB (if any) one time. Under the RUN mode, the program will repeat the execution of OB. The CPU of Siemens S7-1200 is free of physical switch for changing the working mode (STOP or RUN). Please change the working mode by pressing the toolbar button of programming software "Start CPU" // and "STOP CPU" //.

7. 几个常用的位逻辑指令

7. Several commonly-used bit logic instructions

常用的位逻辑指令有常开触点、常闭触点、线圈和置位、复位输出，见表 1.1.2。

The commonly-used bit logic instructions include the normally open contact, normally closed contact, coil, set and reset output, as shown in Table 1.1.2.

几个常用的位逻辑指令　　　　　　　　　　　　　　　　　　　表 1.1.2
Several commonly-used bit logic instructions　　　　　　　　　Table 1.1.2

指令 Instructions	描述 Description	指令 Instructions	描述 Description
┤├	常开触点 Normally open contact	─(R)─	复位输出 Reset output
┤/├	常闭触点 Normally close contact	─(SET_BF)─	置位区域 Set area
┤├	线圈 Coil	─(RESET_BF)─	复位区域 Reset area
─(S)─	置位输出 Set output		

常开触点在指定的位为 1 状态（TRUE）时闭合，为 0 状态（FALSE）时断开。常闭触点在指定的位为 1 状态时断开，为 0 状态时闭合。两个触点串联将进行"与"运算，两个触点并联将进行"或"运算。

The normally open contact is closed when its designated bit is 1 (TRUE), and open when its designated bit is 0 (FALSE). The normally closed contact is open when its designated bit is 1, and open when its designated bit is 0. The series connection of both contacts will execute the "And" operation, and their parallel connection will execute the "Or" operation.

线圈将输入的逻辑运算结果（RLO）的信号状态写入指定的地址，线圈通电（RLO 的状态为"1"）时写入 1，断电时写入 0。

The coil will write the signal status of ROL input in the designated address, 1 will be wrote in at the time of coil power-on (the status of RLO is 1) and 0 will be written in at the time of power-off.

S（Set，置位输出）指令将指定的为操作数置位（变为 1 状态并保持）。

S (Set, set output) instruction is used for setting the designated bit operation number (to status 1 and maintain).

R（Reset，复位输出）指令将指定的位操作数复位（变为 0 状态并保持）。

R (Reset, reset output) instruction is used for resetting the designed bit operation number (to status 0 and maintain).

SET _ BF（置位区域）指令将指定的地址开始的连续的若干个位地址置位（变为 1 状态并保持）。

SET _ BF (set area) instruction is used for setting the several consecutive bit addresses at the beginning of designated address (to status 1 and maintain).

RESET _ BF（复位区域）指令将指定的地址开始的连续若干个地址复位（变为 0 状态并保持）。

RESET _ BF (reset area) instruction is used for resetting the several consecutive bit addresses at the beginning of designated address (to status 0 and maintain).

（二）用 MCGS 嵌入版完成一个人机界面工程

（Ⅱ）To complete one HMI engineering design with embedded MCGS

首先必须在 MCGS 嵌入版的组态环境下进行工程的生成工作。新建工程后，弹出"工作台"窗口，如图 1.1.7 所示。MCGS 嵌入版用"工作台"窗口来管理构成用户应用系统的五个部分，即工作台上的五个标签：主控窗口、设备窗口、用户窗口、实时数据库和运行策略，对应于五个不同的窗口页面，每一个页面负责管理用户应用系统的一个部分，用鼠标单击不同的标签可选取不同窗口页面，对应用系统的相应部分进行组态操作。一般来说，一个人机界面工程不需要对这五个部分全部进行组态，需要根据控制要求来决定是否需要对该部分进行组态。

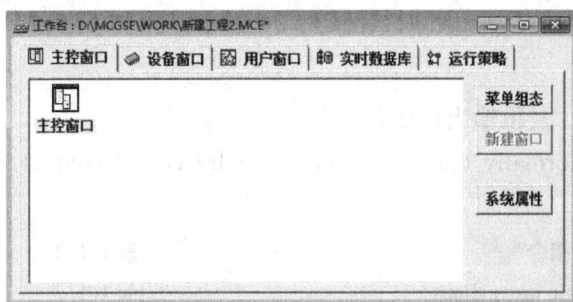

图 1.1.7　工作台窗口

Figure 1.1.7　Workbench window

Firstly, the engineering design generation work must be done under the configuration environment of embedded MCGS. After the establishment of new engineering, the "Workbench" window will pop up as shown in Figure 1.1.7. The embedded MCGS manages the five constituent parts of user application system, i. e. five tags on the workbench; Master control window, device window, user window, real-time database and operation strategies, which are respectively corresponding to these five different window interfaces, each window is responsible for managing one part of user application system, and you can select the different window interface by clicking on different tags and perform the configuration operation corresponding to each part of application system. Generally speaking, the configurations of all these five parts are not required for a HMI engineering design, and you should determine the part has to be configured based on the control requirements.

1. 构造实时数据库

1. To create one real-time database

需要建立一个数据库来实现人机界面数据处理、数据交换和数据的可视化。定义数据对象的过程，就是构造实时数据库的过程。在 MCGS 嵌入版中，数据对象有开关型、数值型、字符型、事件型和组对象等五种类型。通过对这些数据的读写，完成人机交互的过程。

It is necessary to create one database to realize the visualization of data processing, data exchange and data on the HMI. The definition process of data object is the process to create the real-time database. In the embedded MCGS, the data objects cover five types such switch, numeric, character, event and group object. The human-machine interaction process is completed by the read-write operation of these data.

2. 组态用户窗口

2. To configure the user window

用户窗口是由用户来定义、构成 MCGS 嵌入版图形界面的窗口。用户窗口相当于一个"容器"，用来放置图元、图符和动画构件等各种图形对象，通过对图形对象的组态设置，建立与实时数据库的连接，来完成图形界面的设计工作。图形对象是组成用户应用系统图形界面的最小单元。MCGS 嵌入版中的图形对象包括图元对象、图符对象和动画构件三种类型，不同类型的图形对象有不同的属性，所能完成的功能也各不相同。图形对象可以从 MCGS 嵌入版提供的绘图工具箱和常用图符工具箱中选取，如图 1.1.8 所示。

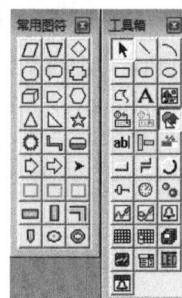

图 1.1.8　绘图工具箱和常用图符工具箱

Figure 1.1.8　Mapping toolbox and commonly-used icon toolbox

The user window is one window defined by the user to create the graphical interface of embedded MCGS. The user window can be equivalent to one "container" which is used for placing the graphic primitives, icons, animation and other various graphic objects, and can establish the connection with the real-time database and complete the design of graphic interface through the configurations of graphic objects. The graphic object is the smallest element constituting the graphic interface of user application system. The graphic objects involved in the embedded MCGS include three types of graphic primitives, icons, animation, the different graphic objects have different attributes and their intended functions are also different. The graphic objects can be selected from the mapping toolbox and commonly-used icon toolbox provided by the embedded MCGS, as shown in Figure 1.1.8.

动画构件本身是一个独立的实体，它比图元和图符包含有更多的特性和功能，它不能和其他图形对象一起构成新的图符。MCGS 嵌入版目前提供的动画构件有：输入框构件、标签构件、流动块构件、百分比填充构件、标准按钮构件、动画按钮构件、旋钮输入构件、滑动输入器构件、旋转仪表构件、动画显示构件、实时曲线构件、历史曲线构件、报警显示构件、自由表格构件、历史表格构件、存盘数据浏览构件、组合框构件。

The animation is an independent entity in itself, which have more features and functions compared to the graphic primitive and icon, and can't form a new icon together with other graphic object. The embedded MCGS can provide the following animation widgets currently: Input box, tag, flow block, percentage-based filler, standard button, animated button, rotary knob input unit, slider input, rotary instrument, animation display, real-time curve component, historic curve component, alarm display component, free form component, historic form component, saved data browse component, and combo box component.

3. 组态设备窗口

3. To configure the device window

设备窗口是 MCGS 嵌入版系统的重要组成部分，在设备窗口中建立系统与外部硬件设备的连接关系，使系统能够从外部设备读取数据并控制外部设备的工作状态，实现对工业过程的实时监控。在 MCGS 嵌入版中，实现设备驱动的基本方法是：在设备窗口内配置不同类型的设备构件，并根据外部设备的类型和特征，设置相关的属性，将设备的操作方法，如硬件参数配置、数据转换、设备调试等都封装在构件之中，以对象的形式与外部设备建立数据的传输通道连接，如图 1.1.9 所示。

The device window is an important part of embedded MCGS, and the connection relation between the system and external hardware is established in the device window to enable the system to read data from the external devices, control the working status of external devices, and realize the real-time monitoring of industrial process. In the embedded MCGS, the device can be driven by the following basic methods: Configure different types of device components in the device window, set the relevant attributes based on the type and features of external device, encapsulate the operation methods of device such as the hardware parameter configuration, data conversion and device debugging in the component, and establish the data transmission channel connection with external device based on the object type, as shown in Figure 1.1.9.

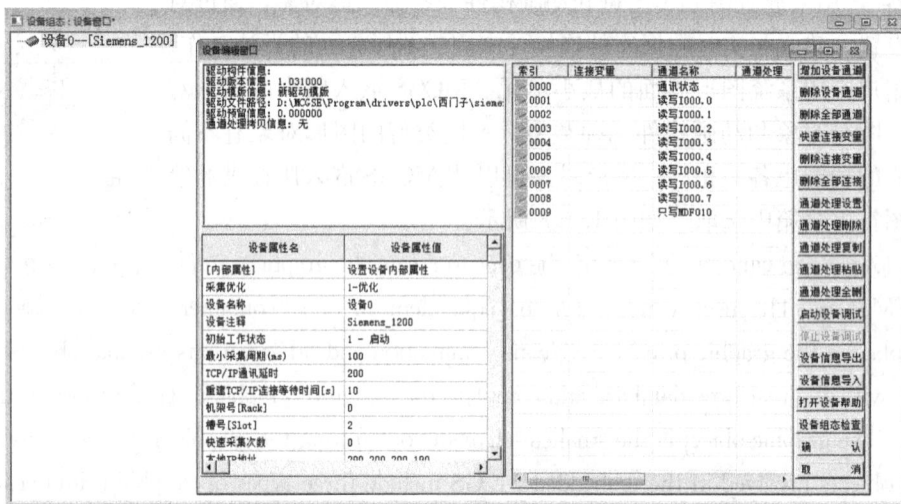

图 1.1.9 设备组态和设备编辑窗口

Figure 1.1.9 Device configurations and device edit window

4. 组态策略窗口

4. To configure the strategy window

组态策略窗口的目的是对系统运行流程的自由控制，使系统能按照设定的顺序和条件，进行操作实时数据库，控制用户窗口的打开、关闭以及控制设备构件的工作状态等一系列工作，从而实现对系统工作过程的精确控制及有序的调度管理。MCGS 提供七种策略类型供用户选择，分别是启动策略、退出策略、循环策略、用户策略、报警策略、事件策略和热键策略，其中除策略的启动方式各自不同之外，其功能本质上没有差别。每种策略都可完成一项特定的功能，而每一项功能的实现又以满足指定的条件为前提。每一个"条件—功能"实体构成策略中的一行，称为策略行，每种策略由多个策略行构成，如图 1.1.10 所示。运行策略的这种结构形式类似于 PLC 系统的梯形图编程语言，但更

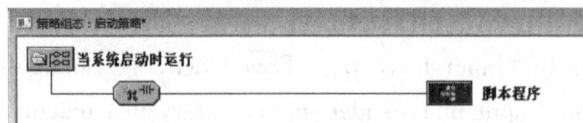

图 1.1.10 策略组态窗口

Figure 1.1.10 Strategy configuration window

加图形化，面向对象性更强，所包含的功能比较复杂，实现过程却相当简单。

The configuration of strategy window is intended to freely control the system running process，and enable the system to operate the real-time database and control the open and close of user window，the working status of device，and other a series of work based on the predetermined sequence and conditions，so as to realize the precise control and orderly dispatching management of the system running process. MCGS offers seven strategy types to user for selection，including the starting strategy, exit strategy, circular strategy, user strategy, alarm strategy, event strategy and hot key strategy，and except for the startup mode，there is no difference in their functions in nature. Each strategy can complete a certain function，and further each function is completed under the premise that the specified conditions are satisfied. Each "Condition-Function" entity forms one line in the strategies，called as the strategy line，and each strategy is composed of multiple strategy lines，as shown in Figure 1.1.10. The structure form of running strategy is similar to the ladder diagram programming language of PLC system，but the former is more graphical and more oriented to object，the functions included are also relatively complex，and the realization process is very simple.

5. 组态主控窗口

5. To configure the master control window

MCGS嵌入版的主控窗口是组态工程的主窗口，是所有设备窗口和用户窗口的父窗口，它相当于一个大的容器，可以放置一个设备窗口和多个用户窗口，负责这些窗口的管理和调度，并调度用户策略的运行。同时，主控窗口又是组态工程结构的主框架，可在主控窗口内设置系统运行流程及特征参数，方便用户的操作。

The master control window of embedded MCGS is the main window of configuration engineering and the parent window of all device windows and user windows，which is equivalent to a big container，and can place one device window and many user windows，take charge of the management and dispatching of these windows，and dispatch the running of user strategy. Meanwhile，the master control window is also the main framework to configure the engineering structure，and can set the running process and feature parameters of system for convenience of user.

在 MCGS嵌入版中，一个应用系统只允许有一个主控窗口。主控窗口是作为一个独立的对象存在的，其强大的功能和复杂的操作都被封装在对象的内部，组态时只需对主控窗口的属性进行正确地设置即可。

In the embedded MCGS，one application system can only accommodate one master control window，the powerful functions and complex operations of master control window as one independent object are encapsulated in the object，thus it is only needed to correctly set the attributes of master control window with respect to the configurations.

三、任务实施

Ⅲ. Task Implementation

（一）构建实时数据库

（Ⅰ）To create one real-time database

分析控制要求可知，需要新建两个开关型变量控制电机的启停，同时还需要新建两个开关型变量监控电机和热继电器的状态，其中，热继电器需要添加报警属性以触发报警策略，见表1.1.3。

It can be known from the analysis of control requirements that it is necessary to create two new switch variables to control the start and stop of motor，and another two new switch variables to monitor the states of motor and thermal relay，of which the thermal relay is assigned with the alarm attributes to trigger the alarm strategy，as shown in Table 1.1.3.

变量名 Variable name	类型 Type	报警 Alarm
HMI 启动 HMI Startup	开关型 Switch type	
HMI 停止 HMI Stop	开关型 Switch type	
热继电器 Thermal relay	开关型 Switch type	开关量报警：报警值为 1 Switch value alarm；Alarm value 1
电机状态 Motor state	开关型 Switch type	

数据变量表 表 1.1.3

Data variables Table 1.1.3

在工作台上的"实时数据库"选项卡中点击"新增对象"，在左边的列表里就会自动新增加一个数据对象，如图 1.1.11 所示。然后双击这个数据对象，弹出数据对象属性设置窗口，如图 1.1.12 所示。根据表 1.1.3 修改数据的名称和类型。

A new data object will be automatically added into the left list after clicking the "Add a new object" in the "Real-time Database" tab on the workbench，as shown in Figure 1.1.11. Then attribute setting window of data object will pop up after double clicking the data object，as shown in Figure 1.1.12. The name and type of data are changed based on Table 1.1.3.

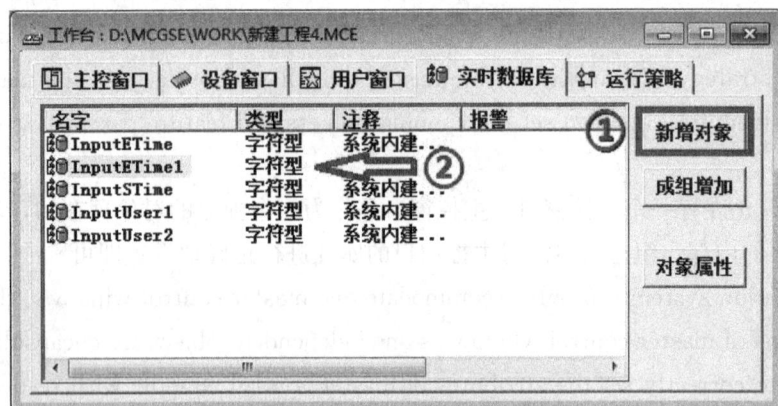
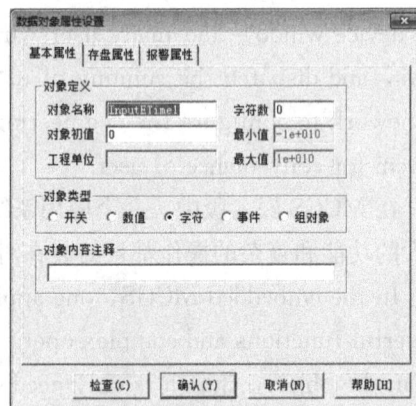

图 1.1.11 实时数据库标签

Figure 1.1.11 Real-time database tags

图 1.1.12 数据对象属性设置窗口

Figure 1.1.12 Attribute setting window of data object

对于"热继电器"变量，需要设置报警属性。点击数据对象属性设置窗口中的"报警属性"选项卡，先勾选"允许进行报警处理"选项，然后在下面的"报警设置"列表中选择报警的方式。在这里当"热继电器"的数值变为 1 时，即触发报警，因此选择"开关量报警"。之后在右边的"报警值"输入框中输入报警值 1。操作过程如图 1.1.13 所示。

For the "thermal relay" variable，it is necessary to set the alarm attributes. Click on the "Alarm Attributes" tab in the attribute setting window of data object，firstly check the "Allow the alarm processing" tab，and secondly select the alarm mode from the below "Alarm Setting" list. Here，when the value of "Thermal Relay" is 1，the alarm is triggered，thus the "Switch Value Alarm" will be selected. Then，the alarm value 1 is entered into the right "Alarm Value" input box. The operating process is shown in Figure 1.1.13.

（二）组态用户窗口

（Ⅱ）To configure the user window

1. 新建用户窗口

1. Create a new user window

在工作台上的"用户窗口"选项卡中点击"新建窗口"按钮，在左边的列表框中就出现了一个名为"窗口 0"的用户窗口，如图 1.1.14 所示。右键点击这个图标，可设置该窗口为启动窗口。按照此方法，再新建"窗口 1"，用于制作报警界面。

A user window named "Window 0" will appear in the left list box after clicking the "Create a New Window" in the "User Window" tab on the workbench，as shown in Figure 1.1.14. Then the window can be set to a startup window by right clicking this icon. In such way，another new "Window 1" is created to prepare the alarm interface.

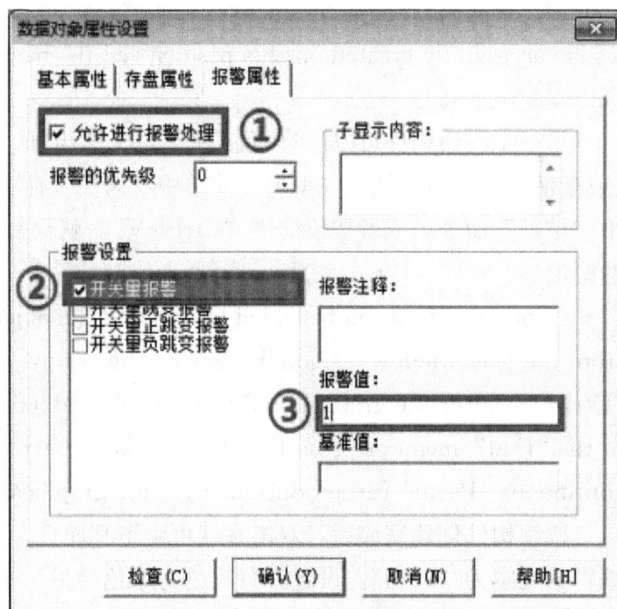

图 1.1.13　报警属性设置

Figure 1.1.13　Alarm attribute setting

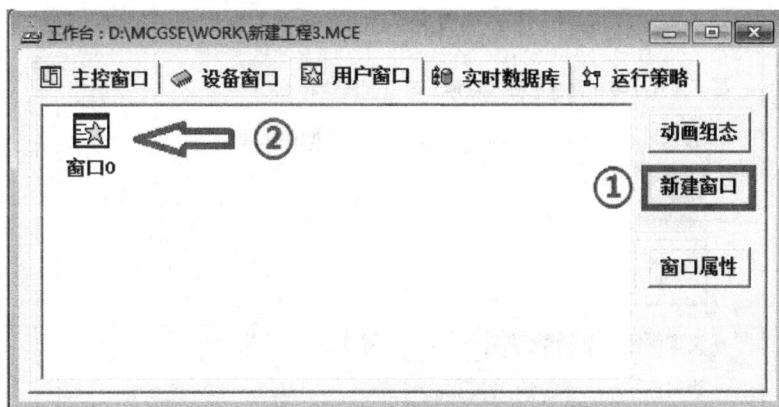

图 1.1.14　新建用户窗口

Figure 1.1.14　Create a new user window

2. 创建和编辑图形对象

2. Create and edit the graphic object

双击"窗口 0"图标，打开用户窗口，便出现了一个空的用户窗口和一个绘图工具箱。在本任务中，需要在用户窗口中添加两个按钮构件和一个电机图形对象。在绘图工具箱中选择标准按钮构件◻，之后把鼠标移到用户窗口内，此时鼠标光标变为十字形，按下鼠标左键不放，在窗口内拖动鼠标到适当的位置，然后松开鼠标左键，就在该位置建立了所需的图形，此时鼠标光标恢复为箭头形状。这样就创建了一个按钮构件。

One blank user window and one mapping toolbox will appear after double clicking the "Window 0" icon to open the user window. In this task，it is necessary to add two buttons and one motor graphic object into the user window. After selecting the standard button ◻ from the mapping toolbox，the mouse cursor is moved to the user window and becomes cruciform structure at this moment，the left mouse button is held down to drag the mouse cursor to a proper position inside the window and then released, one required graphics is successfully created on this position and the mouse cursor recovers to arrow structure. In such way, a new button is created.

另一个按钮构件可以通过复制对象的操作添加。首先，鼠标单击用户窗口内要复制的图形对象，选中（或激活）对象后，执行"编辑"菜单中"拷贝"命令，或者按快捷键"Ctrl＋C"，然后，执行"编辑"菜单中"粘贴"命令，或者按快捷键"Ctrl＋V"，就会复制出一个新的图形，连续"粘贴"，可复制出多个相同的图形。

Another button can be added by copying the object. You can firstly select the graphic object to be copied from the user window, secondly execute the "Copy" instruction in the "Edit" menu or press the shortcut key "Ctrl＋C" after the graphic object being selected (or activated) and thirdly execute the "Paste" instruction in the "Edit" menu or press the shortcut key "Ctrl＋V", finally a new graphics will be produced, and the continuous "Paste" can produce many same graphics.

按钮构件创建完成后，双击它就可编辑其属性，如图 1.1.15 所示。将其中的一个按钮设置为启动按钮，编辑其文本为"启动"，设置其背景色为绿色。另一个按钮设置为停止按钮，编辑其文本为"停止"，设置其背景色为红色。

After creating the button, its attributes can be edited by double clicking it, as shown in Figure 1.1.15. You can set one button to the startup button, edit its text "Startup", and set the background color to green. Then you set another button to the stop button, edit the text "Stop", and set the background color to red.

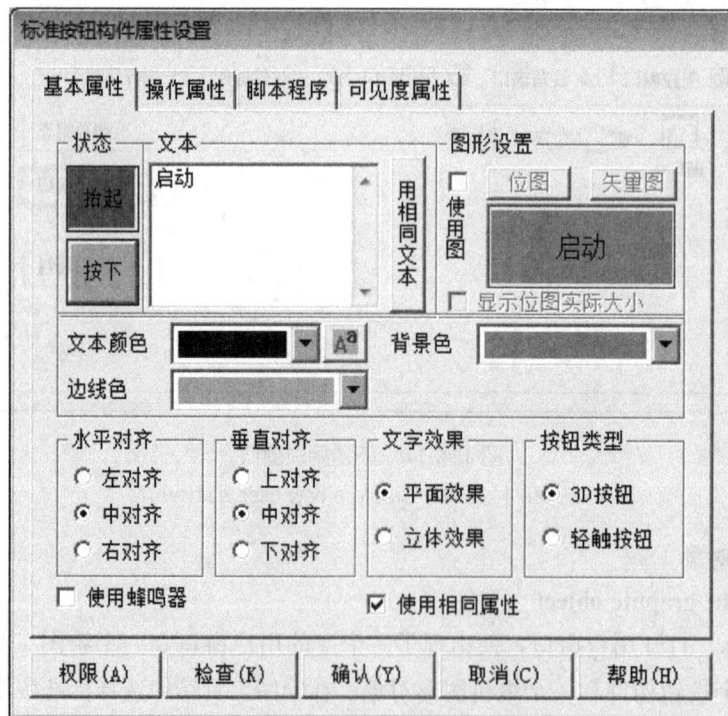

图 1.1.15 设置标准按钮控件基本属性

Figure 1.1.15 Set the essential attributes of standard button control

创建电机图形对象可以这样操作：选择绘图工具箱上的插入元件按钮 ▣，弹出对象元件库管理窗口。在左边的对象元件列表中选择"马达"，右边的列表框中就出现了很多电机的图形，选择其中一个，点击确定即可，如图 1.1.16 所示。这时用户窗口中出现了刚才选择的电机图形，把它拖动到窗口中合适的位置即可。

The motor graphic object can be created by the following ways: The object component library management window will pop up after selecting the button- Insert Component from the mapping toolbox 🖳. The right list box will appear many motor graphics after selecting the "Motor" from the left list of object components, and then you can select one of graphics and click on the confirmation button, as shown in Figure 1.1.16. At this moment, the user window shows the selected motor graphics, which is only required to be dragged to the proper position in the window.

完成以上步骤后，窗口 0 中所有的对象都已创建完成，组成的画面如图 1.1.17 所示.

After the above steps, all objects in the Window 0 have been created successfully, and the final graphics is shown in Figure 1.1.17.

用同样的方法，在窗口 1 中用绘图工具箱中的标签图形对象**A**，创建报警提示语标签.

Create the alarm prompt message tag in the Window 1 with the tag graphic object **A** from the mapping toolbox by the same method.

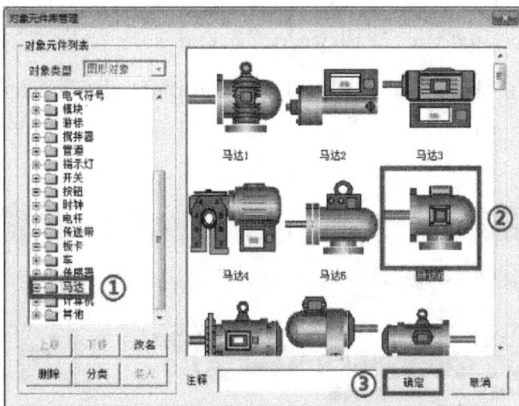

图 1.1.16　创建电机图形对象

Figure 1.1.16　Create the motor graphic object

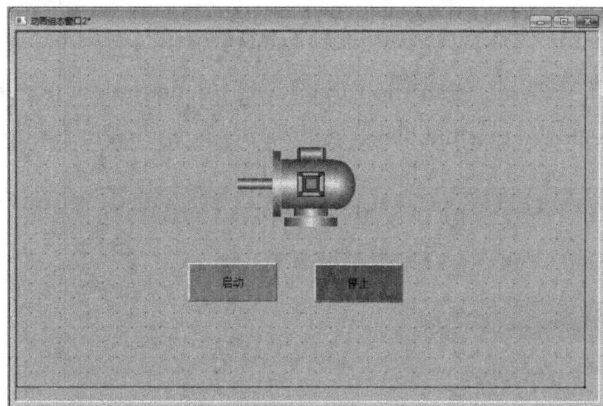

图 1.1.17　窗口 0 画面

Figure 1.1.17　Window 0 view

3. 定义动画连接

3. To define the animation connection

MCGS 嵌入版实现图形动画设计的主要方法是将用户窗口中的图形对象与实时数据库中的数据对象建立相关性连接，并设置相应的动画属性，这样在系统运行过程中，图形对象的外观和状态特征就会由数据对象的实时采集结果进行驱动，从而实现图形的动画效果。

The main methods for the graphic animation design with the embedded MCGS is to establish the correlative connection between the graphic object in the user window and the data object in the real-time database and set the corresponding animation attributes, and in such way, the appearance and state characteristics of graphic object will be driven the real-time results of data object in the system running process, so as to realize the animation effect of graphics.

在本任务中，"启动"按钮控件应与"HMI 启动"变量相关联，当按下"启动"按钮时，"HMI 启动"应设置为 1，松开按钮时，"HMI 启动"应清零。双击"启动"按钮构件，选择属性设置窗口的"操作属性"选项卡。在按钮构件的抬起功能中，勾选数据对象值操作，在其后的操作类型中选择"按 1 松 0"来模拟按钮对数据的操作。之后点击最右边的 ? 按钮，在弹出的变量选择窗口中选择"HMI 启动"变量。最后点击确定完成控件与数据的连接。如图 1.1.18 所示。停止按钮也采用同样的方法操作。

In this task, the "Startup" button control shall be correlated to "HMI Startup" variable, "HMI Starting" should be set to 1 when the "Startup" button is pressed, and should be reset when released. Double click on the "Startup" button, and select the "Operation Attributes" tab from the attribute setting window. Check the data object value in the lifting function of button control, and select "Press 1 Release 0" to stimulate the button operation towards the data in the subsequent operation types. Click on the rightmost ⌐?⌐ button, and select the "HMI Startup" variable from the variable selection window popped up. Click OK to complete the connection between control and data. As shown in Figure 1.1.18. The method is also applicable to Stop Button.

电机图形对象中要显示电机的运行状态，因此应与"电机状态"变量相关联。双击电机图形对象，打开单元属性设置窗口。选择"数据对象"选项卡，点击"填充颜色"右侧的 ？ 按钮，在弹出的变量选择窗口中选择"电机状态"变量。最后点击确定完成图形对象与数据的连接，如图1.1.19所示。

The motor graphic object will show the operating state of motor, therefore it should be correlated to the "Motor State" variable. Double click on the motor graphic object to open the cell attribute setting window. Select the "Data Object" tab, click on the ？ button following the "Fill Color", and select the "Motor State" variable from the variable selection window popped up. Finally, click OK to complete the connection between the graphic object and data, as shown in Figure 1.1.19.

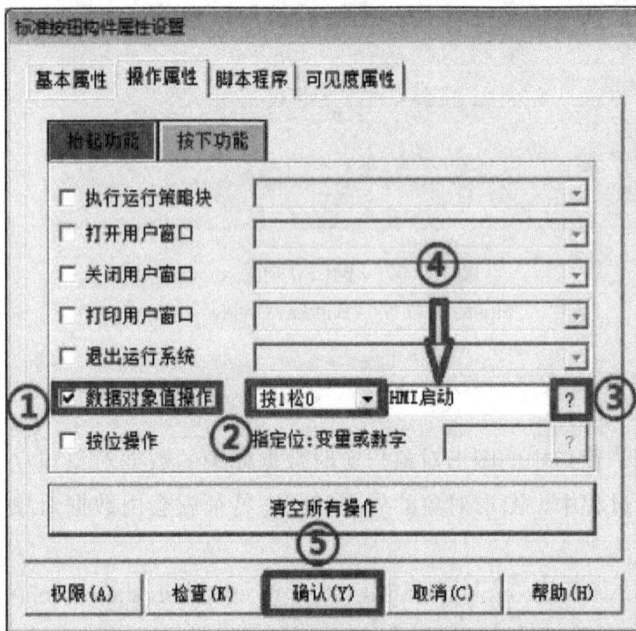

图 1.1.18　在标准按钮控件的属性中关联变量
Figure 1.1.18　Correlate the variables in the attributes of standard button control

图 1.1.19　在电机图形对象的属性中关联变量
Figure 1.1.19　Correlate the variables in the attributes of motor graphic object

当电机过载时，报警窗口弹出，提示电机过载。此时需要在窗口1中的标签图形对象的动画组态属性设置窗口中输入报警提示语，如图1.1.20所示。除此之外，还可以设置标签的填充颜色、字符颜色、边线颜色、对齐方式等属性。

When the motor is overloaded, the alarm window will pop up to prompt the motor overload. At this moment, it is necessary to enter the alarm prompt message into the attribute setting window of animation configurations related to the tab graphic object of Window 1, as shown in Figure 1.1.20. Besides, the filling color, character color, edge color, alignment method and other attributes of tab can also be set.

（三）组态设备窗口

（Ⅲ）To configure the device window

不但用户窗口中的对象与实时数据库需要进行数据关联，数据库中的变量与 PLC 中的变量也要进行关联。当用户操作窗口中的控件时，PLC 才能进行响应，同样，PLC 程序执行的结果才能反映到用户窗口中。在本任务中，启动按钮和停止按钮的数据需要写入 PLC，而电机的运行状态和热继电器的状态需要从 PLC 中读出。因此，需要创建 PLC 和实时数据库变量的通道连接见表 1.1.4。

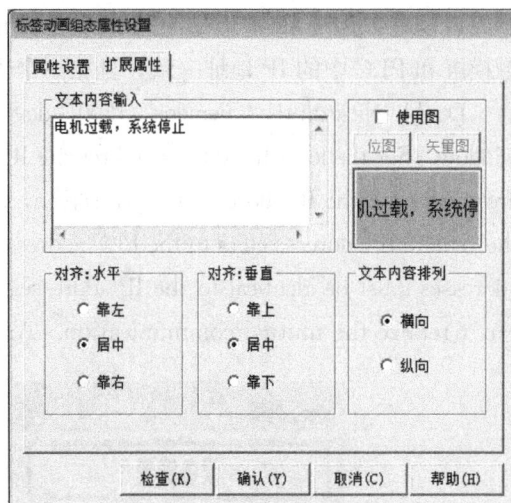

图 1.1.20 标签动画组态属性设置

Figure 1.1.20 Attribute setting of tab animation configurations

Except that the objects in the user window are correlated to the data in the real-time database, the variables in the database should be correlated to these variables in PLC. PLC can render response only when user operating the control in the user window. As well, the results of PLC program execution can be displayed in the user window. In this task, the data of Startup Button and Stop Button should be written in PLC, while the motor running state and thermal relay state should be read out from PLC Therefore, it is necessary to create the channel connection between PLC and real-time database, as shown in Table 1.1.4.

设备通道和连接变量   表 1.1.4

Device channels and connection variables   Table 1.1.4

连接变量 Connection variables	通道名称 Channel name
HMI 启动 HMI Startup	只写 M0.0 Write M0.0 only
HMI 停止 HMI Stop	只写 M0.1 Write M0.1 only
热继电器 Thermal relay	只读 I0.2 Read I0.2 only
电机状态 Motor state	只读 Q0.0 Read Q0.0 only

在设备组态窗口中点击右键，打开设备工具箱，如图 1.1.21 所示。双击列表中的"Siemens_1200"，这样就在设备组态窗口中添加了西门子 1200 的驱动构件，如图 1.1.22 所示。如果列表中没有该设备的驱动构件，则点击"设备管理"按钮，在弹出的设备管理窗口中添加。

Right click in the device configuration window to open the device toolbox, as shown in Figure 1.1.21. Double click on the "Siemens_1200" in the list such that the driving components of Siemens 1200 are added to the device configuration window, as shown in Figure 1.1.22. Click on the "device management" button, and add the driving component in the device management window popped up in the event of no driving component applicable to the device.

在设备组态窗口中双击刚才添加的"Siemens_1200"设备，打开设备编辑窗口。在左下方的设备属性中，设置本地 IP 地址和远端 IP 地址。本地 IP 地址指的是触摸屏的 IP 地址，需要在触摸屏的系统设置中查看或者设置。远端 IP 地址指的是 PLC 的地址，在 PROFINET 接口属性中查看或设置。这两个地址必须跟触摸屏和 PLC 中的 IP 地址一致，且在一个子网内，否则二者之间无法通信。如图 1.1.23 所示。

Double click on the "Siemens _ 1200" device just added in the device configuration window to open the device edit window. Set the local IP address and remote IP address in the bottom left device attributes. Of which the local IP address refers to the IP address of touch screen，it should be viewed or set in the system setting of touch screen，and the remote IP address refers to the PLC address，it can be viewed or set in the PROFINET interface attributes. Both addresses must be identical to the IP addresses of touch screen and PLC and within one sub-net，otherwise they can't realize the mutual communication. As shown in Figure 1. 1. 23.

图 1.1.21　设备工具箱

Figure 1. 1. 21　Device toolbox

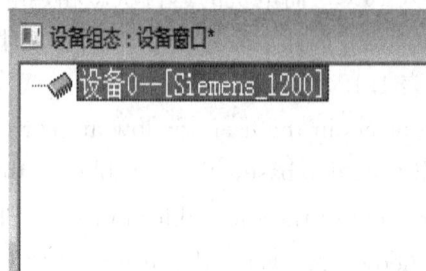

图 1.1.22　添加 Siemens _ 1200 设备

Figure 1. 1. 22　Add siemens _ 1200 device

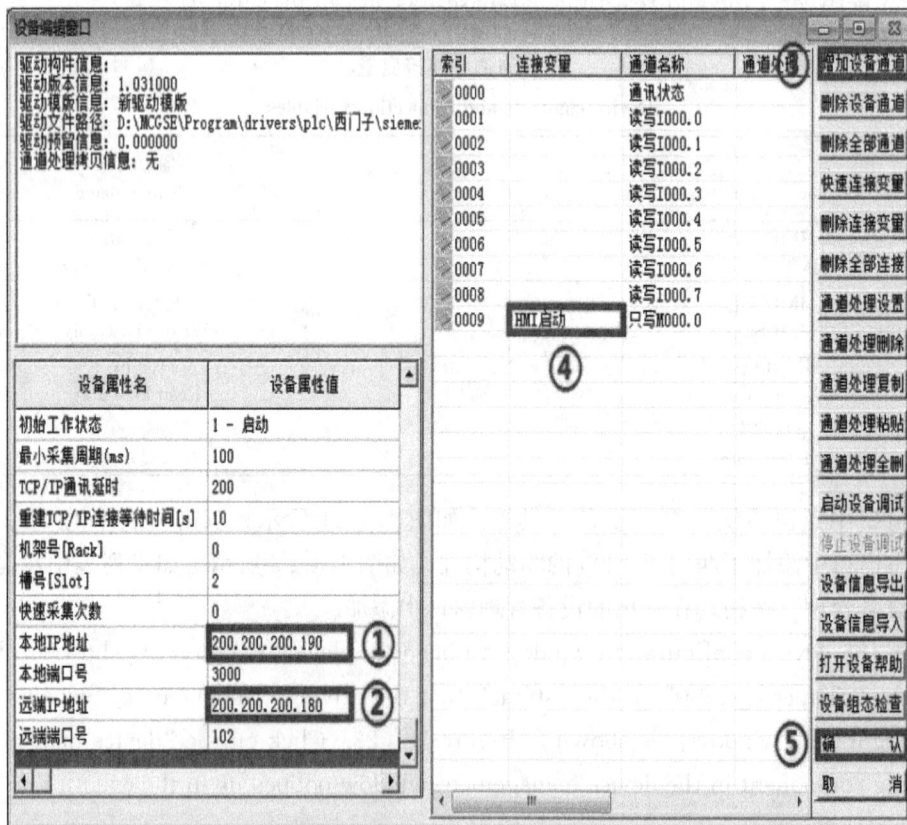

图 1.1.23　设备编辑窗口

Figure 1. 1. 23　Device edit window

点击"增加设备通道"按钮，如图 1.1.23 所示，在弹出的添加设备通道窗口中，按表 1.1.4 逐一添加设备通道。例如添加 M0.0 通道，如图 1.1.24 所示。通道添加完成后，在双击通道名称右边的"连接变量"单元格，在弹出的选择变量窗口中选择变量。最后点击确认完成设备窗口组态。

Click "to add the device channel" button as shown in Figure 1.1.23，and add the device channels one by one in the Add Device Channel Window popped up based on Table 1.1.4. For example the addition of M0.0 channel is shown in Figure 1.1.24. After the addition，double click on the "Connect Variable" cell on the right side of channel name，and then select the variable from the Select Variables Window popped up. Finally，click OK to complete the device window configuration.

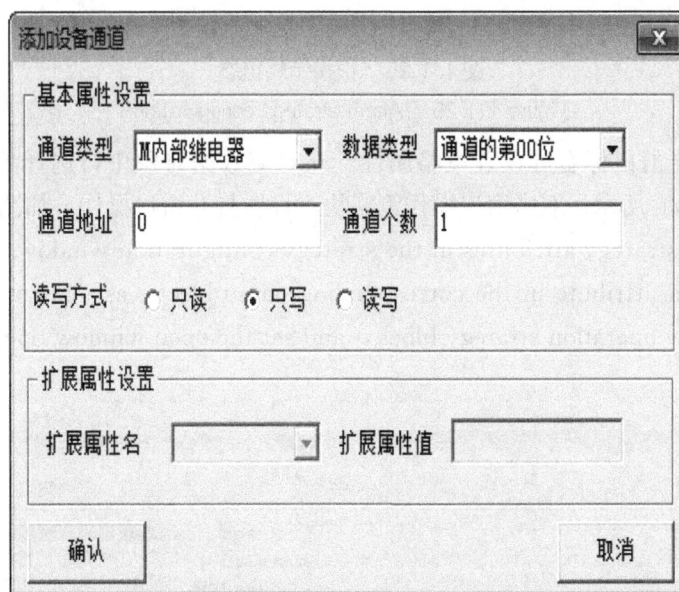

图 1.1.24　添加设备通道窗口

Figure 1.1.24　Add device channel window

（四）组态策略窗口

（Ⅳ）To configure the strategy window

当过载时，热继电器动作，实时数据库中的对应数据发生报警，触发报警策略，打开报警窗口。报警策略由用户在组态时创建，当指定数据对象的某种报警状态产生时，报警策略将被系统自动调用一次。

In case of overload，the thermal relay is enabled，and the corresponding data in the real-time database shows the alarm state and triggers the alarm strategy to open the alarm window. The alarm strategy is created by user in the configurations，when the designated data object appears a certain alarm state，the alarm strategy will be automatically called once by the system.

在工作台的"策略窗口"选项卡中，选择"新建策略"，在弹出的窗口中选择"报警策略"。这样在列表中就增加了一个名为"策略 1"的报警策略。双击它，打开策略组态窗口。在策略组态窗口中，通过点击鼠标右键"新建策略行"和打开"策略工具箱"。再选中策略行最右侧的"策略块"，在策略工具箱中双击选择"窗口操作"，表示当报警策略调用时进行了窗口打开关闭等操作，如图 1.1.25 所示。

You can select to create a new strategy in the "strategy window" tab of workbench，and then select the "alarm strategy" in the window popped up. In such way，there is a new alarm strategy named "Strategy 1" successfully added in the list. Double click it to open the "strategy configuration window". Right click to create a "new strategy line" and open the "strategy toolbox" in the strategy configuration window. Then select the rightmost "strategy block" in the strategy line, and double click to select the "proper operation window" from the strategy toolbox, indicating that the open window, close window and other operations are performed when the alarm strategy is called, as shown in Figure 1.1.25.

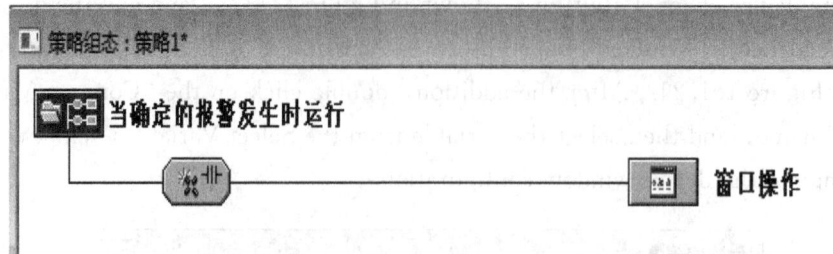

图 1.1.25　报警策略组态

Figure 1.1.25　Alarm strategy configuration

在策略组态窗口中，点击鼠标右键设置策略属性。在对应数据对象中，选择具有报警属性的"热继电器"变量，如图 1.1.26（a）所示。双击窗口操作策略块，设置打开窗口操作，如图 1.1.26（b）所示。

Right click to set the strategy attributes in the strategy configuration window. Select the Thermal Relay variable with the alarm attribute in the corresponding data object, as shown in Figure 1.2.26 （a）. Double click on the window operation strategy block, and set the open window operation, as shown in Figure 1.1.26 （b）.

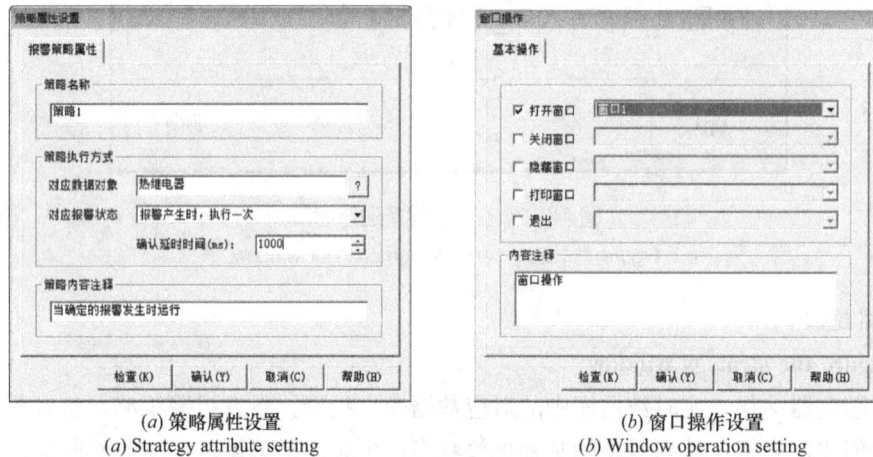

(a) 策略属性设置	(b) 窗口操作设置
(a) Strategy attribute setting	(b) Window operation setting

图 1.1.26　设置策略属性和窗口操作

Figure 1.1.26　Set strategy attribute and window operation

当热继电器复位，报警解除。实时数据库中的对应数据由 1 变 0，产生负跳变，触发事件策略。事件策略由用户在组态时创建，当对应表达式的某种事件状态产生时，事件策略会被系统自动调用一次。如图 1.1.27 所示。

The alarm will be released when the thermal relay is recovered. The corresponding data in the real-time database is changed from 1 to 0, causing the negative transition and trigger the event strategy. The event strategy is created by user in the configurations, and the event strategy will be automatically called once by the system when certain event state corresponding to expression appears. As shown in Figure 1.1.27.

（五）连接 I/O 设备

（Ⅴ）To connect I/O device

按照控制要求，I/O 接线图如图 1.1.28 所示。

Based on the control requirements, the I/O wiring diagram is shown in Figure 1.1.28.

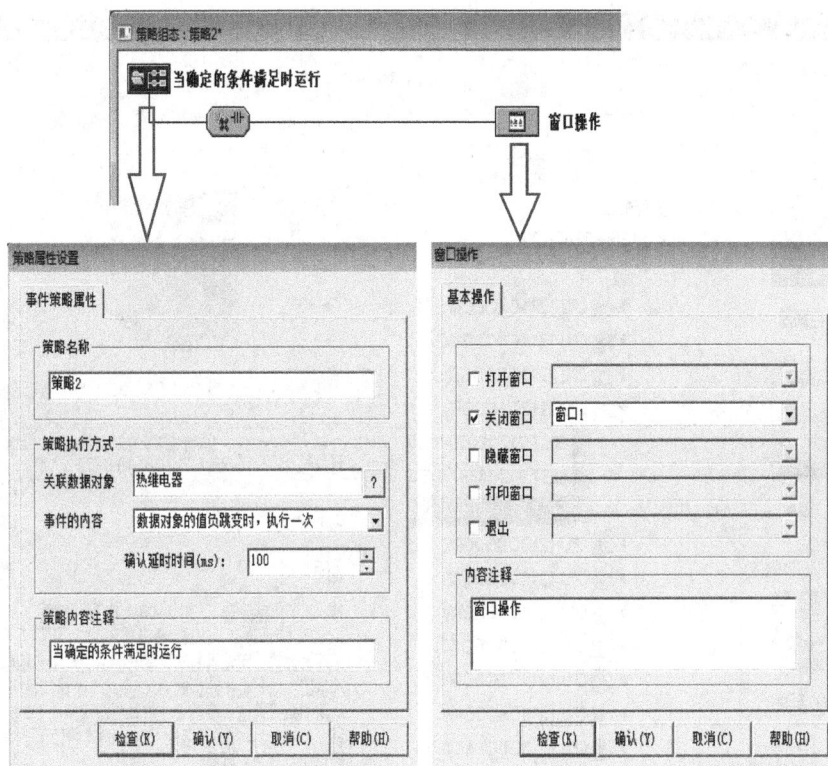

图 1.1.27 事件策略组态及属性设置

Figure 1.1.27 Event strategy configuration and attribute setting

（六）在 TIA 博途中新建项目和设备组态

（Ⅵ）To create a new item and device configuration in TIA Portal

打开 TIA 博途编程软件，点击左下角的链接切换到项目视图。在菜单栏上的"项目"下拉框中选择"新建"，设置好项目名称和路径后就新建了一个项目。

Open the programming software of TIA Portal, click on the bottom left link to switch to the item view. Select the "Create New Item" from the "Item" drop-down box on the menu bar, set the item name and patch to create a new item.

在左边的项目树中点击"添加新设备"，弹出"添加新设备窗口"。选择实际使用的 PLC 的型号和订货号，再点击确认完成设备添加，如图 1.1.29 所示。

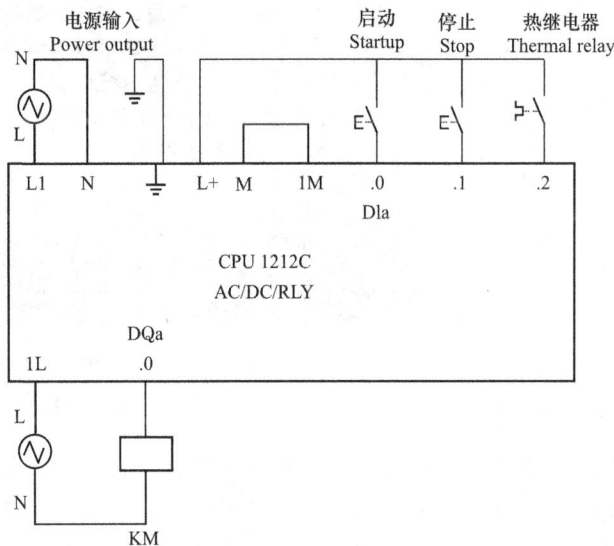

图 1.1.28 I/O 接线图

Figure 1.1.28 I/O wiring diagram

Click on "Add New Device" in the left item tree to pop up the "new device window". Select the actually used PLC model and order number, and then click on OK to complete the device addition, as shown in Figure 1.1.29.

回到项目视图中，可以看到左边的项目树中多了一个名为"PLC_1"的文件夹，如图 1.1.30 所示。点击 ▶ 按钮可以展开它，看到里面各层次的列表。在编程之前，点击列表中的"设备组态"完成设备组态。在设备视图中，可以看到刚才添加的"设备 PLC_1"，它是一个 CPU 模块，位于机架 Rack_0 的 1 号槽位。

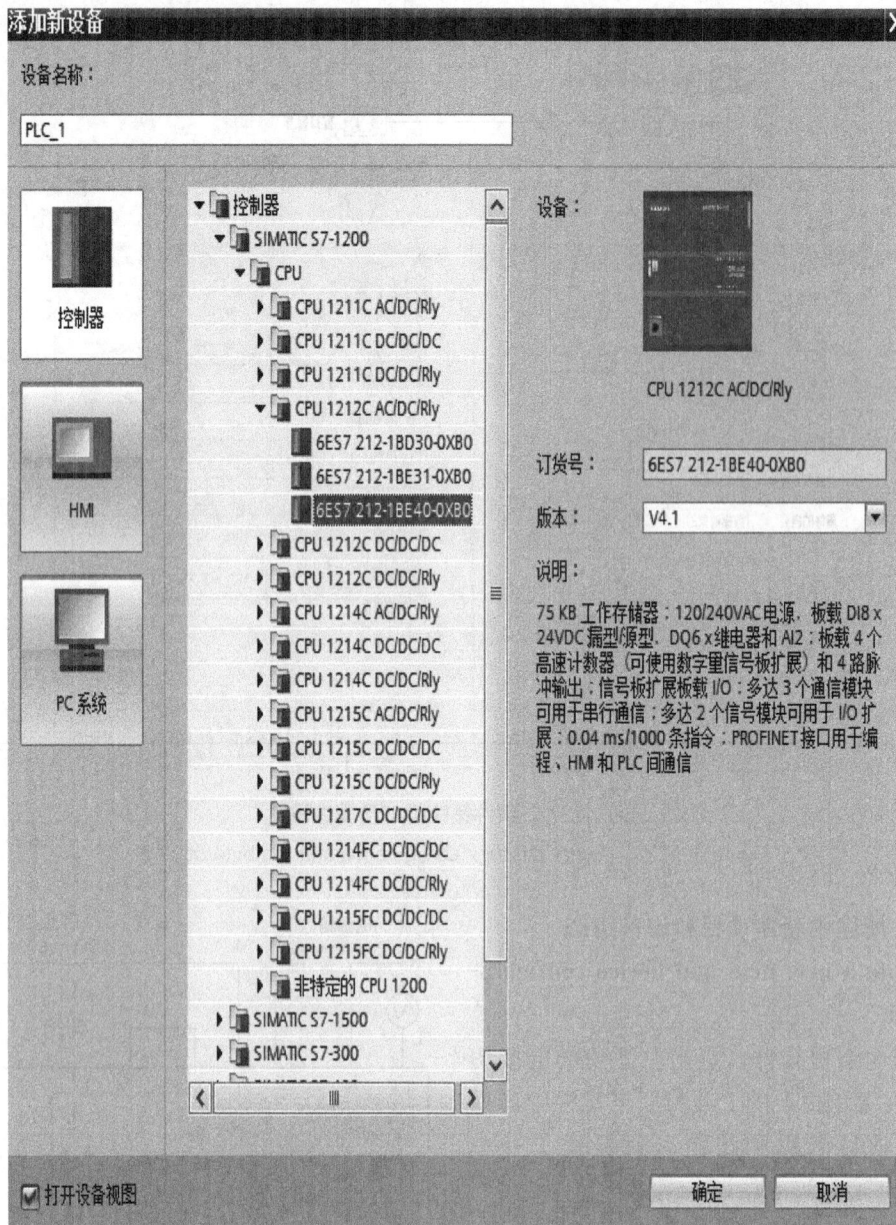

图 1.1.29　添加新设备

Figure 1.1.29　Add a new device

其左边的101、102、103 槽位用于扩展通信模块，右边的2号槽位和3号槽位用于扩展信号模块，CPU模块中间的方框可以扩展信号板。扩展模块的型号和订货号可以在右边的硬件目录中添加。设备组态窗口中的设备要与实际使用的设备一致。在设备视图中选中"PLC_1"，下方的巡视窗口中的属性＞常规选项卡可设置其各项属性。首先在 PROFINET 接口中设置以太网地址，这个地址要与在 MCGS 嵌入版的设备组态窗口中设置的远端 IP 地址相同。

After returning to the item view，you can find out there is a file named "PLC_1" in the left item tree，as shown in 1.1.30. It can be expanded by clicking on the ▶ button，displaying the lists at all levels. Before the programming，the "Device Configuration" in the list is clicked to complete the device configuration. You can find out the "device PLC_1" just added in the device view，which is a CPU module located at No.1 slot position of Rack_0. The left No.101，102 and 103 slot positions are used for extending the communication modules，the right No. 2 and 3 slot positions are used for extending the signal module，and the square box in the middle of the CPU module is used for extending the signal board. The model of extended module and order number can be added in the right Hardware Catalog. The device in the device configuration window should be identical to the actually used device. After "PLC _ 1" being selected in the device view，its attributes can be set via the Attributes>General Tabs in the blow inspector window. Firstly，the Ethernet address is set in the PROFINET interface，the address must be identical to the remote IP address set in the device configuration window of embedded MCGS.

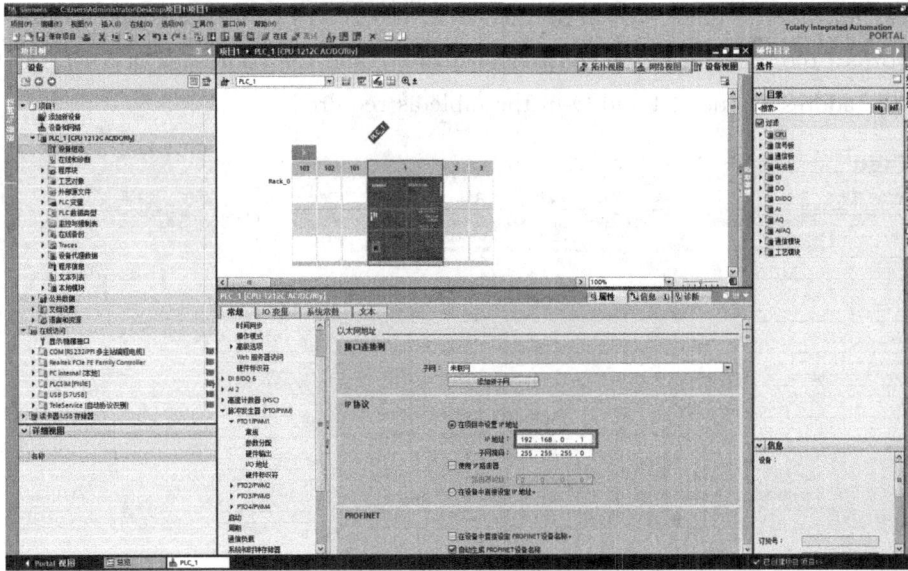

图 1. 1. 30　设备组态

Figure 1. 1. 30　Device configuration

由于 PLC 要与触摸屏进行通信，因此允许触摸屏访问 PLC。在"保护"选项中勾选"允许远程伙伴（PLC、HMI、OPC）使用 PUT/GET 通信访问"。如图 1. 1. 31 所示。

Because PLC will be communicated with touch screen，the touch screen is allowed to access PLC. It is only needed to check "Allow remote partners (PLC，HMI and OPC) to use PUT/GET for communication access" in the "Protection" tabs. As shown in Figure 1. 1. 31.

图 1. 1. 31　设置允许触摸屏访问

Figure 1. 1. 31　Setting of allowing touch screen to access

在设备视图中点击右侧的指向左的小箭头，出现了“设备概览”窗口，如图 1.1.32 所示。这里列出了 PLC 中所有可用的输入输出资源，及给它们分配的地址。在第一行我们可以看到我们所用的 PLC 有 8 个 DI 和 6 个 DQ，它们的地址都是 0，也就是说，8 个 DI 的地址是 I0.0～I0.7，6 个 DQ 的地址是 Q0.0～Q0.5。用户可以根据需要，在表格中修改 I 和 Q 的地址值。

The device overview window will appear after clicking on the right small arrow pointed to left in the device view, as shown in Figure 1.1.32. There are all input and output resources available to PLC and their assigned addresses. In the first line, we can find out that the used PLC has 8 DIs and 6 DQs, their addresses are 0, in other words, the addresses of 8 DIs is I0.0 to I0.7 and the addresses of 6 DQs is Q0.0 to Q0.5. User can modify the address value of I and Q in the table as required.

设备概览

模块	插槽	I 地址	Q 地址	类型	订货号	固件	注释
▼ PLC_1	1			CPU 1212C AC/DC/Rly	6ES7 212-1BE40-0XB0	V4.1	
DI 8/DQ 6_1	1 1	0	0	DI 8/DQ 6			
AI 2_1	1 2	64...67		AI 2			
	1 3						
HSC_1	1 16	1000...10...		HSC			
HSC_2	1 17	1004...10...		HSC			
HSC_3	1 18	1008...10...		HSC			
HSC_4	1 19	1012...10...		HSC			
HSC_5	1 20	1016...10...		HSC			
HSC_6	1 21	1020...10...		HSC			
Pulse_1	1 32		1000...10...	脉冲发生器 (PTO/PWM)			
Pulse_2	1 33		1002...10...	脉冲发生器 (PTO/PWM)			
Pulse_3	1 34		1004...10...	脉冲发生器 (PTO/PWM)			
Pulse_4	1 35		1006...10...	脉冲发生器 (PTO/PWM)			
▸ PROFINET 接口_1	1 X1			PROFINET 接口			
	2						

图 1.1.32 设备概览

Figure 1.1.32 Device overview

（七）建立变量表

（Ⅶ）To create the variable table

打开项目树的文件夹“PLC 变量”，双击其中的“默认变量表”，打开变量编辑器。根据控制要求，将 PLC 所用到的所有地址都定义成变量。如图 1.1.33 所示。

Open the folder “PLC Variables” in the item tree, double click on the “Default Variable Table”, and open the variable editor. Define all available addresses of PLC as the variables based on the control requirements. As shown in Figure 1.1.33.

PLC 变量

	名称	变量表	数据类型	地址	保持	在 H...	可从 ...	注释
1	启动	默认变量表	Bool	%I0.0	☐	☑	☑	
2	停止	默认变量表	Bool	%I0.1	☐	☑	☑	
3	热继电器	默认变量表	Bool	%I0.2	☐	☑	☑	
4	HMI启动	默认变量表	Bool	%M0.0	☐	☑	☑	
5	HMI停止	默认变量表	Bool	%M0.1	☐	☑	☑	
6	KM	默认变量表	Bool	%Q0.0	☐	☑	☑	
7	<添加>				☐	☑	☑	

图 1.1.33 变量表

Figure 1.1.33 Variable table

（八）编写梯形图程序

（Ⅷ）To write the ladder logic program

PLC 程序的目的是要实现电机的启动和停止，也就是起保停程序。由于按钮和触摸屏都能够实现电机的启停，因此需要将启动的触点并联，停止的触点串联。

PLC program is intended to realize the startup and stop of motor, thus it is called as the startup, holding and stop program. Because the button and touch screen can realize the startup and stop of motor, it is needed to connect the contacts to be started in parallel and the contacts to be stopped in series.

双击项目树列表中的"程序块"下的"OB1",打开主程序,即可利用右边指令窗口中的指令编写梯形图程序,如图 1.1.34 所示。

You can double click on "OB1" belonging to the "program block" in the item tree to open the main program, i. e. the instructions in the right instruction window can be used for writing the ladder logic program, as shown in Figure 1.1.34.

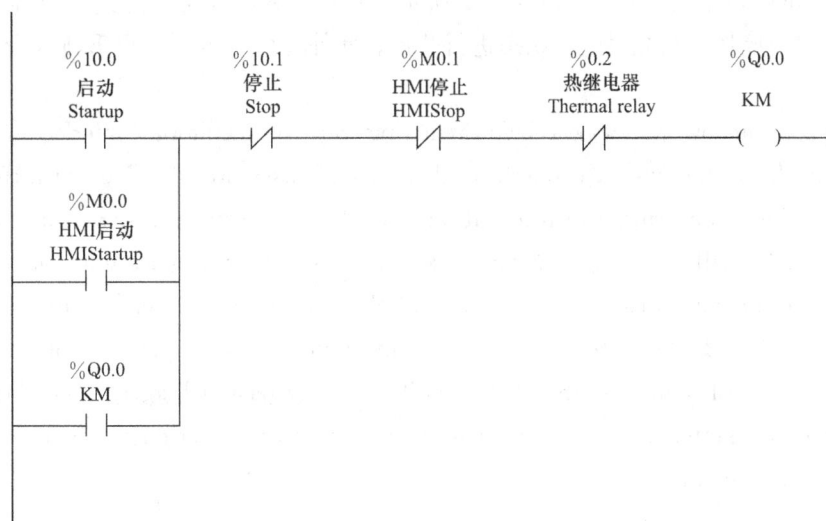

图 1.1.34 梯形图程序
Figure 1.1.34 Ladder logic program

(九) 调试
(Ⅸ) Debugging

完成以上步骤后,将编辑好的 MCGS 嵌入版工程和 TIA 博途项目分别下载至触摸屏和 PLC,再用编程线将二者连接起来,这样三相电机启停的远程控制和运行监控系统就建立起来了。按照控制要求逐项对系统进行调试,直到所有功能都实现为止。

After the above steps, the embedded MCGS engineering and TIA Portal items have been edited are respectively downloaded to the touch screen and PLC, the both are connected with the programming line such that the remote startup and stop control and running monitoring system of three-phase motor is finally created. The system is debugged item by item according to the control requirements until that all functions are realized.

任务二　用触摸屏建立一个三相电机正反转的远程控制和运行监控系统

Task Ⅱ　To Build a Remote Positive and Negative Rotation Control and Operation Monitoring System for Three-phase Motor via the Touch Screen

一、任务分析
Ⅰ. Task Analysis

用人机界面（HMI）、PLC 及相关电气器件组态一个电机正反转控制的监控系统,要求系统具有以下功能：

Assemble the human-machine interface（HMI），PLC and relevant electric devices into the motor positive and negative rotation monitoring system，which should have the following functions：

（1）用按钮可以控制电机的正反转和停止。

（1）The positive and negative rotation，and stop of motor can be controlled with buttons.

（2）在进入监控画面前，弹出用户登录窗口，如图 1.2.1 所示。登录时可以选择以操作员身份登录，密码为"123"，也可以选择以工程师身份登录，密码为"456"。输入正确后，进入监控画面。如输入有错，画面上弹出"密码输入错误，是否重新输入"的提示框，点击提示框中的"是"，用户可重新登录，点击提示框中的"否"，用户也可进入监控画面，如图 1.2.2 所示。只有工程师才可以通过人机界面远程操作电机的正反转和停止，未登录或操作员只能对电机状态进行监控。此外还可点击"用户重新登录"返回用户登录窗口重新登录。

（2）The user logon window will appear before entering into the monitoring interface，as shown in Figure 1.2.1. You can log in the system with the operator identity and password "123" or with the engineer identity and password "456". The monitoring interface will appear after entering the correct password. If the password is wrong，the system will pop up the prompt box "the password is wrong，enter again or not"，if clicks on "Yes"，user can log in again，otherwise if clicks on "No"，user can also enter into the monitoring interface as shown in Figure 1.2.2. Only the user with engineer identity can remotely operate the positive and negative rotation，and stop of motor via the HMI，while the user hasn't logged in or with operator identity only can monitor the motor state. Additionally，user can click on the "user enters again" to return to the user logon window and enter again.

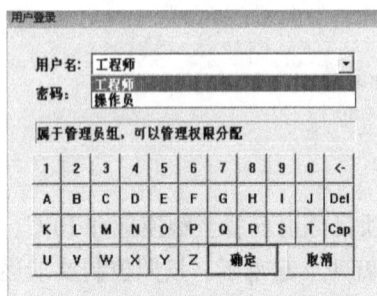

图 1.2.1 用户登录窗口
Figure 1.2.1 User logon window

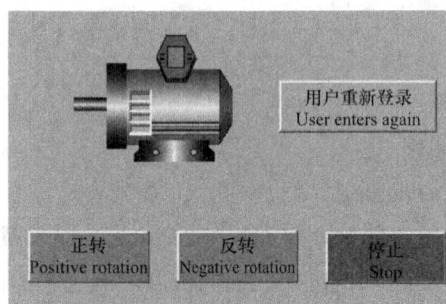

图 1.2.2 监控画面
Figure 1.2.2 Monitoring interface

（3）人机界面能通过电机元件上的指示灯实时监控到电机的运行或停止状态，并在用工程师身份登录时可以远程操作电机的正反转和停止。

（3）HMI should be capable of monitoring the running or stop state of motor in real time via the indicators on the motor components，and user who logged in the system with the engineer identity can remotely operate the positive and negative rotation，and stop of motor.

（4）若电机过载，热继电器动作，则自动弹出报警画面"电机过载，系统停止"，电机停止运行。报警画面弹出 3 秒后自动关闭。

（4）When the motor is overloaded，the thermal relay is actuated，the alarm dialog box "Motor Overloaded，System stopped" will automatically appear，and then the motor will be stopped. The alarm page will automatically disappear after 3 seconds.

为了完成本任务，首先需要对涉及的设备进行选型，根据现有条件，人机界面选择昆仑通态的 MCGS 触摸屏，PLC选择西门子S7-1200系列PLC。通过完成本任务，需要在完成上一个任务的基础上，学习如何管理MCGS嵌入版的用户权限，学会表达式和基本语句的使用以及用户登录函数和定时器函数的使用。

For the purpose of this task, it is firstly necessary to select the required types of devices involved, of which MCGS touch screen and Siemens S7-1200 PLC will be selected under the current conditions. In this task, you should learn the user authority management of embedded MCGS, the use of expression and basic statements, and the use of user logon function and timer function on the basis that the previous task has been completed.

二、相关知识技能
II. Relevant Knowledge & Skills
（一）MCGS嵌入版的用户管理及操作权限
（I）User management and operation authority of embedded MCGS

MCGS嵌入版组态软件提供了一套完善的安全机制，用户能够自由组态控制按钮和退出系统的操作权限，只允许有操作权限的操作员才能对某些功能进行操作。MCGS嵌入版采用用户组和用户的概念来进行操作权限的控制。在MCGS嵌入版中可以定义多个用户组，每个用户组中可以包含多个用户，同一个用户可以隶属于多个用户组。操作权限的分配是以用户组为单位来进行的，即某种功能的操作哪些用户组有权限，而某个用户能否对这个功能进行操作取决于该用户所在的用户组是否具备对应的操作权限。

The configuration software of embedded MCGS provides a complete set of perfect security mechanism, user can freely configure the control button, log off the system's operation authority, and only the authorized operator can operate some functions. The embedded MCGS controls the operation authority based on the concepts of user group and user. There can be many user groups defined in the embedded MCGS, each user group can contain many users, and each user can belong to many user groups. The operation authority is assigned in unit of user group, i. e. which user groups can be authorized with certain function, and the authority of certain user with respect to this function depends on whether the user group it belongs to has the corresponding operation authority.

在MCGS嵌入版组态环境中，选取"工具"菜单中的"用户权限管理"菜单项，弹出用户管理器窗口如图1.2.3所示。在MCGS嵌入版中，固定有一个名为"管理员组"的用户组和一个名为"负责人"的用户，双击它们可以对它们的属性进行设置。管理员组不可删除，其中的用户有权利在运行时管理所有的权限分配工作，管理员组的这些特性是由MCGS嵌入版系统决定的，其他所有用户组都没有这些权利。

You can select the "User Authority Management" item from the "Tool" menu under the configuration environment of the embedded MCGS, and then the user administrator window will pop up, as shown in Figure 1.2.3. There are user group named "Administrator Group" and user named "Responsible Person" in the embedded MCGS, and you can double click on them to set their attributes. The Administrator Group can't be deleted, their users have the right to manage all authority assignments in the system running process, these features of Administrator Group are decided by the embedded MCGS, and other all user groups haven't these rights.

将鼠标选中"负责人"用户，单击"新增用户"按钮，可以添加新的用户名。选中一个用户时，点击"属性"或双击该用户，会出现用户属性设置窗口，如图1.2.4所示，在该窗口中，可以设置和修改用户密码和所属的用户组。

A new user name can be added by selecting the user "Responsible Person" and clicking on the "user addition" button. When one user is selected, the user attribute setting window will appear by clicking on the "attributes" or double clicking on the user, as shown in Figure 1.2.4. In the window, you can set and modify the user password and the user group it belongs to.

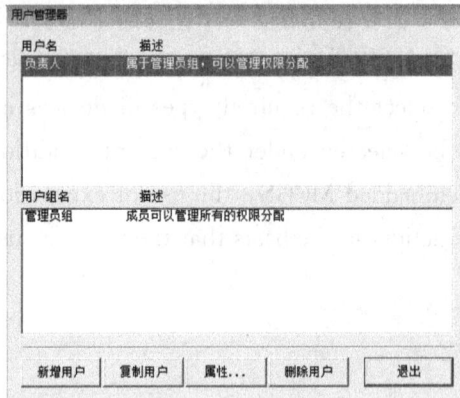

图 1.2.3　用户管理器

Figure 1.2.3　User administrator

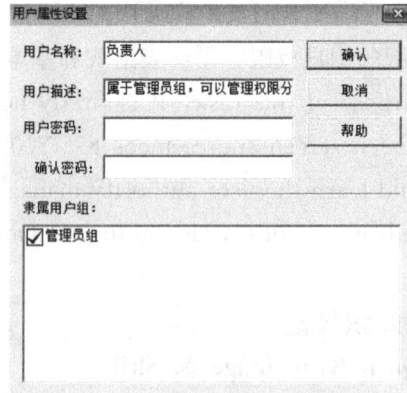

图 1.2.4　用户属性设置窗口

Figure 1.2.4　User attribute setting window

将鼠标选中"管理员组"用户，单击"新增用户"组按钮，可以添加新的用户组。选中一个用户组时，点击"属性"或双击该用户组，会出现用户组属性设置窗口，如图 1.2.5 所示，在该窗口中，可以选择该用户组包括哪些用户，点击"登录时间"按钮，可以设置用户组可以登录的时间，如图 1.2.6 所示。

A new user group can be added by selecting the user group "Administrator Group" and clicking on the "user group addition" button. When one user group is selected，the "user group attribute" setting window will appear by clicking on the "attributes" or double clicking on the user group，as shown in Figure 1.2.5. In the window，you can select the users belonging to the user group，and click on the Logon Time button to set the logon time corresponding to the user group，as shown in Figure 1.2.6.

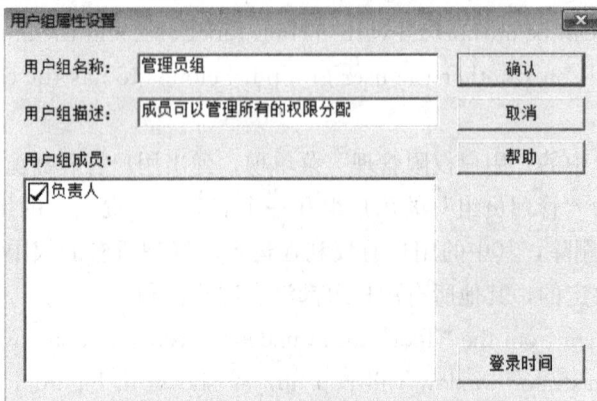

图 1.2.5　用户组属性设置窗口

Figure 1.2.5　User group attribvte setting window

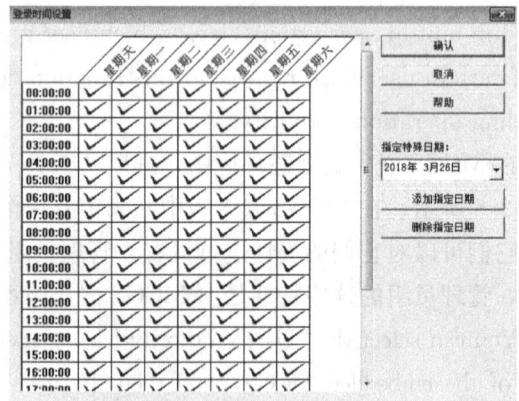

图 1.2.6　登录设置窗口

Figure 1.2.6　Logon setting window

为了更好地保证工程运行的安全、稳定可靠，防止与工程系统无关的人员进入或退出工程系统，MCGS 嵌入版系统提供了对工程运行时进入和退出工程的权限管理。在 MCGS 嵌入版主控窗口中设置"系统属性"，打开窗口如图 1.2.7 所示。点击"权限设置"，可设置工程系统的运行权限，同时可设置系统进入和退出时是否需要用户登录，共有四种组合："进入不登录，退出登录""进入登录，退出不登录""进入不登录，退出不登录"以及"进入登录，退出登录"。

In order to better to ensure the safe，stable and reliable engineering operation and prevent the personnel unrelated to the engineering system from entering into or exiting out the engineering system，the embedded MCGS provides the authority management for entering into and exiting out the engineering system at the time of engineering operation. The "System Attributes" are set in the master control window of embedded MCGS, and the window is shown in Figure 1.2.7. You can set the running authority of engineering system by clicking on the "Authority Setting", and also whether the user logon is required for the system enter and exit, and there are four combinations: "Enter without logon, and exit with logon", "enter with logon, and exit without logon", "enter without logon, and exit without logon", and "enter with logon, and exit without logon".

除了对进入系统和退出系统进行用户权限设置之外，还可以对某些控件进行权限设置，也就是说，可以设置允许某些用户组对某个构件进行操作。标准按钮构件就是可以进行权限设置的构件。双击它弹出标准按钮属性设置窗口，点击左下角的"权限"按钮，如图1.2.8所示，可以在弹出的窗口中进行用户权限设置，如图1.2.9所示。运行时，只有具有操作权限的用户登录时，鼠标在图形对象的上面才变成手状，响应鼠标的按键动作。

Except for the user authority with respect to the system enter and exit, the authorities related to some controls are set, in other words, it can be set that certain user groups are allowed to operate certain component. The standard button is just the component for authority setting. You can double click on the standard button to pop up the attribute setting window of standard button, and then click on the bottom left "Authority" button, as shown in Figure 1.2.8. To pop up the window and set the user authority, as shown in Figure 1.2.9. In the running process, only when the user with the operation authority logs in, the mouse cursor over the graphic object will show the hand form to respond the key action of mouse.

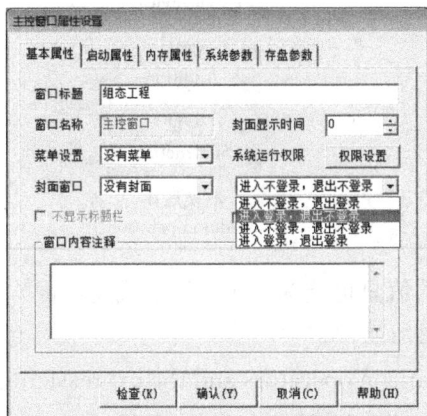

图 1.2.7 主控窗口属性设置窗口

Figure 1.2.7 Attribute setting of master control window

图 1.2.8 标准按钮构件的权限设置

Figure 1.2.8 Authority setting of standard button

(二) MCGS 嵌入版的表达式及基本语句

(Ⅱ) Expressions and basic statements of embedded MCGS

1. MCGS 嵌入版的表达式

1. Expressions of embedded MCGS

由数据对象（包括设计者在实时数据库中定义的数据对象、系统内部数据对象和系统函数）、括号和各种运算符组成的运算式称为表达式，表达式的计算结果称为表达式的值。常量或数据对象是狭义的表达式，这些单个量的值即为表达式的值。运算符包括算术运算符、逻辑运算符和比较运算符。

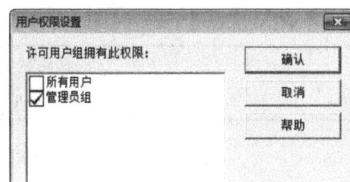

图 1.2.9 用户权限设置窗口

Figure 1.2.9 User authority setting window

The working equations composed of data objects (including the data objects defined by designer in the real-time database, the system's internal data objects and system functions), brackets and various operators are called as the expressions, and the result of expression is called as the expression value. The constants or data objects are the expressions in a narrow sense, and the individual value is just the value of expression. The operators include the arithmetic operators, logical operators and comparison operators.

当表达式中只包含算术运算符，表达式的运算结果为具体的数值时，这类表达式称为算术表达式。算术运算符见表 1.2.1。

When one expression only contains the arithmetic operators and its result is the specific value, such expression is further called as the arithmetic expression. The arithmetic operators are shown in Table 1.2.1

算术运算符   表 1.2.1
Arithmetic operators   **Table 1.2.1**

算术运算符 Arithmetic operators	功能 Function
∧	乘方 Exponentiation
*	乘法 Multiplication
/	除法 Division
\	整除 Exact division
+	加法 Addition
—	减法 Subtraction
Mod	取模运算 Modulo operation

当表达式中包含有逻辑运算符或比较运算符时，表达式的值只可能为 0（条件不成立，假）或非 0（条件成立，真），这类表达式称为逻辑表达式。逻辑运算符见表 1.2.2。

When one expression contains the logic operators or comparison operators and the expression value maybe only zero (the condition is invalid, false) or non-zero (the condition is valid, true), such expression is further called as the logic expression. The logic operators are shown in Table 1.2.2.

逻辑运算符   表 1.2.2
Logical operators   **Table 1.2.2**

逻辑运算符 Logical operators	功能 Function
AND	逻辑与 And
NOT	逻辑非 Not
OR	逻辑或 Or
XOR	逻辑异或 Exclusive or

比较运算符见表 1.2.3。表达式值的类型即为表达式的类型，必须是开关型、数值型、字符型三种类型中的一种。

The comparison operations are shown in Table 1.2.3. The type of expression value is just the expression type，and must be one of switch，value，character types.

比较运算符 Comparison operators	表 1.2.3 Table 1.2.3

比较运算符

Comparison operators

比较运算符 Comparison operators	功能 Function
>	大于 Larger than
>=	大于或等于 Larger than or equal to
=	等于（字符串比较需要使用字符串函数！StrCmp，不能直接使用等于运算符） Equal to（string comparison requires the string function! StrCmp，and can't directly use the＝operator）
<=	小于或等于 Smaller than or equal to
<	小于 Smaller than
—	减法 Subtraction
<>	不等于 Not equal

表达式是构成脚本程序的最基本元素，在 MCGS 嵌入版的组态过程中，也常常需要通过表达式来建立实时数据库对象与其他对象的连接关系，正确输入和构造表达式是 MCGS 嵌入版的一项重要工作。

The expression is the most basic element forming the script program，the expressions are often required to establish the connection relation between the real-time database and other object in the configuring process of embedded MCGS，the correct input and constructing expression are important jobs to the embedded MCGS.

2. MCGS 嵌入版的基本语句

2. Basic statements of embedded MCGS

MCGS 嵌入版的脚本语句包括：赋值语句、条件语句、退出语句、注释语句和循环语句。

The basic statements of embedded MCGS include：Assignment statement，conditional statement，exit statement，comment statement and loop statement.

（1）赋值语句的形式为：数据对象＝表达式。赋值号用"＝"表示，它的具体含义是：把"＝"右边表达式的运算值赋给左边的数据对象。赋值号左边必须是能够读写的数据对象，如：开关型数据、数值型数据以及能进行写入操作的内部数据对象。右边为一个表达式，表达式的类型必须与左边数据对象值的类型相符合，否则系统会提示"赋值语句类型不匹配"的错误信息。

（1）The form of assignment statement：Data object＝expression. Assignment symbol "＝" has the following specific meaning：The operational value of expression on the right of "＝" is assigned to its left data object. There must be writable and readable data objects on the left of assignment symbol，such as：switch type data，numeric type data and writable internal data objects. The right is one expression，and its type must be corresponding to that of left data object，otherwise the system will prompt the error information i. e.

"the assignment statement type isn't matched".

（2）条件语句有如下三种形式：

（2）The conditional statement has the following three forms：

If［表达式］Then［赋值语句或退出语句］

If［expression］Then［assignment statement or exit statement］

If［表达式］Then

If［expression］Then

［语句］

［Statement］

EndIf

If［表达式］Then

If［expression］Then

［语句］

［Statement］

Else

［语句］

［Statement］

EndIf

条件语句中的四个关键字"If"、"Then"、"Else"和"Endif"不分大小写。如拼写不正确，检查程序会提示出错信息。

The four keywords "If"、"Then"、"Else" and "Endif" contained in the conditional statement are case insensitive. In case of spelling mistake, the check program will prompt the error information.

条件语句允许多级嵌套，即条件语句中可以包含新的条件语句，MCGS 脚本程序的条件语句最多可以有 8 级嵌套，为编制多分支流程的控制程序提供方便。

The conditional statement allows the multi-stage nest，namely that the conditional statement can contain new conditional statement（s），there are at most 8-stage nested conditional statements allowed in the script program of MCGS to provide convenience for programming the control procedure with many branch flows.

"IF"语句的表达式一般为逻辑表达式，也可以是值为数值型的表达式，当表达式的值为非 0 时，条件成立，执行"Then"后的语句，否则，条件不成立，将不执行该条件块中包含的语句，而是开始执行该条件块后面的语句。值为字符型的表达式不能作为"if"语句中的表达式。

The expression for "IF" statement is generally logic，but can also be numeric type，when the expression value is non-zero，the condition is true，the statements following "Then" can be executed，otherwise the condition is false，the statements following the condition block，rather than contained in the condition block will be executed. The expression with the character type value can't be used as the expression contained in the "if" statement.

（3）循环语句的结构为：

（3）The structure of loop statement：

While［条件表达式］

While［conditional expression］

...

EndWhile

当条件表达式成立时（非零），循环执行 While 和 EndWhile 之间的语句。直到条件表达式不成立（为零），退出。

When the conditional expression is true (non-zero), the loop execution of statements between While and EndWhile will be performed until that the conditional expression is false (zero), the loop execution will exit.

（4）退出语句为"Exit"，用于中断脚本程序的运行，停止执行其后面的语句。一般在条件语句中使用退出语句，以便在某种条件下，停止并退出脚本程序的执行。

（4）The exit statement is "Exit", which is used for terminating the running of script program and stopping the execution of the following statements. In general, the exit statement is used in the conditional statement such that the execution of script program is stopped and exited under certain condition.

（5）以英文字符单引号''开头的语句称为注释语句，注释语句在脚本程序中只起到注释说明的作用，实际运行时，系统不对注释语句作任何处理。

（5）The statement which begins with the English character '' is called as the comment statement, and such statement can only be used as the explanatory note in the script program and will not be processed by the system in the actual running process.

（三）用户登录操作函数

（Ⅲ）User logon operating functions

在脚本程序中使用用户登录操作函数可以在工程运行时改变操作权限，能够更灵活地管理用户和操作权限。

The user logon operating functions used in the script program can change the operation authority in the engineering running process, and more flexibly manage the users and operation authorities.

用户登录操作函数常用的有如下四个函数：用户登录、退出登录、运行时修改用户密码和用户管理。

The commonly-used user logon operating functions include: Four functions, i.e. user logon, logoff, change user password in the running process, and user management.

1. 用户登录函数！LogOn（）

1. User logon function！LogOn（）

在脚本程序中执行该函数，弹出登录窗口，如图 1.2.10 所示。从用户名下拉框中选取要登录的用户名，在密码输入框中输入用户对应的密码，按回车键或确认按钮，如输入正确则登录成功，函数返回值为 0，否则会出现对应的提示信息。按取消按钮停止登录。登录失败时，函数返回值不等于 0。

The function is executed in the script program, and the logon window will appear, as shown in Figure 1.2.10. You can select the user name which is used for logon from the user name drop-down box, enter the password corresponding to the user name in the password box, and press the enter key or OK button. If the password is true, the system logon will be completed, the return value of this function is 0, otherwise, and the corresponding prompt message will appear. You should press Cancel button to stop the system logon. If the system logon fails, the return value of this function isn't 0.

2. 退出登录函数！LogOff（）

2. LogOff function！LogOff（）

在脚本程序中执行该函数弹出提示框，提示是否要退出登录，"是"表示退出，"否"表示不退出。退出登录成功时，函数返回值为 0，不成功时，返回值不等于 0。

The execution of this function in the script program will pop up the prompt box, i.e. logoff confirmation, "Yes" indicates that the system will be logged off, "No" indicates that the system will not be logged off. After the system logoff, the return value of this function is 0, but if the system logoff fails, the return value isn't equal to 0.

3. 修改密码函数！ChangePassword（）

3. Change password function！ChangePassword（）

在脚本程序中执行该函数弹出修改密码窗口，如图1.2.11所示，供当前登录的用户修改密码。修改密码成功时，函数返回值为0，不成功时，返回值不等于0。

The execution of this function in the script program will pop up the change password window, as shown in Figure 1. 2. 11, which is only used for changing the password of current user. After changing the password, the return value of this function is 0, but if fails, the return value isn't equal to 0.

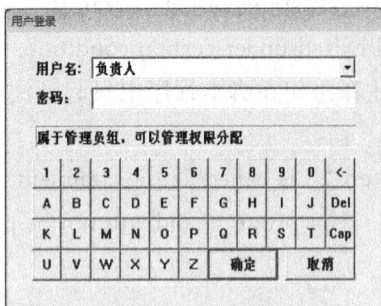

图 1. 2. 10　用户登录窗口

Figure 1. 2. 10　User logon window

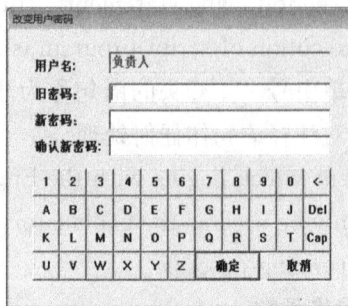

图 1. 2. 11　修改密码窗口

Figure 1. 2. 11　Change password window

4. 用户管理函数！Editusers（）

4. User management function！Editusers（）

在脚本程序中执行该函数弹出用户管理器窗口如前图1.2.3所示，允许在运行时增加删除用户或修改用户的密码和所隶属的用户组。注意：只有在当前登录的用户属于管理员组时，本功能才有效。运行时不能增加、删除或修改用户组的属性。当函数调用成功时，函数返回值为0，不成功时，返回值不等于0。

The execution of this function in the script program will pop up the user administrator window, as shown in Figure 1. 2. 3, which allows adding and deleting user or changing the user password and the user group it belongs to in the running process. Note：This function is valid only if the current user belongs to administrator group. The attributes of user group can't be added, deleted or changed in the running process. If this function is called successfully, its return value is 0, but if fails, the return value isn't equal to 0.

（四）定时器操作函数

（Ⅳ）Timer functions

MCGS嵌入版内置127个系统定时器，即可用的系统定时器范围为1～127。定时器返回时间值为数值型，单位为秒（s）。常用的定时器操作函数有启动定时器函数、停止定时器函数、读取定时器当前值函数和定时器复位函数。

There are 127 built-in timers in the embedded MCGS, i. e. the number of available system timers is between 1 and 127. The time value returned by timers is numeric type in the unit of second（s）. The commonlyused timer functions include the timer run, timer stop, read timer current value and timer reset.

1. 启动定时器函数！TimerRun（定时器号）

1. Timer run function！TimerRun（Timer No.）

函数意义：启动定时器开始工作。

Function meaning：To enable the timer.

返回值：数值型。返回值＝0：调用成功；＜＞0：调用失败。

Return value：Numeric type. Return value＝0：Call success；＜＞0：Call failure.

参数：定时器号。

Parameter：Timer No. .

实例：！TimerRun（1），启动1号定时器工作。

Instance：！TimerRun（1），enable the No. 1 timer.

2. 停止定时器函数！TimerStop（定时器号）

2. Timer stop function！TimerStop（Timer No. ）

函数意义：停止定时器工作。

Function meaning：Disable the timer.

返回值：数值型。返回值＝0：调用成功；＜＞0：调用失败。

Return value：Numeric type. Return value＝0：Call success；＜＞0：Call failure.

参数：定时器号。

Parameter：Timer No. .

实例：！TimerStop（1），停止1号定时器工作。

Instance：！TimerStop（1），disable No. 1 timer.

3. 取定时器当前值函数！TimerValue（定时器号，0）

3. Read timer current value function！TimerValue（Timer No. ，0）

函数意义：取定时器的当前值。

Function meaning：Read the current value of timer.

返回值：将定时器的值以数值型的方式输出（数值格式）。

Return value：Output the current value of timer in the numeric type（numeric format）.

参数：定时器号。

Parameter：Timer No. .

实例：Data3＝！TimerValue（1，0），取定时器1的值给Data3。

Instance：Data3＝！TimerValue（1，0），assign the value of No. 1 timer to Data3.

4. 定时器复位函数！TimerReset（定时器号，数值）

4. Timer reset function！TimerReset（Timer No. ，value）

函数意义：设置定时器的当前值，由第二个参数设定，第二个参数可以是 MCGS 嵌入版变量。

Function meaning：The current value of timer is set by the second parameter which can be the variable of embedded MCGS.

返回值：数值型。返回值＝0：调用成功；＜＞0：调用失败。

Return value：Numeric type. Return value＝0：Call success；＜＞0：Call failure.

参数：定时器号；数值。

Parameter：Timer No. ；Value.

实例：！TimerReset（1，12），设置1号定时器的值为12。

Instance：！TimerReset（1，12），set the value of No. 1 timer to 12.

三、任务实施

Ⅲ. Task Implementation

（一）构建实时数据库

（Ⅰ）To create one real-time database

分析控制要求，可以知道需要新建三个开关型变量控制电机的正反转和停止，同时还需要新建三个开关型变量监控电机的状态和热继电器的状态，其中，热继电器需要添加报警属性以触发报警策略，见表1.2.4。

It can be known from the analysis of control requirements that it is necessary to create three new switch variables to control the positive and negative rotation，and stop of motor，and another three new switch variables to monitor the states of motor and thermal relay，of which the thermal relay should be assigned with the alarm attributes to trigger the alarm strategy，as shown in Table 1. 2. 4.

		数据变量表 Data variables			表 1. 2. 4 Table 1. 2. 4
变量名 Variable name	类型 Type	报警 Alarm
HMI 正转 HMI positive rotation	开关型 Switch type	
HMI 反转 HMI negative rotation	开关型 Switch type	
HMI 停止 HMI Stop	开关型 Switch type	
热继电器 Thermal relay	开关型 Switch type	开关量报警：报警值为 1 Switch value alarm：Alarm value 1
正转状态 State of positive rotation	开关型 Switch type	
正转状态 State of positive rotation	开关型 Switch type	

按照上一任务的操作步骤，根据表 1. 2. 4 构建实时数据库。

The real-time database is created by the steps in the previous task based on Table 1. 2. 4.

（二）组态用户窗口

（Ⅱ）To configure the user window

1. 制作窗口画面和定义动画连接

1. Produce the window page and define the animation connection

按照上一任务的操作步骤，新建监控窗口和报警窗口，并为各窗口创建合适的图形对象，最后定义各对象的动画连接。在这里需要注意的是，电机图形对象只能指示运行和停止两个状态，而电机在正反转时有正转、反转和停止三个状态。这就需要将正转和反转合并为运行状态，也就是说，正转和反转共用一个为 1 的值，电机图形对象就指示为运行状态，二者是或逻辑的关系。那么在设置其属性时，不能直接关联某个数据变量，需要用一个表达式来替代，即正转状态 OR 反转状态，如图 1. 2. 12 所示。

Create the new monitoring window and alarm window by the steps in the previous task，create the proper graphic objects to each window，and finally define the animation connection with each object. Here，it should be noted that the motor graphic object only can indicate the running and stop states，while there are three states i. e. the positive rotation，negative rotation and stop for the positive and negative rotation of motor. The positive rotation and negative rotation should be uniformly considered as the running state，in other words，the positive rotation and negative rotation share the same value 1，the motor graphic object indicates the running state，and there is OR Logic relation between them. Thus，when the attributes are set，certain data variable can't be directly correlated and should be replaced by one expression，i. e. positive rotation state OR negative rotation，as shown in Figure 1. 2. 12.

2. 新建用户组和设置用户权限

2. Create the user group，and set the user authority

根据任务要求，需要添加两个不同的用户，一个是操作员，另一个是工程师，而且因为两个用户的操作权限不同，应该分属两个不同的用户组。

It is necessary to add two different users, operator and engineer based on the task requirements, because the operation authorities of both users are different, they should belong to two different user groups.

在 MCGS 嵌入版的菜单中选择"工具">"用户权限管理"，打开用户管理器。点击下方的用户组列表框的空白处，再点击"新增用户组"按钮，如图 1.2.13 所示。在弹出的用户组属性设置窗口中，新增一个操作员用户组，如图 1.2.14 所示。

Select "Tools" > "User Authority Management" in the menu of embedded MCGS to open the user administrator. Click on the blank inside the below user group list box, and then click on the "Add New User Group" button, as shown in Figure 1.2.13. Add one operator user group the attribute setting window of user group as shown in Figure 1.2.14.

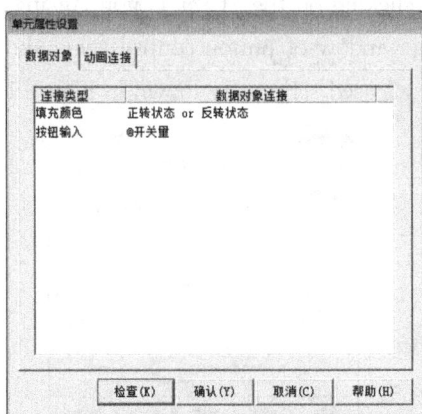

图 1.2.12 电机图形对象的动画连接
Figure 1.2.12 Animation connection of motor graphic object

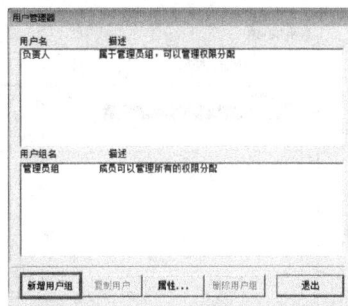

图 1.2.13 在用户管理器中新增用户组
Figure 1.2.13 Add user group in the user administrator

这样，就建立了两个用户组。通过以下步骤在用户组中添加用户：双击"负责人"用户，在弹出的用户属性设置窗口中将其设置为"工程师"用户。按任务要求设置好密码后，设置其隶属于"管理员组"，如图 1.2.15 所示。同样的，点击"新增用户"可新建"操作员"用户，设置其隶属于"操作员组"。

In such way, two user groups are created successfully. The users are added into the user group by the following steps: Double click on the user "Responsible Person", and set the user to "Engineer" in the user attribute setting window. Set the password based on the task requirements, and assign it to the "Administrator Group" as shown in Figure 1.2.15. Similarly, click on "Add User" to create the "Operator" user and assign it to the "Operator Group".

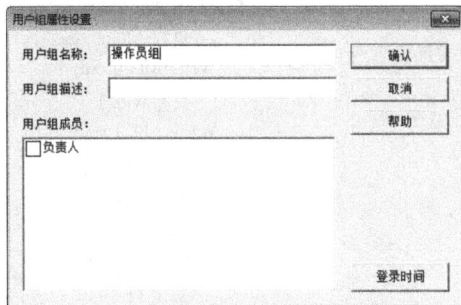

图 1.2.14 设置操作员组属性
Figure 1.2.14 set the attributes of operator group

图 1.2.15 设置工程师用户
Figure 1.2.15 Set the engineer user

只有以工程师身份登录时才可以通过人机界面远程操作电机的正反转和停止，因此需要对远程控制电机的三个按钮构件进行权限设置。在按钮构件的属性设置窗口中，点击左下角的"权限"，在弹出的窗口中对可以操作该按钮的用户组进行设置，如图1.2.16所示。

The user who logged on the system with the engineer identity can remotely operate the positive and negative rotation，and stop of motor，therefore it is necessary to set the user authority of three button controls on the remotely-controlled motor. You should click on the bottom left "Authority" in the attribute setting window of button control，and set the user groups which can operate the button control in the window，as shown in Figure 1.2.16.

点击"用户重新登录"按钮，可以弹出用户登录窗口，重新进行用户登录。这需要使用到用户登录函数。在按钮的属性设置窗口中，选择脚本程序选项卡，在抬起脚本中，输入用户登录函数即可，如图1.2.17所示。

And the user logon window will appear for logon again after clicking on the "User Logon Again" button. The user logon function is needed in this step. In the attribute setting window of button control，you should select the script program tab，and then enter the user logon function in the released script，as shown in Figure 1.2.17.

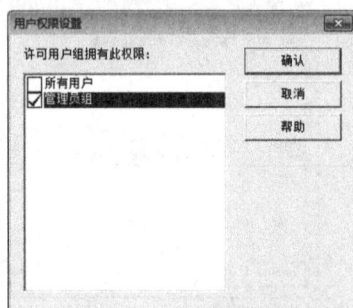

图 1.2.16 按钮构件的用户权限设置

Figure 1.2.16 Set the user authority of button control

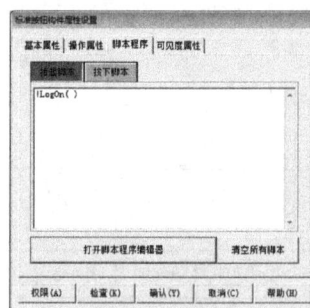

图 1.2.17 编写按钮脚本程序

Figure 1.2.17 Write the script program of button

（三）组态设备窗口

（Ⅲ）To configure the device window

按照上一任务的操作步骤组态设备窗口，按表1.2.5创建PLC和实时数据库变量的通道连接。

Configure the device window by the steps in the previous task，and create the channel connection between PLC and real-time database variable，as shown in Table 1.2.5.

设备通道和连接变量　　　　　　　　　　　　　　　　　　表 1.2.5

Device channels and connection variables　　　　　　　　　　Table 1.2.5

连接变量 Connection variables	通道名称 Channel name
HMI 正转 Positive rotation	只写 M0.0 Write M0.0 only
HMI 反转 Negative rotation	只写 M0.1 Write M0.1 only
HMI 停止 HMI Stop	只写 M0.2 Write M0.2 only
热继电器 Thermal relay	只读 I0.3 Read I0.3 only
正转状态 State of positive rotation	只读 Q0.0 Read Q0.0 only
反转状态 State of negative rotation	只读 Q0.1 Read Q0.1 only

（四）组态策略窗口

（Ⅳ）To configure the strategy window

本任务需要组态一个启动策略，一个报警策略和一个循环策略。由于系统启动时需要进行用户登录，并同时进入监控画面，因此需要组态启动策略，弹出用户登录窗口。启动策略为系统固有策略，在MCGS嵌入版系统开始运行时自动被调用一次。在策略组态窗口中，选择策略行最右侧的策略块为脚本程序，表示当启动策略调用时执行一段脚本程序。在脚本程序编辑栏中，只有写入用户登录函数和打开窗口函数，才能进行用户登录和打开监控画面。如图1.2.18所示。

This task requires configuring one startup strategy, one alarm strategy and one loop strategy. Because the user logon is required for the system startup and meanwhile system will show the monitoring interface, it is necessary to configure the startup strategy to pop up the user logon window. The startup strategy is the inherent strategy of system, and will be automatically called once at the beginning of running the embedded MCGS. If the strategy block on the rightmost of the strategy line in the strategy configuration window is selected as the script program, one section of script program will be executed when the startup strategy is called. The user logon and monitoring interface can be enabled only after writing the user logon and open window functions into the formula bar of script program. As shown in Figure 1.2.18.

当过载时，热继电器动作，实时数据库中的对应数据发生报警，触发报警策略，执行脚本程序。在脚本程序中，需要打开报警窗口和启动定时器。如图1.2.19所示。

In case of overload, the thermal relay is enabled, and the corresponding data in the real-time database shows the alarm state and triggers the alarm strategy to execute the script program. In the script program, it is needed to open the alarm window and enable the timer. As shown in Figure 1.2.19.

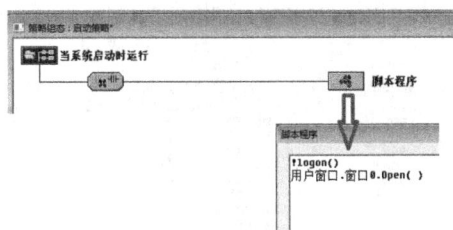

图1.2.18 启动策略组态

Figure 1.2.18 Startup strategy configuration

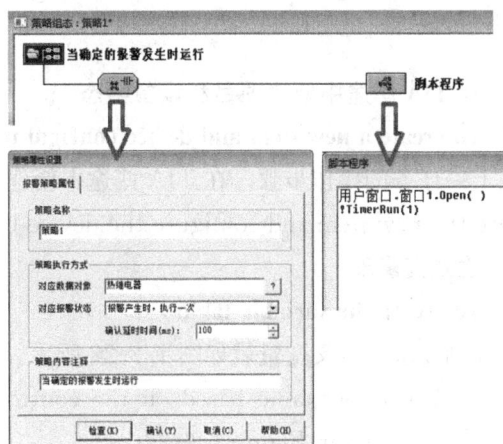

图1.2.19 报警策略组态

Figure 1.2.19 Alarm strategy configuration

当定时器到达指定时间时，报警窗口关闭。判断定时器当前值是否等于设定值依靠循环策略来实现。循环策略为系统固有策略，也可以由用户在组态时创建，在MCGS嵌入版系统运行时按照设定的时间循环运行。在一个应用系统中，用户可以定义多个循环策略，如图1.2.20所示。

When the timer reaches the designated time, the alarm window is closed. The loop strategy can be used for judging whether the current value of timer is equal to the set value. The loop strategy is the inherent strategy of system and can also be created by user in the configurations; the embedded MCGS will circularly run based on the scheduled time. The loop strategy is the inherent strategy of system and can also be created by user in the configurations；the embedded MCGS will circularly run based on the scheduled time. In one application system，user can define many loop strategies，as shown in Figure 1.2.20.

（五）连接 I/O 设备

（Ⅴ）To connect I/O device.

按照控制要求，I/O 接线图如图 1.2.21 所示。

Based on the control requirements，the I/O wiring diagram is shown in Figure 1.2.21.

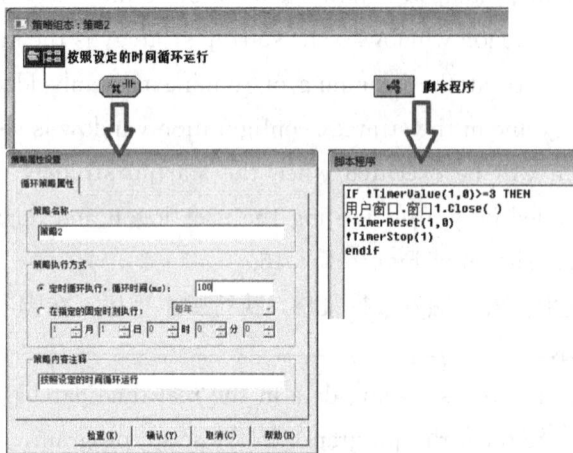

图 1.2.20 循环策略组态

Figure 1.2.20 Loop strategy configuration

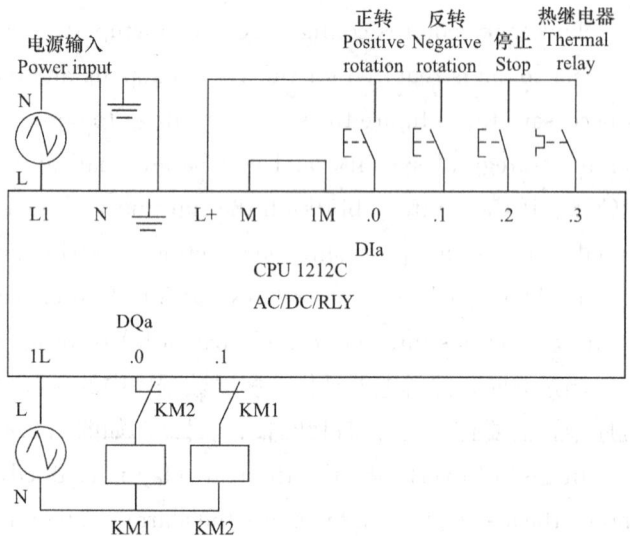

图 1.2.21 I/O 接线图

Figure 1.2.21 I/O wiring diagram

（六）在 TIA 博途中新建项目和设备组态

（Ⅵ）To create a new item and device configuration in TIA Portal

按照上一任务的操作步骤，在 TIA 博途中新建项目和设备组态。

Create the new item and configure the device by the steps in the previous task.

（七）建立变量表

（Ⅶ）To create the variable table

按照控制要求，定义变量表如图 1.2.22 所示。

Define the variable table based on the control requirements，as shown in Figure 1.2.22.

（八）编写梯形图程序

（Ⅷ）To write the ladder logic program

梯形图程序如图 1.2.23 所示。

The ladder diagram program is shown in Figure 1.2.23.

图 1.2.22 变量表

Figure 1.2.22 Variable table

（九）调试

（Ⅸ）Debugging

完成以上步骤后，将编辑好的 MCGS 嵌入版工程和 TIA 博途项目分别下载至触摸屏和 PLC，再用编程线将二者连接起来，这样三相电机启停的远程控制和运行监控系统就建立起来了。按照控制要求逐项对系统进行调试，直到所有功能都实现为止。

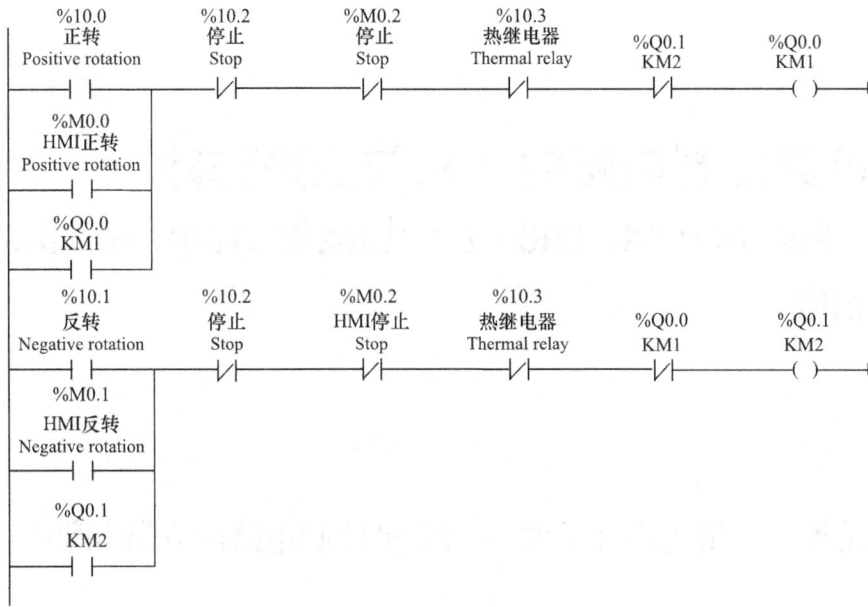

图 1.2.23 梯形图程序

Figure 1.2.23 Ladder logic program

After the above steps, the embedded MCGS engineering and TIA Portal items have been edited are respectively downloaded to the touch screen and PLC, the both are connected with the programming line such that the remote startup and stop control and running monitoring system of three-phase motor is finally created. The system is debugged item by item according to the control requirements until that all functions are realized.

项目一思考题

Questions for Item I

1. S7-1200 的硬件主要由哪些部件组成?

1. What are the main components in the hardware for S7-1200?

2. S7-1200 的 CPU 模块默认的 IP 地址和子网掩码是什么?

2. What is the default IP address and subnet mask for the CPU module of S7-1200?

3. MCGS 嵌入版组态软件中数据对象有哪几种类型?

3. What are the types of data objects in the configuration software of embedded MCGS?

项目二 机床控制电路的认知与故障排除
Item Ⅱ Awareness and Troubleshooting of Machine-tool Control Circuit

任务一 带离心开关的三相电机控制电路的安装与调试

Task Ⅰ Installation and Debugging of Control Circuit for Three-phase Motor with Centrifugal Switch

一、任务分析
Ⅰ. Task Analysis

将图 2.1.1 所示电路图改造为用 PLC 控制的电气系统，要求如下：

To transform the circuit diagram as shown in Figure 2.1.1 into the electrical system controlled by PLC based on the following requirements：

（1）用按钮实现图中电路的反接制动功能。

（1）The function of reverse braking circuit in the figure is realized with the button.

（2）能够记录电机的运行次数，当运行次数达到 5 次时，指示灯 HL1 以 2Hz 闪烁。按下复位按钮后，重新开始计数。

（2）The running times of motor can be recorded，when the number of running times reaches 5，indicator HL1 will flash at the frequency of 2Hz. The counting operation will be restarted after pressing the reset button.

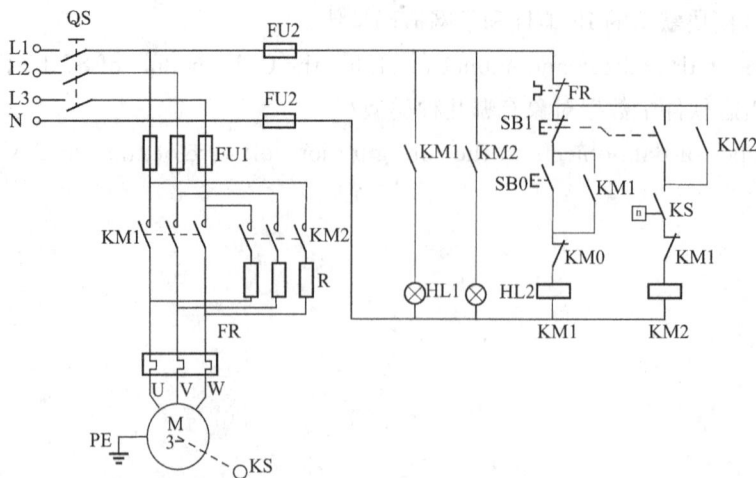

图 2.1.1 带离心开关的反接制动电路

Figure 2.1.1 Reverse braking circuit with centrifugal switch

为了完成本任务，需要了解三相异步电机的几种制动方法和 S7-1200 系统存储器、时钟存储器和计数器的使用方法。

For the purpose of this task，it is firstly necessary to learn about the braking methods of three-phase a-synchronous motor，and the use of S7-1200 system memory，clock memory and timer.

二、相关知识技能
Ⅱ. Relevant Knowledge & Skills

（一）三相异步电机的制动
（Ⅰ） Braking of three-phase asynchronous motor

所谓制动，就是给电机一个与转动方向相反的转矩使它迅速停转（或限制其转速）。制动的方法一般有两类：机械制动和电力制动。

The braking is intended to impose the rotary torque opposite to the rotating direction of the motor to make it quickly stop (or limit the rotating speed of motor). In general, there are two braking methods；Mechanical braking and electrical braking.

1. 机械制动
1. Mechanical braking

机械制动的原理是当电机的定子绕组断电后，利用机械装置使电机立即停转。机械制动的方法有电磁抱闸制动和电磁离合器制动。

The mechanical braking principle is to use the mechanical device to immediately stop the motor after the stator winding of motor is powered off. The mechanical braking methods include the electromagnetic braking and electromagnetic clutch braking.

电磁抱闸的结构如图 2.1.2 所示。

The structure of electromagnetic brake is shown in Figure 2.1.2.

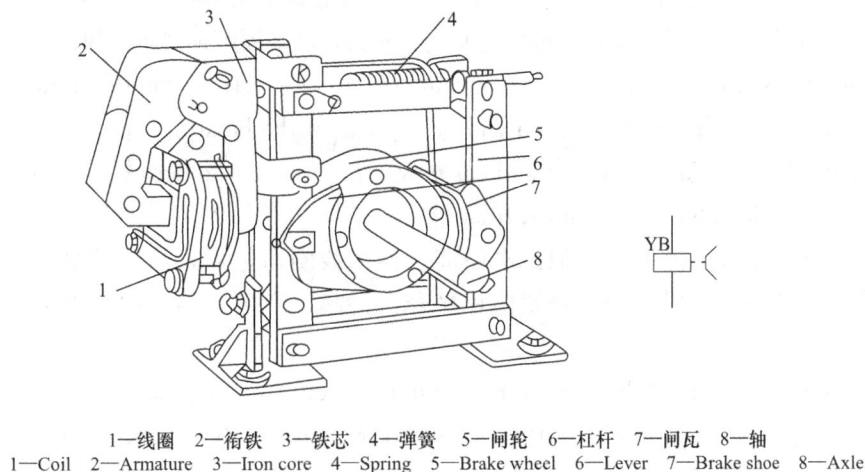

1—线圈　2—衔铁　3—铁芯　4—弹簧　5—闸轮　6—杠杆　7—闸瓦　8—轴
1—Coil　2—Armature　3—Iron core　4—Spring　5—Brake wheel　6—Lever　7—Brake shoe　8—Axle

图 2.1.2　电磁抱闸的结构
Figure 2.1.2　Structure of electromagnetic brake

电磁抱闸分为断电制动型和通电制动型。断电制动的控制电路如图 2.1.3（a）所示，当交流接触器 KM1 失电时，电磁抱闸线圈 YB 失电，闸瓦被弹簧拉下，从而抱住闸轮，电机无法转动，处于制动状态。当 KM1 得电时，YB 得电，衔铁通过杠杆将闸瓦抬起，电机可以转动。通电制动的控制电路如图 2.1.3（b）所示，当按下停止按钮 SB2 时，交流接触器 KM2 得电，电磁抱闸线圈 YB 得电，衔铁将闸瓦拉下抱住闸轮制动；当 YB 线圈失电时，闸瓦被弹簧拉起，与闸轮分开，电机可以转动。

The electromagnetic brake is further divided into the power-off braking and power-on braking. The control circuit of power-off braking is shown in Figure 2. 1. 3 (*a*). When AC contactor KM1 is de-energized，the electromagnetic brake coil YB is also de-energized and the brake shoe is pulled down by the spring，thus contracting the brake wheel and enabling the motor to stop under the braking status. When KM1 is energized，YB is also energized and the brake shoe is lifted by the armature via the lever，thus enabling the motor to rotate. The control circuit of power-on braking is shown in Figure 2. 1. 3 (*b*). When the stop button SB2 is pressed，AC contactor KM2 and electromagnetic brake coil YB is energized，the brake shoe is pulled down by the armature to contract the brake wheel and stop the motor；When YB is de-energized，the brake shoe is pulled up by the spring to separate from the brake wheel，thus enabling the motor to rotate.

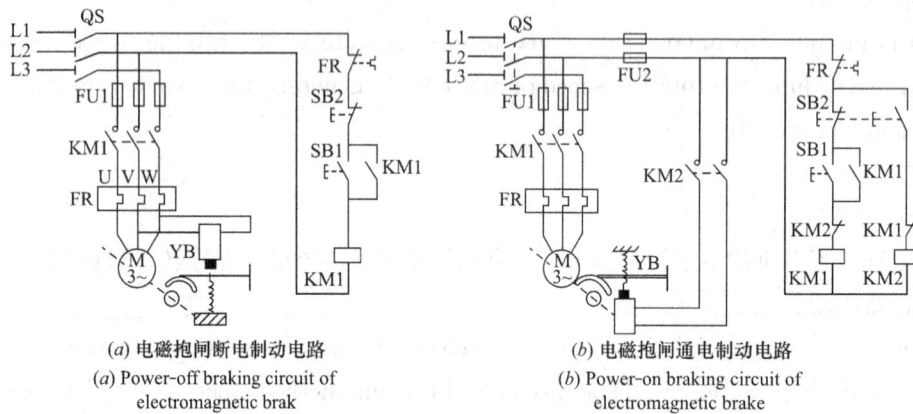

(*a*) 电磁抱闸断电制动电路
(*a*) Power-off braking circuit of electromagnetic brak

(*b*) 电磁抱闸通电制动电路
(*b*) Power-on braking circuit of electromagnetic brake

图 2. 1. 3　电磁抱闸控制电路

Figure 2. 1. 3　Control circuit of electromagnetic brake

电磁抱闸制动的优点：电磁抱闸制动，制动力强，特别是断电制动型抱闸，广泛应用在起重设备上。它安全可靠，不会因突然断电而发生事故。缺点：电磁抱闸体积较大，制动器磨损严重，快速制动时会产生振动。

The electromagnetic brake has the following advantages；The electromagnetic brake has strong braking force，particularly the power-off brake has been widely applied on the hoisting equipment. It is safe and reliable，and no accident will happen due to the sudden power failure. Disadvantages；Large volume，serious brake wear，and vibration in the event of quick braking.

电磁离合器制动的原理与电磁抱闸类似。电磁离合器的线圈通电后，衔铁被吸合，动、静摩擦片分开，电机可以转动。当线圈断电时，动、静摩擦片接合在一起，产生足够大的摩擦力使电机断电后立即制动。

The principle of electromagnetic clutch braking is similar to electromagnetic brake. After the coil of electromagnetic clutch being energized，the armature is picked up and the dynamic and static friction plates are separated，thus enabling the motor to rotate. After the coil being de-energized，the dynamic and static friction plates are contacted to generate the enough friction force to immediately stop the motor in the event of power interruption.

2. 电力制动

2. Electric braking

电力制动的原理是电机产生一个和电机实际旋转方向相反的电磁转矩，使电机迅速停转。电力制动常用的方法有：反接制动、能耗制动、电容制动和回馈制动。

The principle of electrical braking is to enable the motor to produce an electromagnetic torque opposite to its actual rotating direction to quickly stop rotation. The commonly-used electrical braking methods include: Reverse braking, dynamic braking, capacitor braking and feedback braking.

（1）反接制动原理是在停车时，把电机接成反向运行，则其定子旋转磁场便反向旋转，在转子上产生的电磁转矩亦随之变为反向，成为制动转矩。当电机转速接近零值时，应立即切断电机电源。否则电机将反转。为此，在反接制动设施中，为保证电机的转速被制动到接近零值时，能迅速切断电源，防止反向启动，常利用速度继电器来自动并及时地切断电源。离心开关就是一种速度继电器。

（1）The reverse braking principle is to connect the motor to make it rotate in the reverse direction at the time of parking, in such case its stator's rotating magnetic field will rotate in the reverse direction and the direction of electromagnetic torque produced on its rotor will also be reversed, thus producing the braking torque. When the rotating speed of motor is close to zero, the power supply should be immediately cut off. Otherwise, the motor will rotate in the reverse direction. Therefore, the reverse braking facilities often use the speed relay to automatically and timely cut off the power supply in order to ensure the power supply can be quickly cut off when the rotating sped of motor is close to zero and prevent the motor from reverse startup. The centrifugal switch is a kind of speed relay.

反接制动的控制电路如图2.1.4所示。当按下停止按钮SB2时，正向运行接触器KM1失电，反向运行接触器KM2得电，电机反向运行，进行制动。当速度接近零时，离心开关KS的常开触点断开，KM2失电，切断反转电路。在这里需要注意的是，在电机处于正转运行状态时，离心开关KS动作，常开触点闭合，也就是说，在SB2按下时，KM2是可以得电的。由于电路图中的触点状态为断电时的状态，因此在图中画的是常开触点。

The control circuit of reverse braking is shown in Figure 2.1.4. When the stop button SB2 is pressed, the KM1 for positive rotation contactor is de-energized and KM2 for negative rotation contactor is energized, and the motor will run in the reverse direction for braking. When the speed is close to zero, the normally open contact of centrifugal switch KS is disconnected, KM2 is de-energized, and the reverse rotation circuit is cut off. Here, it should be noted that when the motor is running in the positive direction, the centrifugal switch KS is actuated and the normally open contact is closed, in other words, KM2 can be energized when SB2 is pressed. Because the contact status as shown in the circuit diagram is corresponding to the status of power failure, the contact in the circuit diagram is a normally open contact.

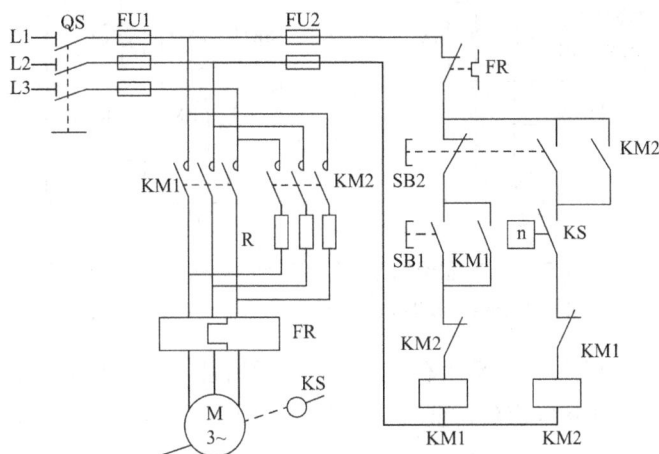

图2.1.4　反接制动控制电路

Figure 2.1.4　Control circuit of reverse braking

反接制动的优点是制动力强、停转迅速、无需直流电源；缺点是制动准确性差，冲击力强，电能消耗多。因此适用于10kW以下小容量的电机，制动要求迅速、系统惯性不大，不经常启动与制动的设备，如铣床、镗床、中型车床等主轴的制动控制。

The reverse braking has the advantages of strong braking force, quick deceleration and no DC power supply; its disadvantages include poor braking accuracy, strong impact, and large energy consumption. Therefore, it is applicable to the braking control of shafts used in this equipment with the need for quickly braking small-capacity motors below 10kW, smaller system inertia, and infrequent startup and braking, such as miller, boring machine and medium-sized lathe.

（2）能耗制动的原理是电机切断交流电源后，立即在定子线组的任意两相中通入直流电，利用转子感应电流受静止磁场的作用以达到制动目的。典型的能耗制动控制电路有无变压器半波整流能耗制动电路和有变压器的全波整流能耗制动电路。

（2）The principle of dynamic braking is that DC power supply is immediately connected to any two phases of stator winding after cutting off AC power supply to the motor, and then the braking is realized by the rotor induced current affected by the stationary magnetic field. The typical control circuits of dynamic braking include the half-wave rectification dynamic braking circuit without the transformer and the full-wave rectification dynamic braking circuit with the transformer.

无变压器半波整流单向启动能耗制动控制电路如图2.1.5所示。该线路采用单只晶体管半波整流器作为直流电源，所用附加设备较少，线路简单，成本低，常用于10kW以下小容量电机，且对制动要求不高的场合。

The half-wave rectification one-way startup dynamic braking circuit without the transformer is shown in Figure 2.1.5. The control circuit is connected to single-transistor half-wave rectifier as its DC power supply, and often used for small-capacity motors below 10kW and situations with no strict braking requirement due to the less additional devices, simple circuit and low cost.

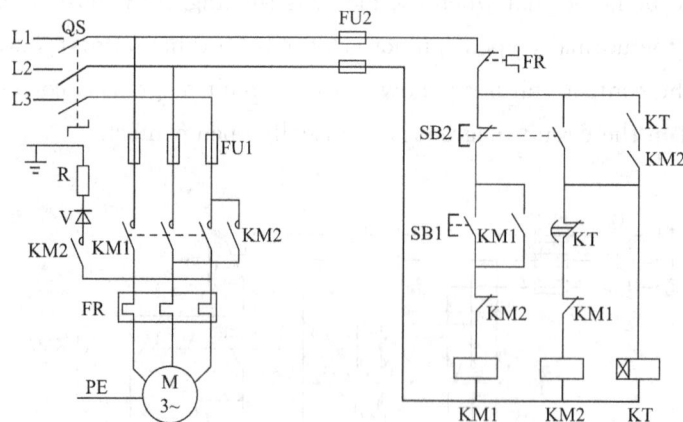

图 2.1.5　无变压器半波整流单向启动能耗制动控制电路

Figure 2.1.5　Control circuit of half-wave rectification one-way startup dynamic braking without the transformer

对于10kW以上容量较大的电机，多采用有变压器全波整流能耗制动控制电路，如图2.1.6所示。其中直流电源由单相桥式整流器VC供给，TC是整流变压器，电阻R是用来调节直流电流的，从而调节制动强度。整流变压器原边与整流器的直流侧同时进行切换，有利于提高触头的使用寿命。

For the large-capacity motors above 10kW, the control circuit of full-wave rectification dynamic braking with the transformer is often used, as shown in Figure 2.1.6. Of which DC power supply is offered by single-phase bridge rectifier VC，TC is rectifier transformer，resistance R is used for regulating DC current，so as to regulate the braking strength． The primary winding of rectifier transformer is switched over together with the DC side of rectifier，which is conductive to extend the useful life of contact．

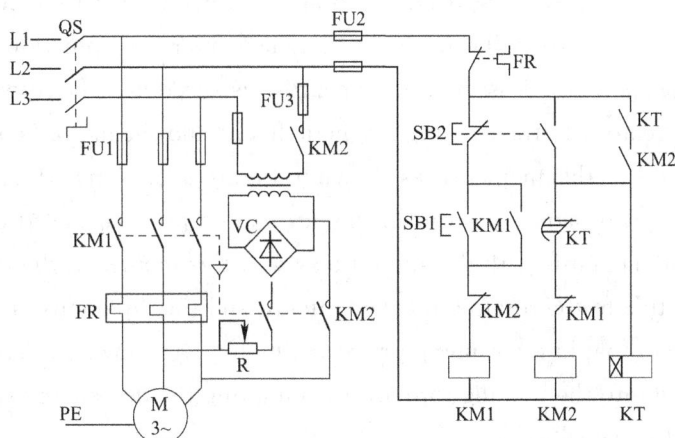

图 2.1.6 有变压器全波整流能耗制动控制电路

Figure 2.1.6 Control circuit of full-wave rectification dynamic braking with the transformer

能耗制动的优点：制动准确、平稳，且能量消耗小。缺点：需要直流电源，设备费用较高，制动力较弱，低速时制动力矩小。适用于要求制动准确、平稳的场合。

Advantages of dynamic braking：Accurate braking，stability，and low energy consumption． Disadvantages：DC power supply，higher equipment costs，relatively poor braking force，and small braking torque at low speed． Applicable to the motors with accurate and stable braking．

（3）电容制动

（3）Capacitor braking

电容制动的原理是当电机切断电源后，通过立即在电机定子绕组的出线端接入电容器迫使电机迅速停转。当旋转着的电机断开交流电源时，转子内仍有剩磁，随着转子的惯性转动，有个随着转子转动的旋转磁场，这个磁场切割定子绕组产生感生电动势，并通过电容器回路形成感生电流，该电流产生的磁场与转子绕组中感生电流互相作用，产生一个制动力矩，使得电机受制动停转。电容制动控制电路如图 2.1.7 所示。

The principle of capacitor braking is that the capacitor is immediately connected to the outgoing terminal of stator winding and force to quickly stop the motor after cutting off the power supply to the motor． When the AC power supply to the motor is cut off in the running process，there is still residual magnetism in the rotor，and the rotating magnetic field will cut the stator winding to produce the induced electromotive force and generate the induced current along the capacitor circuit with the inertia rotation of the rotor． The magnetic field caused by such induced current interacts with the induced current in the rotor winding to generate the braking torque and stop the motor under the braking force． The control circuit of capacitor braking is shown in Figure 2.1.7．

（4）回馈制动

（4）Feedback braking

回馈制动又叫再生发电制动，主要用在起重机械和多速异步电机上。当起重机在高处开始下放重物时，电机转速 n 小于同步转速 n_1，转子相对于旋转磁场逆时针切割磁感线，这时电机处于电动运行状态，转差率 $0<s<1$，如图 2.1.8（a）。但由于重力作用，在重物下放过程中，会使电机的转速大于同步转速 n_1，这时电

机处于发电运行状态，转差率 $s<0$，转子相对于旋转磁场顺时针切割磁感线，其转子电流和电磁转矩的方向都与电动运行时相反，如图 2.1.8（b）所示。电磁力矩变为制动力矩，从而限制了重物的下降速度，保证了设备和人身安全。

The feedback braking also called as regenerative braking is mainly used on the hoisting equipment and multispeed asynchronous motor. When the hoisting equipment starts to put down weights from the high place，the motor's rotating speed n is less than the synchronous speed n_1，the rotor cuts the magnetic induction lines counterclockwise relative to the rotating magnetic field，and the motor is in the inching running status with the slip ratio $0<s<1$ at this moment，as shown in Figure 2.1.8 (a). But the rotating speed of motor may be larger than the synchronous speed n_1 in the descending process due to the action of gravity，the motor is the generative running status with the slip ratio $s<0$ at this moment，the rotor cuts the magnetic induction lines clockwise relative to the rotating magnetic field，and the directions of rotor current and electromagnetic torque is opposite to that in the running process of motor，as shown in Figure 2.1.8 (b). The electromagnetic torque is turned into the braking torque，thus limiting the descending speed of weights and guaranteeing the equipment and personal safety.

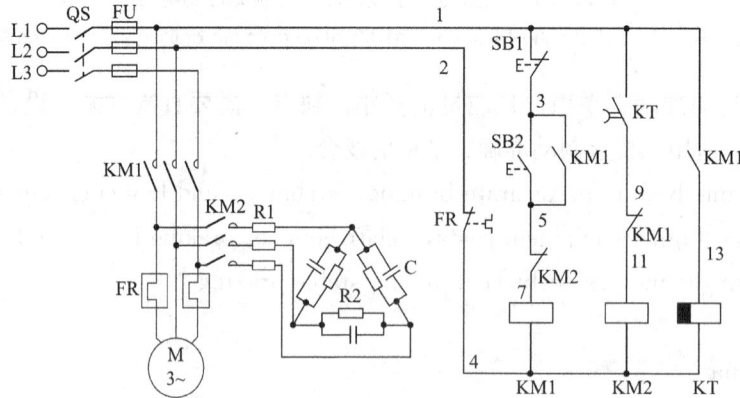

图 2.1.7　电容制动控制电路

Figure 2.1.7　Control circuit of capacitor braking

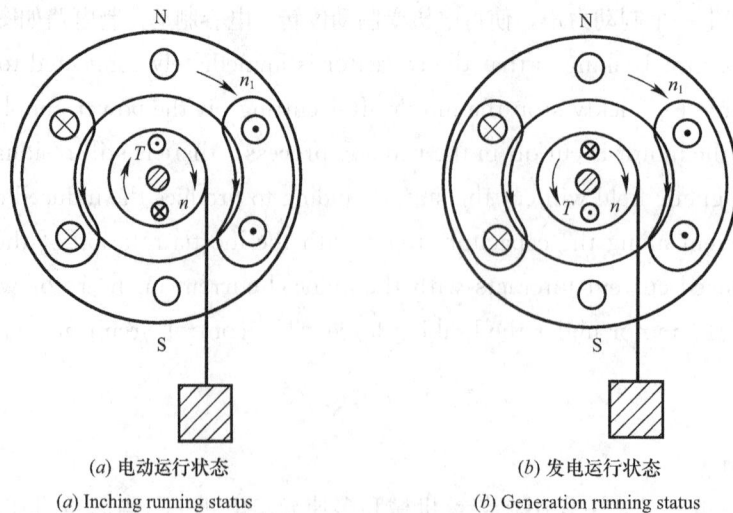

(a) 电动运行状态　　　　　　　　　(b) 发电运行状态

(a) Inching running status　　　　　　(b) Generation running status

图 2.1.8　回馈制动原理

Figure 2.1.8　Principle of feedback braking

（二）西门子 S7-1200 的系统存储器和时钟存储器

（Ⅱ）System memory and clock memory of Siemens S7-1200

在进行 S7-1200 的设备组态时，勾选 CPU 模块的属性中"系统和时钟存储器"的可启用系统存储器和时钟存储器的相应字节，方便编程中使用，如图 2.1.9 所示。系统存储器组态了一个字节，其中的每个字节会在发生特定事件时启用（值＝1），见表 2.1.1。

The corresponding bytes to be enabled of the system memory and clock memory of "System and Clock Memory" in the attributes of CPU module are selected to be used in the programming process at the time of configuring S7-1200, as shown in Figure 2.1.9. The system memory configures one byte, of which each bit will be enabled in the event of a certain event (value＝1), as shown in Table 2.1.1.

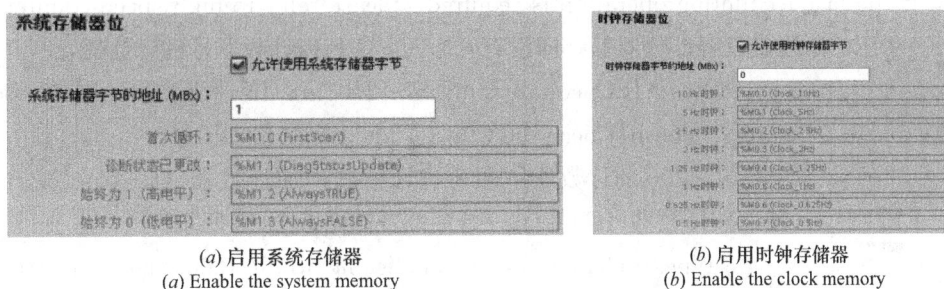

(a) 启用系统存储器
(a) Enable the system memory

(b) 启用时钟存储器
(b) Enable the clock memory

图 2.1.9 启用系统和时钟存储器

Figure 2.1.9 Enable the system and clock memory

系统存储器的功能 表 2.1.1

Functions of system memory Table 2.1.1

位 Bit	7 6 5 4	3	2	1	0
功能 Function	保留值 0 Reserved value 0	始终熄灭值 0 Constant extinguishing value	常开值 1 Normally open value 1	诊断状态指示 Diagnostic status indication 1：变化 1：Change 0：无更改 0：Unchanged	首次扫描指示 First scanning indication 1：启动后首次扫描 1：First scanning after the startup 0：不是首次扫描 0：Not first scanning

时钟存储器组态了一个字节，该字节的各个位分别按固定的时间间隔循环启用和禁用。每个时钟位都会在相应的 M 存储器位产生一个方波脉冲。见表 2.1.2。

The clock memory configures one byte, and it's all bits are enabled and disabled circularly with the fixed time interval. Each clock bit will generate one square-wave pulse at the corresponding M memory bit. As shown in Table 2.1.2.

时钟存储器的功能 表 2.1.2

Functions of clock memory Table 2.1.2

位 Bit	7	6	5	4	3	2	1	0
周期（s） Period（s）	2.0	1.6	1.0	0.8	0.5	0.4	0.2	0.1
频率（Hz） Frequency（Hz）	0.5	0.625	1	1.25	2	2.5	5	10

（三）西门子 S7-1200 的计数器

（Ⅲ）Counter of Siemens S7-1200

S7-1200 的计数器为 IEC 计数器，用户程序中可以使用的计数器数量仅受 CPU 的存储器容量限制。

S7-1200 is equipped with IEC counter，and the number of available counters in the user program is only limited by the memory capacity of CPU.

这里所说的计数器是软件计数器，最大计数速率受所在 OB 的执行速率限制。指令所在 OB 的执行频率必须足够高，以检测输入脉冲的所有变化，如果需要更快的计数操作，请参考高速计数器（HSC）。

The said counter is software counter，and the maximum count rate is limited by the execution rate of OB. The execution rate of OB where the instruction is located must be high enough to detect all changes in the input pulse，if the faster counting operation is required，please refer to high-speed counter.

注：S7-1200 的 IEC 计数器没有计数器号（即没有 C0、C1 这种带计数器号的计数器）。

Notes；IEC counters of S7-1200 are free of counter number（i. e. any counter doesn't bear the C0，C1 and other marks）.

S7-1200 的计数器包含 3 种计数器，加计数器（CTU）、减计数器（CTD）和加减计数器（CTUD）。指令位置如图 2.1.10 所示。

3 types of counters are applicable to S7-1200，i. e. up counter（CTU），down counter（CTD）and up-down counter（CTUD）. The instruction location is shown in Figure 2.1.10.

图 2.1.10　计数器指令的位置

Figure 2.1.10　Location of counter instruction

S7-1200 的计数器是函数块，调用它们时，需要生成保存计数器数据的背景数据块。将计数器直接拖入工作区，自动生成计数器的背景数据块，如图 2.1.11 所示，该块位于"系统块＞程序资源"中。在生成一个计数器时，需要在指令中修改计数值的数据类型。不同数据类型的计数器计数范围不同，见表 2.1.3。

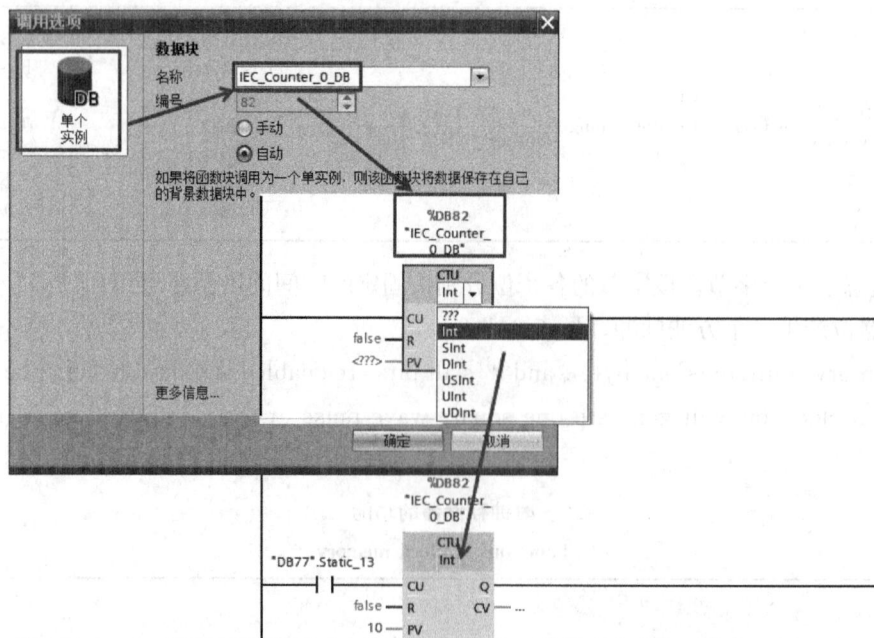

图 2.1.11　生成一个计数器

Figure 2.1.11　Generate one counter

The counters of S7-1200 are function blocks, and when these function blocks are called, it is necessary to generate the background block to save the counter data. The counter is directly dragged to the working area to automatically generate the background data block of counter, as shown in Figure 2.1.11, and the block is located in "system block>program resource". It is necessary to modify the data type of count value when one counter is generated. For different data types, the counting range of counter is also different, as shown in Table 2.1.3.

不同数据类型计数器的计数范围 表 2.1.3

Counting range of counters with different data types Table 2.1.3

数据类型 Data type	计数范围 Counting range
SINT	$-128\sim127$
INT	$-32768\sim32767$
DINT	$-2147483648\sim2147483647$
USINT	$0\sim255$
UINT	$0\sim65535$
UDINT	$0\sim4294967295$

计数器引脚功能见表 2.1.4。

The pin function of counter is shown in Table 2.1.4.

计数器的引脚功能 表 2.1.4

Pin function of counter Table 2.1.4

名称 Name	说明 Description	数据类型 Data type	备注 Remarks
CU	加计数器输入脉冲 Input pulse of CTU	BOOL	仅出现在 CTU, CTUD Only in CTU and CTUD
CD	减计数器输入脉冲 Input pulse of CTD	BOOL	仅出现在 CTD, CTUD Only in CTD and CTUD
R	CV 清 0 CV zero clearing	BOOL	仅出现在 CTU, CTUD Only in CTU and CTUD
LD	CV 设置为 PV Set CV to PV	BOOL	仅出现在 CTD, CTUD Only in CTD and CTUD
PV	预设值 Preset value	整数 Integer	
Q	输出位 Output bit	BOOL	仅出现在 CTU, CTD Only in CTU and CTD
QD	输出位 Output bit	BOOL	仅出现在 CTUD Only in CTUD
QU	输出位 Output bit	BOOL	仅出现在 CTUD Only in CTUD
CV	计数值 Count value	整数 Integer	

加计数器（CTU）的工作原理如图 2.1.12 所示，可归纳总结为：

The working principles of CTU is shown in Figure 2.1.12, and can be summarized as follows:

（1）每当 CU 从 "0" 变为 "1"，CV 增加 1；

(1) CV is increased by 1 whenever CU is changed to "1" from "0";

（2）当 CV＝PV 时，Q 输出"1"，此后每当 CU 从"0"变为"1"，Q 保持输出"1"，CV 继续增加 1 直到达到计数器指定的整数类型的最大值；

（2）When CV is equal to PV，Q outputs "1"，subsequently Q continues outputting "1" and CV is increased by 1 whenever CU is changed to "1" from "0" until to reach the designated maximum integer value of counter；

（3）在任意时刻，只要 R 为"1"时，Q 输出"0"，CV 立即停止计数并回到 0。

（3）Q will output "0"，and CV immediately stops counting and returns to 0 as long as R is "1" at any time.

减计数器（CTD）的工作原理如图 2.1.13 所示，可归纳总结为：

The working principles of CTD is shown in Figure 2.1.13，and can be summarized as follows：

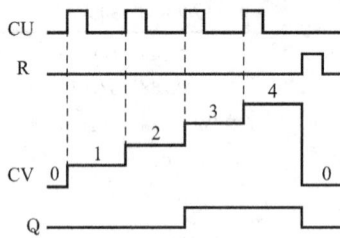

图 2.1.12　加计数器的工作原理
Figure 2.1.12　Working principle of CTU

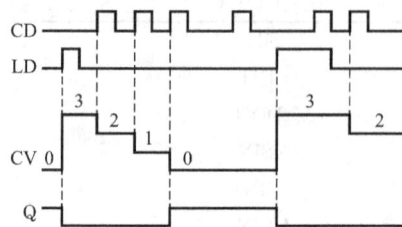

图 2.1.13　减计数器的工作原理
Figure 2.1.13　Working principle of CTD

（1）每当 CD 从"0"变为"1"，CV 减少 1；

（1）CV is decreased by 1 whenever CD is changed to "1" from "0"；

（2）当 CV＝0 时，Q 输出"1"，此后每当 CU 从"0"变为"1"，Q 保持输出"1"，CV 继续减少 1 直到达到计数器指定的整数类型的最小值；

（2）When CV is equal to 0，Q outputs "1"，subsequently Q continues outputting "1" and CV is decreased by 1 whenever CU is changed to "1" from "0" until to reach the designated minimum integer value of counter；

（3）在任意时刻，只要 LD 为"1"时，Q 输出"0"，CV 立即停止计数并回到 PV 值。

（3）Q will output "0"，and CV immediately stops counting and returns to PV value as long as LD is "1" at any time.

加减计数器（CTUD）的工作原理如图 2.1.14 所示，可归纳总结为：

The working principles of CTUD is shown in Figure 2.1.14，and can be summarized as follows：

（1）每当 CU 从"0"变为"1"，CV 增加 1，每当 CD 从"0"变为"1"，CV 减少 1；

（1）CV is increased by 1 whenever CU is changed to "1" from "0"，and CV is decreased by 1 whenever CD is changed to "1" from "0"；

（2）当 CV≥PV 时，QU 输出"1"，当 CV＜PV 时，QU 输出"0"；当 CV≤0 时，QD 输出"1"，当 CV＞0 时，QD 输出"0"；

（2）When CV is larger than or equal to PV，QU outputs "1"，and when CV is less than PV，QU outputs "0"；When CV is less than or equal to 0，QD outputs "1"，and when CV is larger than 0，QD outputs "0"；

（3）CV 的上下限取决于计数器指定的整数类型的最大值与最小值。

（3）The upper and lower limits of CV depends on the designated maximum and minimum integer value of counter.

（4）在任意时刻，只要 R 为"1"时，QU 输出"0"，CV 立即停止计数并回到 0；只要 LD 为"1"时，QD 输出"0"，CV 立即停止计数并回到 PV 值。

（4）QU will output "0"，and CV immediately stops counting and returns to 0 as long as R is "1"；OD outputs "0"，and CV immediately stops counting and returns to PV value as long as LD is "1".

三、任务实施

Ⅲ．Task Implementation

（一）连接 I/O 设备。

（Ⅰ）To connect I/O device.

按照控制要求，I/O 接线图如图 2.1.15 所示。

Based on the control requirements，the I/O wiring diagram is shown in Figure 2.1.15.

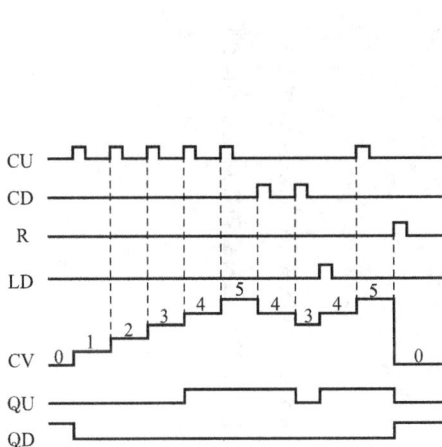

图 2.1.14 加减计数器的工作原理

Figure 2.1.14 Working principle of CTUD

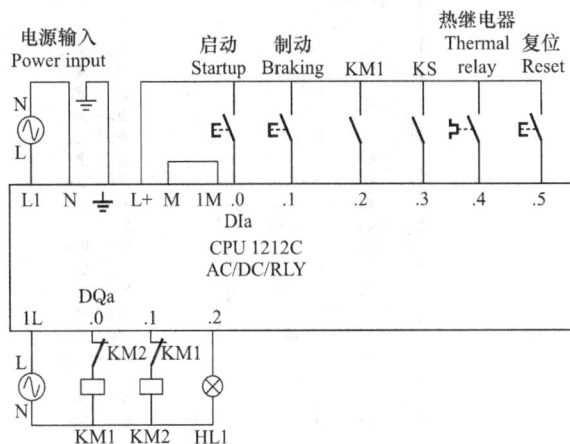

图 2.1.15 I/O 接线图

Figure 2.1.15 I/O wiring diagram

（二）在 TIA 博途中新建项目和设备组态

（Ⅱ）To create a new item and device configuration in TIA Portal

由于本任务无需连接触摸屏，因此不需要在设备组态时在 CPU 模块的"保护"选项中勾选"允许远程伙伴（PLC、HMI、OPC）使用 PUT/GET 通信访问"。但是，由于在这里需要用到 2Hz 的脉冲信号，因此需要在"系统和时钟存储器"中勾选"启用时钟存储器字节"，存储器的字节地址设为 0。

Because the connection to touch screen isn't required in executing this task，user will not check "Allow the remote partners (PLC，HMI and OPC) to use PUT/GET communication access" in the "Protection" option of CPU module at the time of device configuration. However，because 2Hz pulse signal will be used in this task，it is necessary to check "Enable the bytes of clock memory" in the "System and clock memory"，and set the byte address of memory to 0.

（三）建立变量表

（Ⅲ）To create the variable table

按照控制要求，定义变量表如图 2.1.16 所示。

Define the variable table as shown in Figure 2.1.16 based on the control requirements.

（四）编写梯形图程序

（Ⅳ）To write the ladder logic program

KM1 在电机启动运行时得电并保持通电，按下制动时失电，是一个起保停程序。KM2 在按下制动按钮时得电开始制动，在离心开关 KS 常开触点断开时失电，也是一个起保停程序。由于 KM1 和 KM2 控制电机的正转和反转，因此不能同时得电，需要在程序中互锁。用 KM1 的常开触点对电机启动的次数进行计数，当计数器达到设定值 5 时，输出 Q 变为 1，再串联一个 2Hz 脉冲信号的常开触点以达到信号灯闪烁的目的。

KM1 is energized and held at the time of motor startup，and de-energized at the time of pressing braking button，which is a startup，holding and stop program. KM2 is energized to execute the braking program at the time of pressing the braking button，and de-energized when the normally open contact of centrifugal switch KS is disconnected，which is also a startup，holding and stop program. Because KM1 and KM2 are used for controlling the positive rotation and negative rotation of motor，they can't be energized at the same time and should be interlocked in the program. The normally open contact of KM1 is used for counting the frequency of motor startup，when the counter reaches the set value 5，output Q is turned into 1，another normally open contact of 2Hz pulse signal will be connected to KM1 in series to realize the signal indicator flashing.

图 2.1.16 变量表

Figure 2.1.16 Variable table

梯形图程序如图 2.1.17 所示。

The ladder diagram program is shown in Figure 2.1.17.

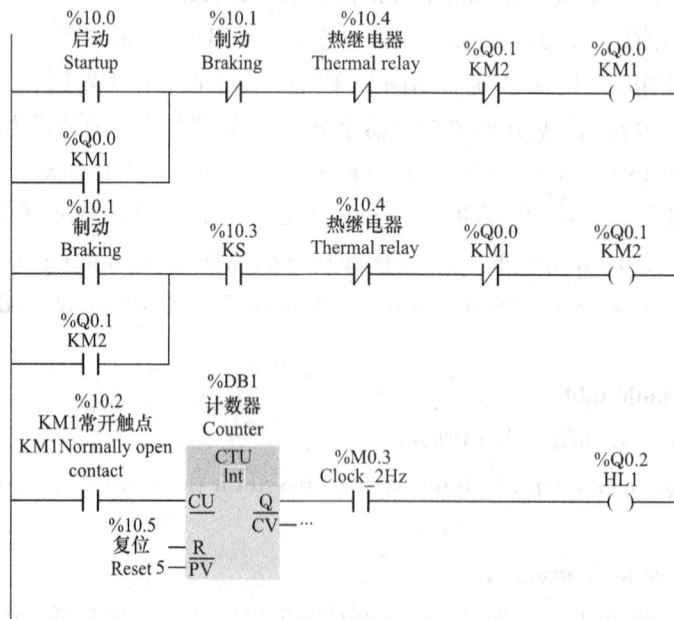

图 2.1.17 梯形图程序

Figure 2.1.7 Ladder logic program

（五）调试

（Ⅴ）Debugging

完成以上步骤后，将编辑好的 TIA 博途项目下载至 PLC，按照控制要求逐项对系统进行调试，直到所有功能都实现为止。

After the above steps，the edited TIA Portal items are downloaded to the PLC，and then the system will be debugged item by item based on the control requirements until that all functions are realized.

任务二 星形-三角形降压启动控制电路的安装与调试

Task Ⅱ Installation and Debugging of Control Circuit for Star-like-triangle Reduced-voltage Startup

一、任务分析

Ⅰ. Task Analysis

将图 2.2.1 所示电路图改造为用 PLC 控制的电气系统，要求如下：

To transform the circuit diagram as shown in Figure 2.2.1 into the electrical system controlled by PLC based on the following requirements：

（1）能实现图中电路的降压启动功能。

（1）The reduced-voltage startup function of circuit as shown in the figure can be realized.

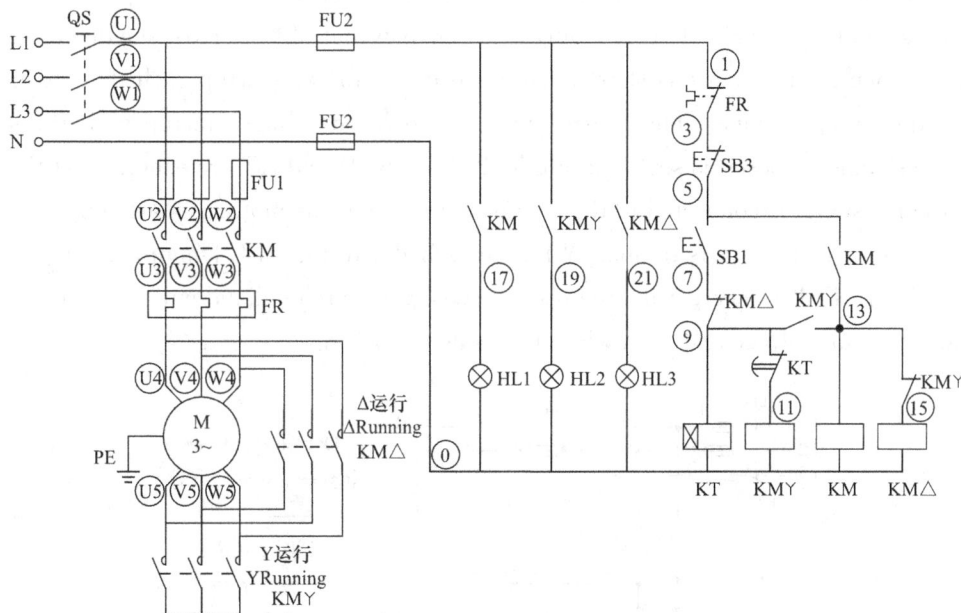

图 2.2.1 星形-三角形降压启动控制电路

Figure 2.2.1 Control circuit of star-like-triangle reduced-voltage startup

（2）为了完成本任务，需要了解三相异步电机几种降压启动的方法和 S7-1200 定时器的使用方法。

（2）For the purpose of this task，it is necessary to learn about several reduced-voltage startup methods of three-phase asynchronous motor and the use of S7-1200 timer.

二、相关知识技能

Ⅱ. Relevant Knowledge & Skills

（一）三相异步电机的降压启动控制线路

（Ⅰ）Reduced-voltage startup control circuit of three-phase asynchronous motor

降压启动就是利用启动设备将电压适当降低后，加到电机的定子绕组上进行启动，待电机启动运转后，再使其电压恢复到额定电压正常运转。

The reduced-voltage startup is that the voltage is firstly reduced to proper level by the startup device, then the reduced voltage is applied to the stator winding to start up the motor, and finally the reduced voltage will be recovered to the rating voltage to realize the normal running after the startup.

降压启动的方法有定子绕组串接电阻降压启动、自耦变压器降压启动、星形-三角形（Y-△）降压启动。

The reduced-voltage startup methods include the reduced-voltage startup with the stator winding connected to resistance in series, the reduced-voltage startup with auto-transformer, and star-triangle（Y-△）reduced-voltage startup.

1. 定子绕组串接电阻降压启动

1. Reduced-voltage startup with the stator winding connected to resistance in series

定子绕组串接电阻降压启动的方法是电机启动时，在电机的定子绕组上串联电阻，由于电阻的分压作用，使加在电机的定子绕组上的电压低于电源电压，待电机启动完成后，再将电阻短接，电机便在额定电压下正常运行。定子绕组串接电阻启动的控制电路如图 2.2.2 所示。按下 SB1，KM1 与 KT 同时得电，电阻串联接入电路中，电机降压启动，时间继电器 KT 开始定时。当 KT 定时时间到，其延时常开触点闭合，KM2 得电，电阻被短路，失去作用，电机全压运行，同时 KT 失电。

Its working principle is that the stator winding is connected to the resistance in series at the time of motor startup, the voltage on the motor's stator winding is lower than that of power voltage due to the voltage sharing of resistance, then the resistance is short-circuited after the motor startup such that the motor can normally run under the rating voltage. The control circuit of reduced-voltage startup with the stator winding connected to the resistance in series is shown in Figure 2. 2. 2. KM1 and KT are energized at the same time by pressing SB1, the resistance is connected to the circuit in series and the motor is started up under the reduced voltage, and the time relay KT starts timing. When the scheduled time of KT is up, the normally open time delay contact is closed, KM2 is energized, the resistance is short-circuited and will be out of action, the motor will run under the full voltage and meanwhile KT is de-energized.

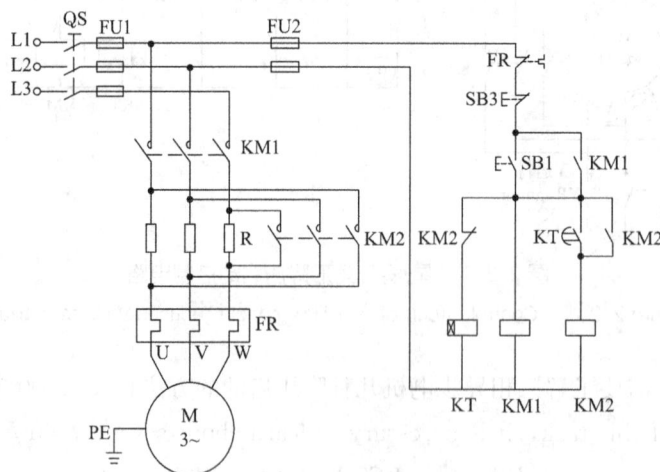

图 2.2.2　定子绕组串接电阻启动控制电路

Figure 2. 2. 2　Control circuit of reduced-voltage startup with the stator winding connected to the resistance in series

这种降压启动的方法由于电阻上有热能损耗，如用电抗器则体积大、成本又较高，因此该方法很少使用。

The reduced-voltage startup method is rarely used due to the loss of thermal energy on the resistance, and large-volume and high-cost electric reactor, if any.

2. 自耦变压器降压启动

2. Reduced-voltage startup with auto-transformer

自耦变压器降压启动的方法是在电机启动时，利用自耦变压器来降低加在电机定子绕组上的启动电压。待电机启动后，再使电机与自耦变压器脱离，从而在全压下正常运行。自耦变压器降压启动的控制电路如图2.2.3所示。按下启动按钮SB2后，KM1与KM2得电，三相电源经过自耦变压器降压后接入三相绕组，电机降压启动。同时，时间继电器KT得电，开始定时。当定时时间到，KT的延时常开触点闭合，KM3得电，三相电源直接接入三相绕组，电机全压运行，同时KM1、KM2与KT失电，自耦变压器从电路中断开。

Its working principle is that the auto-transformer is used to reduce the startup voltage to the stator winding when the motor is started up. Then the motor is separated from the auto-transformer after the startup, thus enable it to run under the full voltage. The control circuit of reduced-voltage startup with auto-transformer is shown in Figure 2.2.3. KM1 and KM2 are energized by pressing the startup button SB2, the three-phase power supply is connected to the three-phase winding after the voltage reduction by the auto-transformer, and the motor is started up under the reduced voltage. Meanwhile, the time relay KT is energized to start timing. When the scheduled time is up, the normally open time delay contact of KT is closed, KM3 is energized, three-phase power supply is directly connected to three-phase winding, the motor will run under the full voltage, meanwhile KM1, KM2 and KT are de-energized, and the auto-transformer is disconnected from the circuit.

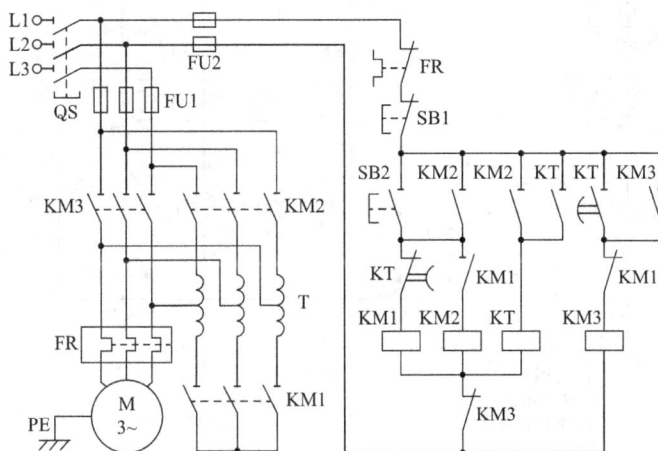

图2.2.3 自耦变压器降压启动控制电路

Figure 2.2.3 Control circuit of reduced-voltage startup with auto-transformer

自耦变压器降压启动的优点是，可以按允许的启动电流和所需的启动转矩来选择自耦变压器副边的不同抽头实现降压启动，而且不论电机定子绕组采用星形接法或三角形接法都可以使用。缺点是设备的体积较大，因而成本较高。

The advantages of reduced-voltage startup with auto-transformer are that the different taps on the secondary side of auto-transformer can be selected based on the allowable startup current and required startup torque to realize the reduced-voltage startup regardless of star or triangle connected motor stator winding. The disadvantages are that large volume and high cost.

3. 星形-三角形（Y-△）降压启动

3. Star-triangle（Y-△）reduced-voltage startup

Y-△降压启动的方法是电机启动时，把电机的定子绕组接成星形，电机定子绕组电压低于电源电压启动，启动即将完毕时再恢复成三角形，电机便在额定电压下正常运行。此时启动电流只有全电压启动电流的 1/3，但启动力矩也只有全电压启动力矩的 1/3。Y-△降压启动控制电路如图 2.2.4 所示。按下启动按钮 SB1 后，KT、KMY 和 KM 得电，电动机绕组接成星形启动，同时时间继电器 KT 开始定时。当定时时间到，KT 的延时常闭触点断开，KMY 失电，KM△得电，KT 失电。电机绕组接成三角形全电压运行。

Its working principle is that the stators are connected in star when the motor is started up, the voltage of stator winding is lower than the power supply voltage, the star connection is turned into triangle connection immediately before the startup, and finally the motor will normally run under the rating voltage. At this moment，the startup current is only 1/3 of that under the full voltage, however the startup torque is only 1/3 of that under the full voltage. The control circuit of Y-△ reduced-voltage startup is shown in Figure 2.2.4. KT，KMY and KM are energized by pressing the startup button SB1，the motor windings are connected in star and meanwhile the time relay KT starts timing. When the scheduled time is up，the normally closed time delay contact is disconnected，KMY is de-energized，KM△ is energized, and KT is de-energized. The motor windings are connected in triangle to run under the full voltage.

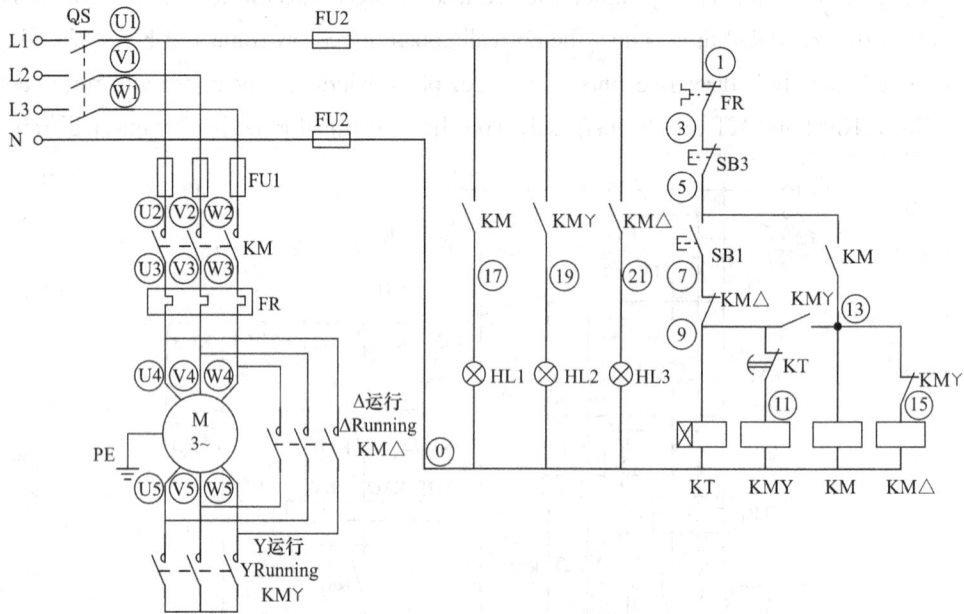

图 2.2.4　Y-△降压启动控制电路

Figure 2.2.4　Control circuit of Y-△ reduced-voltage startup

　　Y-△ 降压启动适用于对电机启动力矩无严格要求但要限制电机启动电流且电机满足 380V/△的接线条件。

　　Y-△ reduced voltage startup is applicable under the circumstances that there is no strict requirement for the startup torque of motor，the startup current of motor should be limited，and the motor satisfies the 380V/△ wiring condition.

（二）西门子 S7-1200 的定时器

（Ⅱ）Timer of Siemens S7-1200

S7-1200 的定时器为 IEC 定时器，用户程序中可以使用的定时器数量仅受 CPU 的存储器容量限制。

S7-1200 is equipped with IEC timer, and the number of available counters in the user program is only limited by the memory capacity of CPU.

　　使用定时器，需要使用定时器相关的背景数据块或者数据类型为 IEC_TIMER（或 TP_TIME、TON_

TIME、TOF＿TIME、TONR＿TIME）的 DB 块变量，上述的不同变量代表着不同的定时器。

To use the timer，the background data block or DB block variables with the data type IEC＿TIMER（or TP＿TIME，TON＿TIME，TOF＿TIME and TONR＿TIME）related to the timer should be used，and the above different variables represent different timers.

注：S7-1200 的 IEC 定时器没有定时器号（即没有 T0、T37 这种带定时器号的定时器）。

Notes：IEC timers of S7-1200 are free of timer number（i. e. any timer doesn't bear the T0，T37 and other marks）.

S7-1200 包含四种定时器：生成脉冲定时器（TP）、接通延时定时器（TON）、关断延时定时器（TOF）、时间累加器（TONR）。指令位置如图 2.2.5 所示。

S7-1200 is equipped with four types of timers：Pulse generation timer（TP），on-delay timer（TON），off-delay timer（TOF），and time accumulator（TONR）. The instruction location is shown in Figure 2.2.5.

S7-1200 的定时器是函数块，调用它们时，需要生成保存定时器数据的背景数据块。将定时器的功能框指令直接拖入工作区中，自动生成定时器的背景数据块，如图 2.2.6 所示，该块位于"系统块＞程序资源"中。

The timers of S7-1200 are function blocks，and when these function blocks are called，it is necessary to generate the background block to save the timer data. The timer is directly dragged to the working area to automatically generate the background data block of timer（see Figure 2.2.6），and the block is located in "system block ＞ program resource".

图 2.2.5 定时器指令的位置

Figure 2.2.5 Location of timer instruction

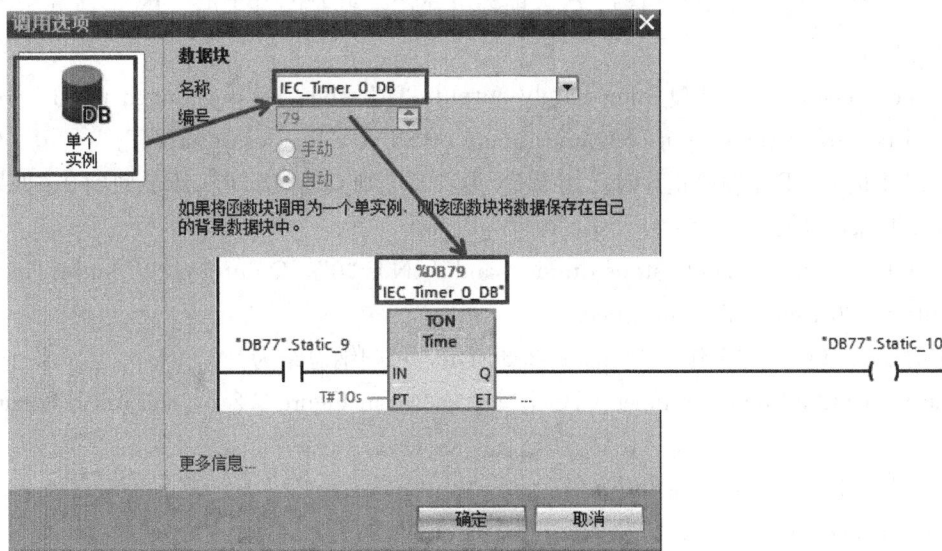

图 2.2.6 生成一个定时器

Figure 2.2.6 Generate one timer

定时器引脚功能见表 2.2.1。表中的 PT 为预设时间值，ET 为定时开始后经过的时间，称为当前时间值，它们的数据类型为 32 位的 Time，单位为毫秒（ms），最大定时时间为 T＃24D＿20H＿31M＿23S＿647MS，其中"D""H""M""S""MS"分别为"日""小时""分""秒"和"毫秒"。可以不分配给 Q 和

59

ET 指定地址。

The pin function of timer is shown in Table 2.2.1. In the table, PT is the default time value, ET is the elapsed time (called as the current time value) after start timing, their data type is 32-bit Time in the unit of millisecond (ms), the maximum scheduled time is $T\sharp 24D\_20H\_31M\_23S\_647MS$, and "D" "H" "M" "S" and "MS" are "day" "hour" "minute" "second" and "millisecond" respectively. It is allowed that the designated address will not be assigned to output Q and ET.

定时器的引脚功能   表 2.2.1

Pin function of timer   Table 2.2.1

名称 Name	说明 Description	数据类型 Data type	备注 Remarks
IN	输入位 Input bits	BOOL	TP、TON、TONR：0＝禁用定时器，1＝启用定时器 TP, TON and TONR：0＝disable timer，1＝enable timer TOF：0＝启用定时器，1＝禁用定时器 TOF：0＝enable timer，1＝disable timer
PT	设定的时间输入 Set time input	TIME	
R	复位 Reset	BOOL	仅出现在 TONR 指令 Only in the TONR instruction
Q	输出位 Output bit	BOOL	
ET	已计时的时间 Time counted	TIME	

生成脉冲定时器（TP）的工作原理如图 2.2.7 所示，可归纳总结为：

The working principles of pulse generation timer (TP) is shown in Figure 2.2.7, and can be summarized as follows：

（1）IN 从 "0" 变为 "1"，定时器启动，Q 立即输出 "1"；当 ET＜PT 时，IN 的改变不影响 Q 的输出和 ET 的计时；

（1）The timer is enabled, and Q immediately outputs "1" when IN is changed to "1" from "0"；The changes in IN will not affect the output of Q and timing of ET if ET is less than PT；

（2）当 ET＝PT 时，ET 立即停止计时，如果 IN 为 "0"，则 Q 输出 "0"，ET 回到 0；如果 IN 为 "1"，则 Q 输出 "0"，ET 保持计时。

（2）If ET＝PT, ET immediately stops timing, and if IN is "0", Q outputs "0" and ET returns to 0；If IN is "1", Q outputs "0" and ET is remained.

接通延时定时器（TON）的工作原理如图 2.2.8 所示，可归纳总结为：

The working principles of on-delay timer (TON) are shown in Figure 2.2.8, and can be summarized as follows：

（1）IN 从 "0" 变为 "1"，定时器启动；

（1）If IN is changed to "1" from "0", the timer is enabled；

（2）当 ET＝PT 时，Q 立即输出 "1"，ET 立即停止计时并保持最终计时数值；

（2）If ET＝, Q immediately outputs "1", and ET immediately stops timing and remains the final time value；

（3）在任意时刻，只要 IN 变为 "0"，ET 立即停止计时并回到 0，Q 输出 "0"。

（3）ET immediately stops timing and returns to 0, and Q outputs "0" as long as IN is changed to "0" at any time.

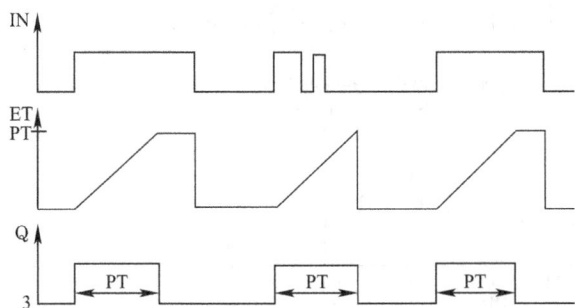

图 2.2.7 生成脉冲定时器的工作原理
Figure 2.2.7 Working principles of pulse generation timer

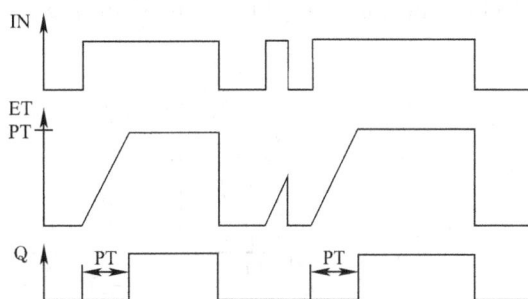

图 2.2.8 接通延时定时器的工作原理
Figure 2.2.8 Working principles of on-delay timer

关断延时定时器（TOF）的工作原理如图 2.2.9 所示，可归纳总结为：

The working principles of off-delay timer（TOF）are shown in Figure 2.2.9，and can be summarized as follows：

（1）只要 IN 为"1"时，Q 即输出为"1"；

（1）Q outputs "1" as long as IN is "1"；

（2）IN 从"1"变为"0"，定时器启动；

（2）If IN is changed to "0" from "1"，the timer is enabled；

（3）当 ET＝PT 时，Q 立即输出"0"，ET 立即停止计时并保持最终计时数值；

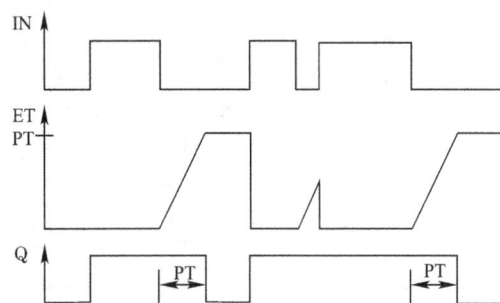

图 2.2.9 关断延时定时器的工作原理
Figure 2.2.9 Working principles of off-delay timer

（3）If ET＝PT，Q immediately outputs "0"，and ET immediately stops timing and remains the final time value；

（4）在任意时刻，只要 IN 变为"1"，ET 立即停止计时并回到 0。

（4）ET immediately stops timing and returns to 0 as long as IN is changed to "1" at any time.

三、任务实施
Ⅲ. Task Implementation
（一）连接 I/O 设备
（Ⅰ）To connect I/O device
按照控制要求，I/O 接线图如图 2.2.10 所示。

Based on the control requirements，the I/O wiring diagram is shown in Figure 2.2.10.

（二）在 TIA 博途中新建项目和设备组态
（Ⅱ）To create a new item and device configuration in TIA Portal
由于本任务无需连接触摸屏，因此不需要在设备组态时在 CPU 模块的"保护"选项中勾选"允许远程伙伴（PLC、HMI、OPC）使用 PUT/GET 通信访问"。也不需要在"系统和时钟存储器"中勾选"启用时钟存储器字节"。

Because the connection to touch screen isn't required in executing this task，user will not check "Allow the remote partners（PLC，HMI and OPC）to use PUT/GET communication access" in the "Protection" option of CPU module at the time of device configuration. There is also no need for checking "Enable the bytes of clock timer" in "System and clock memory".

（三）建立变量表
（Ⅲ）To create the variable table
按照控制要求，定义变量表如图 2.2.11 所示。

Define the variable table as shown in Figure 2. 2. 11 based on the control requirements.

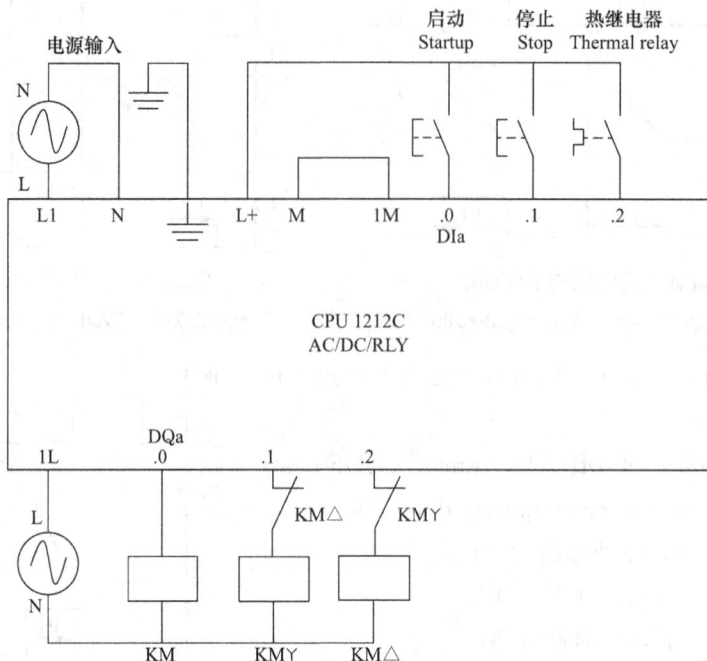

图 2. 2. 10　I/O 接线图

Figure 2. 2. 10　I/O wiring diagram

图 2. 2. 11　变量表

Figure 2. 2. 11　Variable table

（四）编写梯形图程序

（Ⅳ）To write the ladder logic program

按下启动按钮后，"KMㄚ"和"KM"得电，同时启动定时器开始定时。当定时时间到，定时器的常闭触点断开，"KMㄚ"失电，常开触点闭合，"KM△"得电。按下停止按钮后，"KM"失电，定时器复位，所有输出复位。

The "KMㄚ" and "KM" are energized and the timer is enabled to start timing at the same time by pressing the startup button. When the scheduled time is up, the normally closed contact of timer is open, "KMㄚ" is de-energized, the normally open contact is contacted, and "KM△" is energized. The "KM" is de-energized, and the timer and all outputs are reset by pressing the stop button.

梯形图程序如图 2. 2. 12 所示。

The ladder diagram program is shown in Figure 2. 2. 12.

（五）调试

（Ⅴ）Debugging

完成以上步骤后，将编辑好的 TIA 博途项目下载至 PLC，按照控制要求逐项对系统进行调试，直到所有功能都实现为止。

After the above steps，the edited TIA Portal items are downloaded to the PLC，and then the system will be debugged item by item based on the control requirements until that all functions are realized.

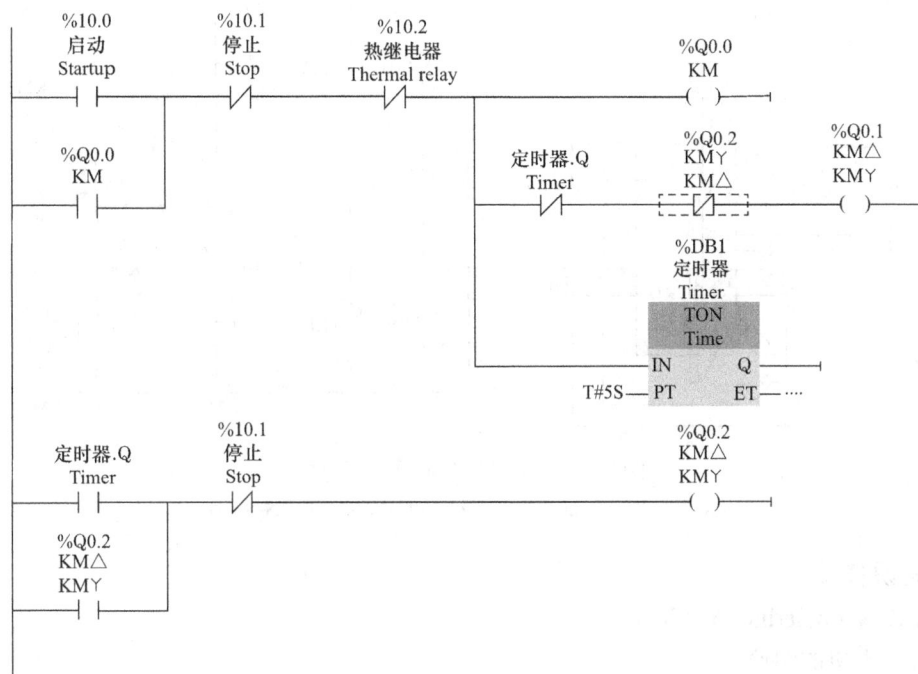

图 2.2.12　梯形图程序

Figure 2.2.12　Ladder logic program

任务三　双速电机调速控制电路的安装与调试

Task Ⅲ　Installation and Debugging of Speed Control Circuit for Double Speed Motor

一、任务分析

Ⅰ. Task Analysis

将图 2.3.1 所示电路图改造为用 PLC 控制的电气系统，要求如下：

To transform the circuit diagram as shown in Figure 2.3.1 into the electrical system controlled by PLC based on the following requirements：

（1）能实现图中电路的电机调速功能；

（1）The motor speed control function in the circuit diagram can be realized；

（2）按下 SB1 按钮，电机以低速运行 4s 后停止，再次按下启动按钮 SB1 后，高速运行 6s 后停止。

（2）The motor will stop after running for 4s at low speed by pressing the button SB1，and then stop after running for 6s at high speed by pressing the startup button SB1 again.

为了完成本任务，需要了解三相异步电机的几种调速方法和 S7-1200 比较指令、扫描操作数信号上升沿/下降沿指令的使用方法。

For the purpose of this task，it is necessary to learn about several speed control methods of three-phase asynchronous motor and the use of S7-1200's comparison instruction，scanning operand signal rising edge/falling edge instructions.

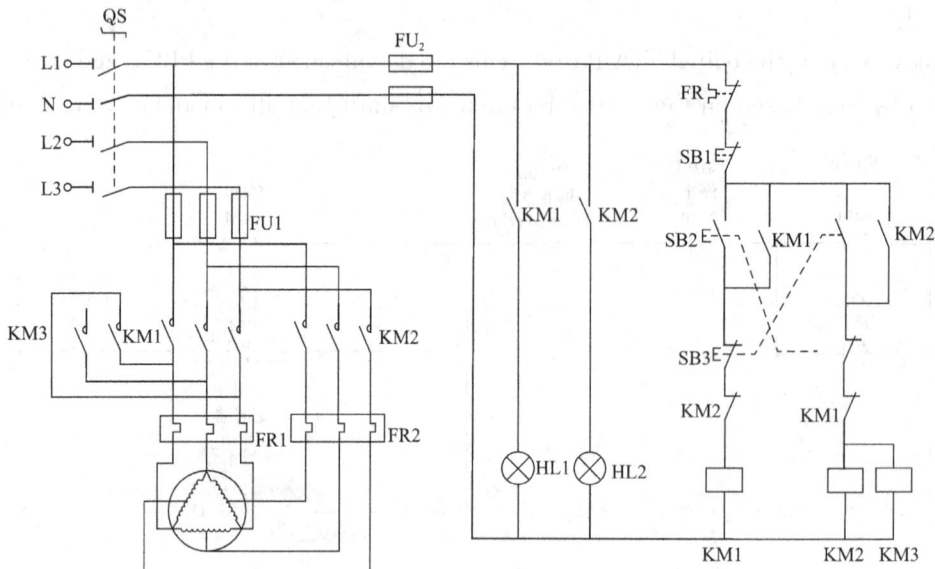

图 2.3.1 双速电机调速控制电路

Figure 2. 3. 1 Speed control circuit of double-speed motor

二、相关知识技能

Ⅱ. Relevant Knowledge & Skills

（一）三相异步电机的调速

（Ⅰ）Speed control of three-phase asynchronous motor

三相异步电机的转速公式：

Three-phase asynchronous motor speed formula：

$$n = (1-s)\frac{60f_1}{p}$$

由公式可知，改变异步电机转速可通过三种方法来实现：一是改变电源频率 f_1；二是改变转差率 s；三是改变磁极对数 p。

It can be seen from the rotating speed formula of three-phase asynchronous motor that the speed of three-phase asynchronous motor can be changed by three methods；Firstly，change the supply frequency f_1；secondly，change the slip ratio s；Thirdly，change the number of pole pairs p.

1. 变频调速

1. Variable frequency speed control

改变异步电机的电源频率调速称为变频调速，主要由变频器来实现，它主要由整流器和逆变器两大部分组成。整流器先将频率为 50Hz 的三相交流电转换为直流电，再由逆变器变换频率 f_1 可调、电压有效值 U1 也可调的三相交流电，供给三相鼠笼式电机。由此可实现电机的无级调速，并具很强的机械特性。

The method to complete the speed control of asynchronous motor by changing the supply frequency is called as variable frequency speed control，and the supply frequency is mainly changed by the frequency converter，which is mainly composed of rectifier and inverter. The rectifier converts the 50Hz three-phase AC power supply into DC power supply，and then the inverter converts DC power supply into three-phase AC power supply with adjustable frequency $f1$ and effective voltage $U1$ before supplying to three-phase squirrel-cage motor. In such way，the stepless speed control of motor can be realized with strong mechanical features.

2. 变极调速

2. Pole changing speed control

改变异步电机的磁极对数调速称为变极调速。变极调速是通过改变定子绕组的连接方式来实现的，它是有级调速，且只适用于鼠笼式异步电机。由于调速时其转速呈跳跃性变化，因而只用在对调速性能要求不高的设备，如铣床、镗床、磨床等机床上。变极调速的其中一种绕组连接方式如图2.3.2所示。当绕组以三角形相接时，每相的两个绕组是串联的方式，这样形成的旋转磁场有两对磁极，电机低速运行，如图2.3.2（a）。当绕组以双星形相接时，每相的两个绕组是并联的方式，这样形成的旋转磁场只有一对磁极，比三角形连接时少了一半，此时电机高速运行，如图2.3.2（b）。

The method to complete the speed control of asynchronous motor by changing the number of pole pairs is called as the pole-changing speed control. The pole changing speed control is realized by changing the connection mode of stator winding, which is the step speed control and only applicable to the squirrel-cage asynchronous motor. Because the rotating speed shows the jumping changes at the time of speed control, it is only applicable to the equipment without strict requirement for the speed control performance, such as miller, boring machine and grinder. One of winding connection types for the pole-changing speed control as shown in Figure 2.3.2. When the windings are connected in triangle, two windings per phase are connected in series such that the rotating magnetic field generated has two pairs of magnetic poles, and the motor runs at low speed, as shown in Figure 2.3.2 (a). When the windings are connected in star, two windings per phase are connected in parallel such that the rotating magnetic field generated only has one magnetic pole, half of that in triangle connection, and motor runs at high speed, as shown in Figure 2.3.2 (b).

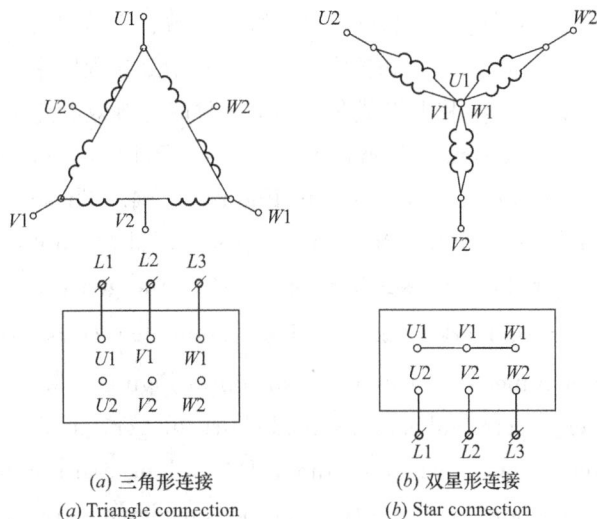

图 2.3.2 双速电机的绕组连接方式

Figure 2.3.2 Winding connection type of double speed motor

3. 变转差率调速

3. Speed control by changing the slip ratio

变转差率调速是绕线式电机特有的一种调速方法。这种调速方式广泛应用于各种起重设备中。变转差率调速的其中一种方法是在转子中串接电阻，如图2.3.3所示，这样的调速方法调速范围不大，特别是负载小时，调速范围更小。另外，由于电阻会消耗一部分功率，因此这种方法调速的效率较低。

The speed control by changing the slip ratio is a unique speed control method for the wound rotor type motor. Such speed control method is widely applied on various lifting and hoisting equipment. One of connection types is that rotor is connected to resistance in series, as shown in Figure 2.3.3, the speed control range of such method is small, especially smaller if the motor is running under the small load. Additionally, its efficiency is relatively low because the resistance may consume partial power.

（二）西门子 S7-1200 的比较指令

（Ⅱ）Comparison instruction of Siemens S7-1200

S7-1200 中的比较指令用来比较两个数据类型相同的数值的大小，指令位置如图2.3.4所示。按照比较类型的不同，分为六种类型：等于、不等于、大于或等于、小于或等于、大于、小于。生成比较指令后，双击指令符号中的问号，再单击出现的▼，可以从出现的下拉菜单中选择数据类型，如图2.3.5所示，支持的数据类型包括：整数、双整数、实数、无符号短整数、无符号整数、无符号长整数、短整数、字符串、字符、时间、DTL 和长实数等。比较指令在程序中只是作为条件来使用，可以将它视为一个等效的触点。当两个数值 IN1 和 IN2 满足给出的条件时，等效触点接通。

The comparison instructions of S7-1200 are used to compare both values with same data type，and the instruction location refers to Figure 2.3.4. These comparison instructions are classified into six types by the differences in the compared types；equal to，not equal to，larger than or equal to，smaller than or equal to，larger than and smaller than. After the generation of comparison instructions，you can double click on the question mark among the instruction characters，then click on ▼，and select the data type from the drop-down menu appeared，as shown in Figure 2.3.5. The supported data types include；integer，double integer，real number，unsigned short integer，unsigned integer，unsigned long integer，short integer，character string，character，time，DTL，long real number，etc. The comparison instructions can only be used as the conditions，and they can be considered as one equivalent contact. When both values IN1 and IN2 satisfy the given condition，the equivalent contact is connected.

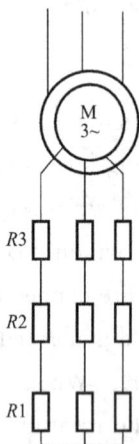

图 2.3.3 转子串接电阻的电路图

Figure 2.3.3 Circuit diagram of rotor connected to resistance in series

图 2.3.4 比较指令的位置

Figure 2.3.4 Location of comparison instructions

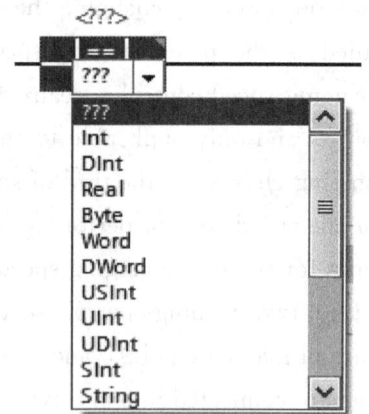

图 2.3.5 数据类型的选择

Figure 2.3.5 Selection of data types

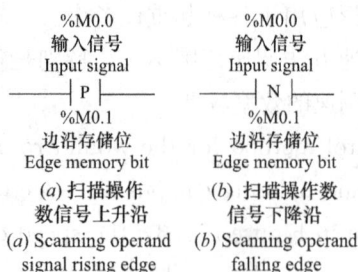

图 2.3.6 扫描操作数信号上升沿/下降沿指令

Figure 2.3.6 Scanning operand signal rising edge/falling edge instructions

（三）扫描操作数信号上升沿/下降沿指令

（Ⅲ）Scanning operand signal rising edge/falling edge instructions

图 2.3.6 (a) 为扫描操作数信号上升沿指令，如果该触点上面的输入信号 M0.0 由 0 状态变为 1 状态（即 M0.0 上升沿到来），则该触点接通一个扫描周期。图 2.3.6 (b) 为扫描操作数信号下降沿指令，如果该触点上面的输入信号 M0.0 由 1 状态变为 0 状态（即 M0.0 下降沿到来），则该触点接通一个扫描周期。边沿检测触点不能放在电路末端。

Figure 2.3.6 (a) shows the scanning operand signal rising edge instruction，if the input signal M0.0 on the contact is changed to 1 from 0 (i. e. the arrival of M0.0 rising edge)，the contact is connected for one scanning cycle. Figure 2.3.6 (b) shows the scanning operand signal falling edge instruction，if the input signal M0.0 on the contact is changed to 0 from 1 (i. e. the arrival of M0.0 falling edge)，the contact is connected for one scanning cycle. The edge detection contact can't be placed on the end of circuit.

图 2.3.6 中的 M0.1 为边沿存储位，用来存储输入信号上一个扫描周期的状态。通过比较输入信号的当前状态和上一个扫描周期的状态，来检测信号的边沿。边沿存储位的地址只能在程序中使用一次，它的状态不能在其他地方被改写。只能用 M、DB 和 FB 的静态局部变量（Static）来作边沿存储位，不能用块的临时局部数据或 I/O 变量来作边沿存储位。

M0.1 in Figure 2.3.6 is the edge memory bit, and used for saving the status of input signal during the last scanning cycle. The signal edge is detected by comparing the current status of input signal and the status during the last scanning cycle. The address of edge memory bit is only used once in the program, its status can't be modified at any other locations. The static local variables (Static) of M, DB and FB can only be used as the edge memory bits, rather than the temporary local data of blocks or I/O variables.

三、任务实施
Ⅲ. Task Implementation
（一）连接 I/O 设备
（Ⅰ）To connect I/O device

按照控制要求，I/O 接线图如图 2.3.7 所示。

Based on the control requirements, the I/O wiring diagram is shown in Figure 2.3.7.

（二）在 TIA 博途中新建项目和设备组态
（Ⅱ）To create a new item and device configuration in TIA Portal

由于本任务无需连接触摸屏，因此不需要在设备组态时在 CPU 模块的"保护"选项中勾选"允许远程伙伴（PLC、HMI、OPC）使用 PUT/GET 通信访问"。也不需要在"系统和时钟存储器"中勾选"启用时钟存储器字节"。

图 2.3.7　I/O 接线图

Figure 2.3.7　I/O wiring diagram

Because the connection to touch screen isn't required in executing this task, user will not check "Allow the remote partners (PLC, HMI and OPC) to use PUT/GET communication access" in the "Protection" option of CPU module at the time of device configuration. There is also no need for checking "Enable the bytes of clock timer" in "System and clock memory".

（三）建立变量表
（Ⅲ）To create the variable table

按照控制要求，定义变量表如图 2.3.8 所示。

Define the variable table as shown in Figure 2.3.8 based on the control requirements.

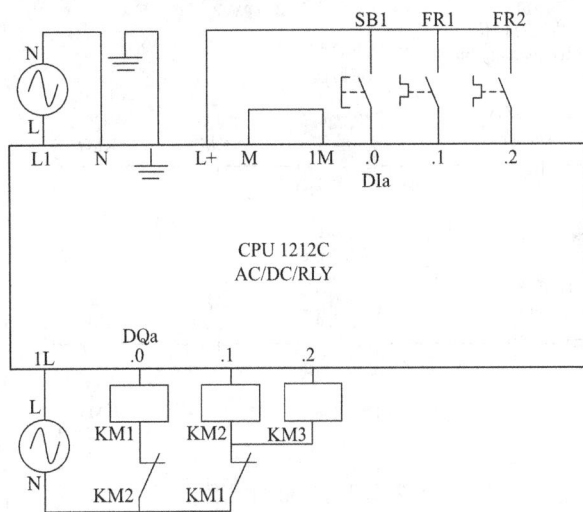

PLC 变量							
	名称	变量表	数据类型	地址	保持	在 H...	可从...	注释
1	SB1	默认变量表	Bool	%I0.0		☑	☑	
2	FR1	默认变量表	Bool	%I0.1		☑	☑	
3	FR2	默认变量表	Bool	%I0.2		☑	☑	
4	KM1	默认变量表	Bool	%Q0.0		☑	☑	
5	KM2	默认变量表	Bool	%Q0.1		☑	☑	
6	KM3	默认变量表	Bool	%Q0.2		☑	☑	
7	边沿存储位	默认变量表	Bool	%M0.1		☑	☑	

图 2.3.8　变量表

Figure 2.3.8　Variable table

（四）编写梯形图程序

（Ⅳ）To write the ladder logic program

对 SB1 按下的次数进行计数。当计数值等于 1 时，启动定时器 1 生成一个 4s 的脉冲，双速电机低速运行 4s 后停止。当计数值等于 2 时，启动定时器 2 生成一个 6s 的脉冲，双速电机高速运行 6s 后停止。当定时器 2 的定时结束后，复位计数器。

To count the number of times SB1 is pressed. When the count value is 1，No. 1 timer is enabled to generate a pulse for 4s，and double speed motor stops after running at low speed for 4s. When the count value is 2，No. 2 timer is enabled to generate a pulse for 6s，and double speed motor stops after running at high speed for 6s. When the scheduled time of No. 2 timer is up，the counter is reset.

梯形图程序如图 2.3.9 所示。

The ladder diagram program is shown in Figure 2.3.9.

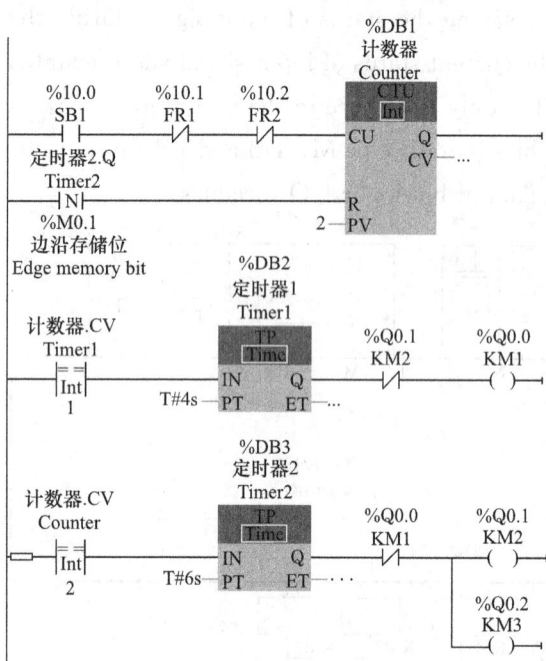

图 2.3.9　梯形图程序

Figure 2.3.9　Ladder logic program

（五）调试

（Ⅴ）Debugging

完成以上步骤后，将编辑好的 TIA 博途项目下载至 PLC，按照控制要求逐项对系统进行调试，直到所有功能都实现为止。

After the above steps，the edited TIA Portal items are downloaded to the PLC，and then the system will be debugged item by item based on the control requirements until that all functions are realized.

任务四　X62W 铣床电气控制单元常见故障检测与排除

Task Ⅳ　Detection and Troubleshooting of Common Faults in the Electrical Control Units of X62W Miller

一、任务分析

Ⅰ. Task Analysis

参考 X62W 铣床的电气原理图，如图 2.4.1 所示，排除 X62W 铣床电路板上所设置的三个故障，使该电路能正常工作，同时完成维修工作票。如图 2.4.2 所示。

Refer to the electrical schematic diagram of X62W miller，as shown in Figure 2.4.1，three faults preset on the circuit board of X62W miller aretroubleshot to recover the normal functioning of circuit，and meanwhile complete the work arrangement. As shown in Figure 2.4.2.

为了完成本任务，需要熟悉 X62W 铣床电路的控制原理，并在此基础上学习机床电路故障的检测和排除方法。

For the purpose of this task，it is necessary to know well the electrical control principles of X62W，and learn the circuit fault detection and troubleshooting of lathes on this basis.

二、相关知识技能

Ⅱ. Relevant Knowledge & Skills

（一）X62W 铣床电路的控制原理

（Ⅰ）Electrical control principles of X62W miller

1. 主轴电机的控制

1. Control of spindle motor

控制线路的启动按钮 SB1 和 SB2 是异地控制按钮，方便操作。SB3 和 SB4 是停止按钮，也是异地控制按钮。KM3 是主轴电机 M1 的启动接触器，KM2 是主轴反接制动接触器，SQ7 是主轴变速冲动开关，KS 是速度继电器。

The startup buttons SB1 and SB2 of control circuit are remotely controlled buttons for convenience. SB3 and SB4 are stop buttons，and also remotely controlled buttons. KM3 is the startup contactor of spindle motor M1，KM2 is the reverse braking contactor of spindle，SQ7 is the variable speed impulse switch of spindle，and KS is the speed relay.

（1）主轴电机的启动

（1）Startup of spindle motor

启动前先合上电源开关 QS，再把主轴转换开关 SA5 扳到所需要的旋转方向，然后按启动按钮 SB1（或 SB2），接触器 KM3 得电动作，其主触点闭合，主轴电机 M1 启动。

You should close the power switch QS before the startup, then turn spindle change-over switch SA5 to the required rotational direction，press the startup button SB1（or SB2），energize and connect the contactor KM3，close its main contact，and finally start up the spindle motor M1.

（2）主轴电机的停车制动

（2）Parking braking of spindle motor

当铣削完毕，需要主轴电动机 M1 停车，此时电机 M1 运转速度在 120 转/分以上时，速度继电器 KS 的常开触点闭合（9 区或 10 区），为停车制动作好准备。当要 M1 停车时，就按下停止按钮 SB3（或 SB4），KM3 断电释放，由于 KM3 主触点断开，电机 M1 断电作惯性运转，紧接着接触器 KM2 线圈得电吸合，电动机 M1 串电阻 R 反接制动。当转速降至 120 转/分以下时，速度继电器 KS 常开触点断开，接触器 KM2 失电断开，停车反接制动结束。

When the spindle motor M1 requires to be stopped after the milling and if the rotating speed of motor M1 is more than 120rpm at this moment，you should close the normally open contact of speed relay KS（area 9 or 10）to be ready for the parking braking. When M1 requires to be stopped，you should press the stop button SB3（or SB4），de-energize and release KM3，enable the motor M1 to rotate under the action of inertia after the power loss，next energize and connect the coils of contactor KM2，and reverse the resistance R connected to the motor M1 in series for braking. When the rotating speed is decreased below 120rpm，you should disconnect the normally open contact of speed relay KS，disconnect the contactor KM2，and complete the reverse braking for parking.

（3）主轴的冲动控制

（3）Impulse control of spindle

当需要主轴冲动时，按下冲动开关 SQ7，SQ7 的常闭触点 SQ7-2 先断开，而后常开触点 SQ7-1 闭合，使接触器 KM2 通电吸合，电机 M1 启动，冲动完成。

When the spindle impulse is needed，you should press the impulse switch SQ7，disconnect the normally closed contact SQ7-2 of SQ7，and then close the normally open contact SQ7-1 to energize and connect the contactor KM2，start up the motor M1，and complete impulse.

2. 工作台进给电机控制

2. Control of workbench feed motor

转换开关 SA1 是控制圆工作台的，在不需要圆工作台运动时，转换开关扳到"断开"位置，此时 SA1-1 闭合，SA1-2 断开，SA1-3 闭合；当需要圆工作台运动时将转换开关扳到"接通"位置，则 SA1-1 断开，SA1-2 闭合，SA1-3 断开。

The change-over switch SA1 is used for controlling the round workbench，the switch is turned to the "Off" position when the motion of round workbench isn't needed，SA1-1 is closed，SA1-2 is disconnected and SA1-3 is closed at this moment；The switch is turned to "On" position when the motion of round workbench is needed，SA1-1 is disconnected，SA1-2 is closed and SA1-3 is disconnected.

（1）工作台纵向进给

（1）Longitudinal feed of workbench

工作台的左右（纵向）运动是由装在床身两侧的转换开关（此开关用于切换不同运动方向的机械传动结构，电路图中未画出）跟行程开关 SQ1 和 SQ2 来完成，需要进给时把转换开关扳到"纵向"位置，拨动进给手柄向右，行程开关 SQ1 被压下，常开触点 SQ1-1 闭合，常闭触点 SQ1-2 断开，接触器 KM4 得电，电机 M2 正转，工作台向右运动；当工作台要向左运动时，拨动进给手柄向右，行程开关 SQ2 被压下，常开触点 SQ2-1 闭合，常闭触点 SQ2-2 断开，接触器 KM5 得电，电机 M2 反转工作台向左运动。在工作台上设置一块挡铁，两边各设置一个限位开关，当工作台纵向运动到极限位置时，挡铁撞到位置开关工作台停止运动，从而实现纵向运动的终端保护。

The left-to-right (longitudinal) motion of workbench is completed by the change-over switches on both sides of lathe (such switches are used for changing over the mechanical transmission mechanisms moving along different directions，which are not indicated on the circuit diagram) and travel switches SQ1 and SQ2，you should turn the change-over switches to the "longitudinal" position，pull the feed handle rightward and press the travel switch SQ1. At this moment，the normally open contact SQ1-1 is closed，the commonly closed contact SQ1-2 is disconnected，the contactor KM4 is energized，the motor M2 rotates clockwise to enable the workbench to move to the right；When the workbench requires moving to the left，you should turn the feed handle to the right. At this moment，the travel switch SQ2 is pressed，and subsequently the normally open contact SQ2-1 is closed，the normally closed contact SQ2-2 is open，the contactor KM5 is energized，and the motor M2 runs in the reverse direction to enable the workbench to the left. There is one stop iron (with one limit switch on both sides) on the workbench，when the workbench longitudinally moves to the limiting position，the stop iron crashes into the limit switch and the workbench stops moving，thus realizing the end protection of longitudinal motion.

（2）工作台升降和横向（前后）进给

（2）Vertical and horizontal (forward-backward) feed of workbench

由于铣床的机构不能完成复杂的机械传动，方向进给只能通过操作装在床身两侧的转换开关跟行程开关 SQ3 和 SQ4 来完成工作台上下和前后运动。在工作台上也分别设置有一块挡铁，两边各设置有一个限位开关，当工作台升降和横向运动到极限位置时，挡铁撞到位置开关工作台停止运动，从而实现纵向运动的终端保护。

Because the transmission mechanism of miller can't complete the complex mechanical transmission，the vertical and horizontal motion of workbench (directional feed) can only be realized by operating the change-over switch and travel switch SQ3 and SQ4 on both sides of lathe. There is also one stop iron (with one limit switch on both sides) on the workbench, when the workbench vertically and horizontally moves to the limiting position, the stop iron crashes into the limit switch and the workbench stops moving, thus realizing the end protection of longitudinal motion.

1) 工作台向上（下）运动

1) Upward and downward motion of workbench

在主轴电机启动后，把装在床身一侧的转换开关扳到"升降"位置再将进给手柄拨向下，行程开关 SQ3 被压下，SQ3 常开触点闭合，SQ3 常闭触点断开，接触器 KM4 得电，电机 M2 正转，工作台向下运动。到达想要的位置时松开按钮工作台停止运动。反之，将进给手柄拨向上，SQ4 被压下，KM5 得电，电 M2 反转，工作台向上运动。

After the startup of spindle motor, if the user turns the change-over switch on one side of lathe to "Up-Down" position and then turns the feed handle downward. At this moment, the travel switch SQ3 is pressed, the normally open contact of SQ3 is closed, the normally closed contact of SQ3 is open, the contactor KM4 is energized, and the motor M2 rotates clockwise to realize the downward motion of workbench. When the workbench reaches the desired position, it stops moving by releasing the button. Conversely, if the feed handle is turned upward, the SQ4 will be pressed, KM5 is energized, the motor M2 rotates counterclockwise to realize the upward motion of workbench.

2) 工作台向前（后）运动

2) Forward (backward) motion of workbench

在主轴电机启动后，把装在床身一侧的转换开关扳到"横向"位置，拨动进给手柄，压下 SQ3 或 SQ4 可操作工作台向前或向后运动。

After the startup of spindle motor, if the user turns the change-over switch on one side of lathe to the "forward-backward" position and toggles the feed handle, the SQ3 or SQ4 will be pressed to realize the forward or backward motion of workbench.

3. 连锁问题

3. Interlocking

铣床在上，下，前，后四个方向进给时，又操作纵向控制这两个方向的进给，将造成铣床重大事故，所以必须连锁保护。当上，下，前，后四个方向进给时，SQ3-2 或 SQ4-2 中的某一个断开，KM4 或 KM5 中的某一个得电，若操作纵向任一方向，SQ1-2 或 SQ2-2 两个触点中的一个断开，接触器 KM4 或 KM5 立刻失电，电机 M2 停转，从而得到保护。

If the feeding operation is controlled along the two longitudinal directions when the miller is executing the feeding operation along the upward, downward, forward and backward directions, the miller will suffer from serious accident, thus the interlocking protection must be provided. When the miller is executing the feeding operation along the upward, downward, forward and backward directions, one of SQ3-2 or SQ4-2 is open, one of KM4 or KM5 is energized, if the motion in any longitudinal direction is controlled, one of SQ1-2 or SQ2-2 contact is open, the contactor KM4 or KM5 is immediately de-energized, and the motor M2 stops running to realize the protection.

同理，当纵向操作时又操作某一方向而选择了向左或向右进给时，SQ1 或 SQ2 被压下，它们的常闭触点 SQ1-2 或 SQ2-2 是断开的，接触器 KM4 或 KM5 都通过 SQ3-2 和 SQ4-2 串联的这条回路接通。若发生误操作，选择了上，下，前，后某一方向的进给，就一定使 SQ3-2 或 SQ4-2 中的某一个断开，使 KM4 或 KM5 断电释放，电机 M2 停止运转，避免了机床事故。

Similarly, if the user operates the miller in a certain direction to make it feed to the left or right at the time of longitudinal operation, SQ1 or SQ2 is pressed, its normally closed contact SQ1-2 or SQ2-2 is open, the contactor KM4 or KM5 is connected via the series loop of SQ3-2 and SQ4-2. In case of false operation enabling the miller to feed along a certain direction i. e. upward, downward, forward or backward, the user must open one of SQ3-2 or SQ4-2, de-energize and release KM4 or KM5, and stop the motor M2 to avoid the machine accident.

（1）进给冲动

(1) Feed impulse

为在改变进给速度时使齿轮顺利啮合。在选择好进给速度将变速盘向里推进时，挡块压动位置开关 SQ6，首先使常闭触点 SQ6-2 断开，然后常开触点 SQ6-1 闭合，接触器 KM4 通电吸合，电机 M2 启动。但它并未转起来，位置开关 SQ6 已复位，首先断开 SQ6-1，而后闭合 SQ6-2。接触器 KM4 失电，电机失电停转。这样一来，使电机接通一下电源，齿轮系统产生一次抖动，使齿轮啮合顺利进行。要冲动时按下冲动开关 SQ6，模拟冲动。

To engage the gear smoothly at the time of changing the feed rate. When the variable gear plate is pushed inward at the selected feed rate, the stop block presses the position switch SQ6 to firstly open the normally closed contact SQ6-2, and then close the normally open contact SQ6-1, energize and engage the contactor KM4, and start up the motor M2. But the motor can't be started up, the position switch SQ6 has been recovered to firstly open SQ6-1 and then close SQ6-2. The contactor KM4 is de-energized, and the motor stops running due to the power loss. So, the user will power on the motor, the gear system will jitter once in order to engage the gear smoothly. If the impulse is required, the user can press the impulse switch SQ6 to simulate the impulse.

（2）工作台的快速移动

(2) Quick movement of workbench

在工作台向某个方向运动时，按下按钮 SB5 或 SB6（两地控制），接触器闭合 KM6 通电吸合，它的常开触点（4 区）闭合，电磁铁 YB 通电（指示灯亮）模拟快速进给。

When the workbench moves along a certain direction, the user presses the button SB5 or SB6 (remote and local control), then the contactor KM6 will be energized and engaged, its normally open contact (area 4) is closed, the electromagnet YB is energized (indicator is on) to simulate the quick feed.

（3）圆工作台的控制

(3) Control of round workbench

把圆工作台控制开关 SA1 扳到"接通"位置，此时 SA1-1 断开，SA1-2 接通，SA1-3 断开，主轴电机启动后，圆工作台即开始工作，其控制电路是：电源进线＞SQ4-2＞SQ3-2＞SQ1-2＞SQ2-2＞SA1-2＞KM4 线圈＞电源出线。接触器 KM4 通电吸合，电机 M2 运转。

The control switch SA1 of round workbench is turned to the "On" position, SA1-1 is disconnected, SA1-2 connected and SA1-3 disconnected at this moment, and the round workbench starts running after the startup of spindle motor, of which its control circuit is shown as follows: Power incoming line ＞SQ4-2＞ SQ3-2＞SQ1-2＞SQ2-2＞SA1-2＞KM4 coil＞power outgoing line. The motor M2 will be started up after the contactor KM4 being energized and engaged.

现实中铣床为了扩大机床的加工能力，可在机床上配备安装圆工作台，这样可以进行圆弧或凸轮的铣削加工。拖动时，所有进给系统均停止工作，只让圆工作台绕轴心回转。该电动带动一根专用轴，使圆工作台绕轴心回转，铣刀铣出圆弧。在圆工作台开动时，其余进给一律不准运动，若有误操作动了某个方向的进给，则必然会使开关 SQ1～SQ4 中的某一个常闭触点断开，使电机停转，从而避免了机床事故的发生。按下主轴停止按钮 SB3 或 SB4，主轴停转，圆工作台也停转。

In order to increase the processing capacity，the real miller can be equipped with a round workbench such that the arc or cam milling can be completed. All feed systems are interrupted at the time of dragging and only the round workbench rotates around the axis. The motor drives one special spindle to make the round workbench rotate around the axis and the milling cutter operate in the arc shape. When the round workbench is enabled，other feed systems must be kept static，but if the feed operation along a certain direction is actuated by the false operation，one of normally closed contacts SQ1 to SQ4 will be certainly open to stop the motor running，thus avoiding the occurrence of machine accident. The user can also press the spindle stop button SB3 or SB4 to stop the spindle rotating，thus stopping the round workbench.

4. 冷却照明控制

4. Cooling and lighting control

要启动冷却泵时开关扳到 SA3，接触器 KM1 通电吸合，电机 M3 运转冷却泵启动。机床照明由变压器 T 供给 36V 电压，工作灯由 SA4 控制。

To enable the cooling pump，the user should turn the switch SA3 to energize and engage the contactor KM1 and start up the motor M3，thus enabling the cooling pump. The machine lighting is enabled under 36V voltage supplied by the transformer T，and the working lamps are controlled by SA4.

（二）机床电路故障的检测和排除方法

（Ⅱ）Detection and troubleshooting of machine circuit fault

1. 故障现象的调查

1. Investigation of fault phenomenon

机床发生故障后，首先应切断机床的电源，并询问机床操作人员，故障发生前后的情况，这样可以还原出故障发生时的情形，有助于判断故障发生的原因。之后应检查熔断器内的熔体是否熔断、导线连接螺钉是否松动、触点的活动是否受限等，如果有此类现象发生，先将这些现象排除后再进行下一步的检查。接下来可以用手触摸电机、变压器等电器元件是否存在发热的现象。如果存在显著发热的现象，应考虑有短路或者过载的情况，重点检查通过电流大、动作频繁的元件。

In case of machine fault，the investigator should firstly power off the machine，inquire the machine operator of the machine condition before and after the fault，in such way the circumstance under which the fault occurs can be restored to facilitate to find out the fault reasons. Then，the investigator should check whether the fuse element is burned out，the conductor jointing screw is loose，the contact action is limited，etc.，if there are such phenomena，they should be eliminated before the further check. Next，the investigator can touch the motor，transformer and other electrical components for overheating by hand. If there is obvious overheating phenomenon，the consideration should be given to the short circuit or overload problem，and the investigator should emphatically check the possibility of faults experienced by these components with large current and frequent action.

2. 通电试车

2. Power on and commissioning

在经过调查研究了解设备的状况，大概确定故障现象后，调查人员会对故障发生的原因有另一个预先的判断。在保证不会造成触电事故、不会损坏设备的前提下，可以通过试车对故障现象进行确认，主要是观察电路设备的动作情况和预先的判断的是否一致，这样就需要结合电路原理图来进行进一步的分析。例如，在对 X62W 铣床进行试车时，应根据前述的电路控制原理的顺序，从主轴运动开始，到进给运动的控制，再到冷却泵的控制，逐一去验证每一项的功能是否能够实现，如果不能实现，则应该给出故障现象更详细的描述。例如，在试车时发现主轴电机不转，这是一个很笼统的故障现象，引起该故障的原因可能是主电路缺相，也可能是控制电路断路。此时要进一步的观察故障现象，看接触器是否得电动作。如果接触器动作，应考虑是主电路的问题；反之，则是控制电路的问题。通过试车，可以得到故障更具体的现象，这是之前的查询和观察无法获得的，接下来就可以通过逻辑分析和判断来确定故障所在。

The investigator will have another beforehand judgment on the fault reasons after knowing the machine condition and roughly determining the fault phenomena through the investigations. Under the premise of guaranteeing that there is no electric shock accident and machine damage, the fault phenomena can be confirmed by commissioning to mainly observe the actions of electrical machine are consistent with the beforehand judgment such that it is necessary to combine with the schematic circuit diagram for the further analysis. For example, the investigator should check the spindle movement, the control of feed motion and the control of cooling pump in the sequence specified on the prescribed schematic circuit diagram at the time of commissioning X62W miller in order to verify that each function can be realized one by one, and give more detailed description of the fault phenomenon if any function isn't realized. For example, if the spindle motor can't run at the time of commissioning, the fault phenomenon is very general and may be attributive to the phase loss of main circuit or the disconnection of control circuit. At this moment, the investigator should further observe the fault phenomenon and check whether the contractor can be actuated after being electrified. If the contactor can be actuated, the consideration should be given to the main circuit fault; conversely, it should be due to the control circuit fault. More specific fault phenomena can be found through the commissioning, which can't be obtained from the previous inquiry and observations, next the investigator can determine the fault position by the logic analysis and judgment.

3. 检测故障点

3. Detection of fault point

通过逻辑分析，能够确定故障的范围，但是要确定故障点，必须利用仪器、仪表对故障的线路检查测量。机床电路故障的排除方法，就是先在一个较大的范围分析故障现象，也就是在"面"的范围内确定故障的位置。然后通过粗略的测量，进一步缩小故障可能存在的范围，确定故障的"线"。最后，再通过细致的测量，确定故障"点"，找到故障。这个方法可以总结成"由面到线、由线到点"，对所有电路以大化小，整个过程都离不开对故障现象的分析和推理。为保证安全，在检测过程中，应全程断电，利用万用表的欧姆挡对电路进行测量。

The range of fault can be determined by the logic analysis, but the fault point is determined only after the check and measurement by the apparatus and instrument. The troubleshooting methods of machine's circuit fault are to firstly analyze the fault phenomena, and then determine the fault position within a larger range, i. e. the "plane" range. Then the rough measurement is carried out to further reduce the possible range of fault and determine the "line" of fault. Finally, the fine measurement is carried out to determine the "point" of fault and find out the fault. The method can be summarized to minimize all circuits from "plane to line and line to point", and the entire process can't be separated from the analysis and inference of fault phenomena. In order to ensure the safety, the circuit should be measured in the ohms range by the multi-meter without the electricity in the entire process.

4. 排除故障，通电试车

4. Troubleshooting, and power-on commissioning

找到故障点后，接着就是排除故障。如果是电器元件有损坏，就更换电气元件。如果是导线松脱，就紧固导线。故障排除后，通电试车复查，完成故障排除任务。

The troubleshooting is carried out immediately after finding out the fault point. If the electrical component is damaged, the damaged should be replaced by a new one. If the conductor becomes loose, it should be tightened. After the troubleshooting, the machine should be energized and commissioned to finally complete the troubleshooting task.

三、任务实施

Ⅲ. Task Implementation

以 X62W 铣床照明灯不亮为例，说明排除机床电路故障的方法。

The method to troubleshoot the circuit fault of machine will be explained with an example that X62W miller's lighting lamp is off.

（1）通过查询和断电观察，发现故障的现象是 X62W 铣床照明灯 EL 不亮。电路元件均未有烧坏迹象，熔断器也没有熔断。电机及变压器的温度也正常。初步判定为电路中有断路，有可能断路出现在电源部分，导致整个照明和显示电路没有供电。这只是在一个比较大的范围内确定了故障的位置。

（1）It can be learned from the inquiry and power-off observation that the fault phenomenon is X62W miller's lighting lamp EL is off. The electrical components are free of burnout indication, and the fuse is also intact. The temperature of motor and transformer is also normal. It can be preliminarily determined that the fault is attributive to the open circuit, the open circuit possibly exists in the power supply, resulting in the entire lighting and display circuit isn't energized. The fault position is only determined with a larger range.

（2）通电试车，发现除了照明灯 EL 不亮之外，其他指示灯都亮。这说明断路没有发生在电源和变压器上，而是发生在了 EL 所在的回路中。具体来说就是从 184 到 192 之间有断路。这样就进一步将故障范围缩小了。

（2）It is found from the power-on commissioning that other indicators are on except the lighting lamp EL. It indicates that the open circuit exists in the circuit where EL is located, rather than the power supply and transformer. More specially, there is open circuit from 184 to 192. Therefore, the range of fault is further reduced.

（3）在确定具体的故障点时，需要用到万用表对电路进行测量。先将铣床电路的电源断开，将万用表打到欧姆挡，将开关 SA4 接通，再依次测量 184 到 187、187 到 188、188 到 189、189 到 190、190 到 191、191 到 192 之间的电阻，如果电阻不为无穷大，即说明该段电路没有断路。如果电阻为无穷大，就说明该段电路发生了断路，应将该段导线重新连接。测试结果表明 184 到 187 之间电阻为无穷大，发生了断路，将导线重新连接即可。

（3）The multi-meter will be used for the circuit measurement in order to determine the specific fault point. The tester should firstly disconnect the power supply to the miller, turn the multi-meter to the ohms range, switch on SA4, and then successively measure the resistance from 184 to 187, 187 to 188, 188 to 189, 189 to 190, 190 to 191, and 191 to 192. If the resistance isn't infinitely great, it indicates that the section isn't open. If the resistance is infinitely great, it indicates that the section is open, and the conductor should be reconnected. The test results indicate that the resistance from 184 to 187 is infinitely great, the section is open, and the conductor should be reconnected.

（4）再次通电试车，照明灯亮，故障排除。

（4）The lighting lamp is on in the power-on commissioning, indicating that the fault has been troubleshot.

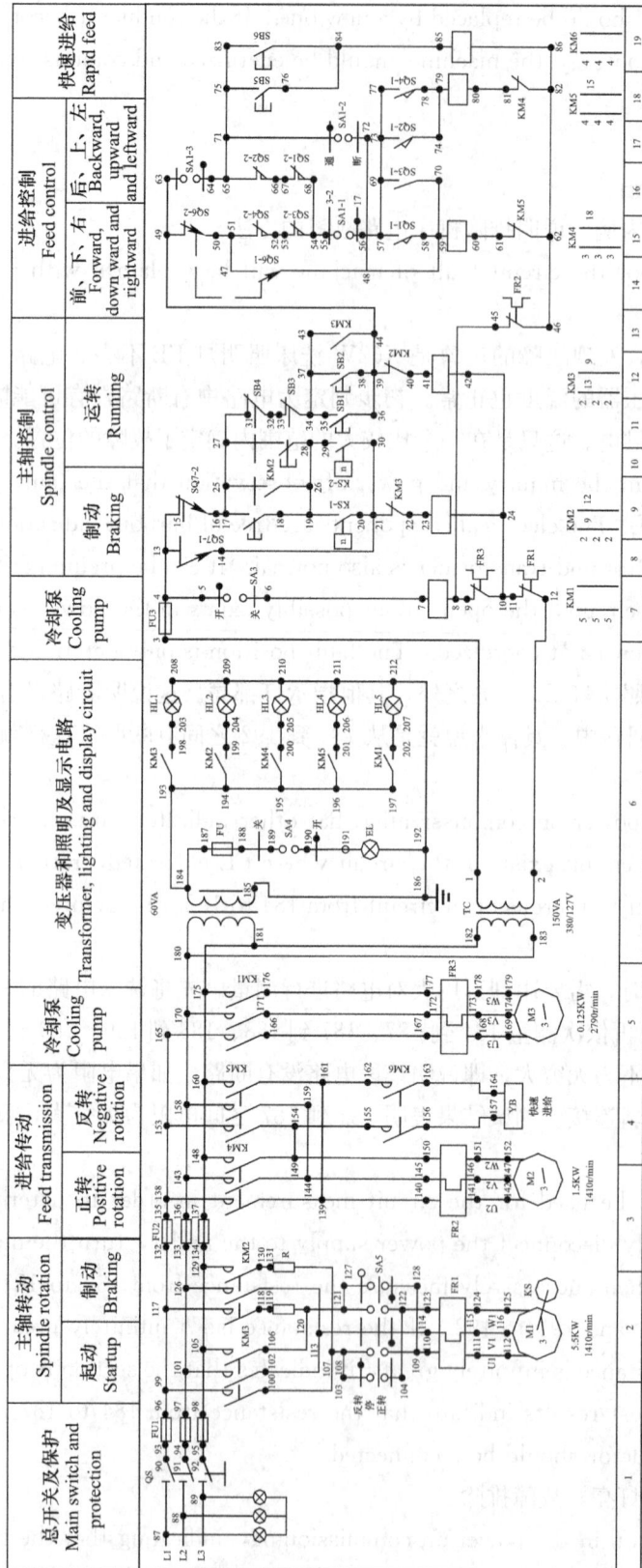

图2.4.1 X62W铣床电气原理图

Figure 2.4.1 Electrical schematic diagram of X62W miller

维修工作票
Maintenance arrangement

工作票编号 No.：_____
Work Arrangement No.：_____
发单日期：20 年　月　日
Issuance of date ×× ××，20××

工位号 **Station No.**	
工作任务 **Work task**	X62W 铣床电气线路故障检测与排除 Electrical circuit fault detection and troubleshooting of X62W miller
工作时间 **Working hours**	自__年__月__日__时__分至__年__月__日__时__分 From Hour Minute on /MM/DD/YY to Hour Minute on /MM/DD/YY
工作条件 **Work conditions**	登录学号：（即两位数的工位号，如：01、10、20 等） Logon number：（two-digit station number，for example：01，10 and 20） 登录密码：无 Logon password；None 观察故障现象和排除故障后试机通电；检测及排故过程停电 Observe the fault phenomena，and carry out the power-on commissioning after the troubleshooting；Detect and power off in the troubleshooting process
工作许可人签名 **Signature of the work approver**	
维修要求 **Maintenance requirements**	1. 在工作许可人签名后方可进行检修 1. The maintenance can be performed with the signature of work approver 2. 对电气线路进行检测，确定线路的故障点并排除调试填写下列表格 2. Detect the electrical circuit，determine and eliminate the fault point in the circuit，and fill in the below form 3. 严格遵守电工操作安全规程 3. Strictly abide by the safety procedures for electrical operation 4. 不得擅自改变原线路接线，不得更改电路和元件位置 4. Don't change the original wiring，and the position of circuit and electrical components without the authorization 5. 完成检修后能恢复该铣床各项功能 5. Restore all functions of miller after the detection and maintenance
故障现象描述 **Description of fault phenomena**			
故障检测和排除过程 **Fault detection and troubleshooting process**			
故障点描述 **Description of fault point**			

图 2.4.2　维修工作票

Figure 2.4.2　Maintenance arrangement

项目二思考题

Questions for Item Ⅱ

1. 时钟存储器字节的哪一位的时钟脉冲周期为 500ms？

1. Which bit of clock memory bytes has the clock period is 500ms?
2. RLO 是什么意思？
2. What does RLO mean?
3. 铣床电路中的三相电机分别起什么作用？
3. What are the functions of three-phase motor in the miller circuit?

项目三 三相电机和直流电机的调速控制
Item III Speed Control of Three-phase Motor and DC Motor

任务一 三相电机的多段调速控制
Task I Multi-segment Speed Control of Three-phase Motor

一、任务分析
I. Task Analysis

用西门子 S7-1200 系列 PLC 与 G120C 变频器对三相电机进行多段调速控制。变频器参数设置为第一段速度为 150rpm，第二段速度为 250rpm，第三段速度为 350rpm、第四段速度为 500rpm，加速时间 0.6s，减速时间 0.4s。

To perform the multi-segment speed control of three-phase motor by Siemens S7-1200 series PLC and G120C frequency converter. The parameters of frequency converter are set to 150rpm in the first segment, 250rpm in the second segment, 350rpm in the third segment, 500rpm in the fourth segment, 0.6s acceleration time and 0.4s deceleration time.

要求系统具有以下功能：

The system has the following functions：

（1）按下启动按钮 SB1，电机以 250rpm 正转启动；

（1）The motor can rotate clockwise at the speed of 250rpm by pressing the startup button SB1；

（2）再次按下 SB1 按钮，电机以 350rpm 反转运行；

（2）The motor can rotate counterclockwise at the speed of 350rpm by pressing the startup button SB1 again；

（3）再次按下 SB1 按钮，电机以 150rpm 正转运行；

（3）The motor can rotate clockwise at the speed of 150rpm by pressing the startup button SB1 again；

（4）再次按下 SB1 按钮，电机以 500rpm 正转运行；

（4）The motor can rotate clockwise at the speed of 500rpm by pressing the startup button SB1 again；

（5）按下停止按钮 SB2，电机停止。

（5）The motor can stop the rotation by pressing the stop button SB2.

为了完成本任务，需要在掌握西门子 G120C 变频器的基本安装和调试方法的基础上学习固定转速设定值调速的设定方法。

For the purpose of this task，it is necessary to learn about the methods of speed control at the speed setting value of Siemens G120C frequency converter on the basis of mastering the basic installation and commissioning methods.

二、相关知识技能

II. Relevant Knowledge & Skills

（一）西门子 **G120C** 变频器的基本安装和调试方法

（Ⅰ）**Basic installation and commissioning methods of Siemens G120C frequency converter**

1. 将变频器及其组件接入电网

1. To connect the frequency converter and its components to the power grid

根据变频器的使用环境，选择适当的组件并将其接入电网，如图 3.1.1 所示。通常与变频器一起使用的组件有输入电抗器、电源滤波器、输出电抗器等，这些组件起到保护电机或变频器的作用。

The proper components are selected and connected to the power grid based on the service environment of frequency converter，as shown in Figure 3.1.1. Generally，the components combined with the frequency converter include the input reactor，power filter，output reactor，etc.，these components can protect the motor or frequency converter.

变频器的底部有电源、电机和制动电阻的接口，如图 3.1.2 所示。注意，电源接口和电机接口不可互换，否则会烧毁变频器。

There are interfaces for connecting to power，motor and braking resistance on the bottom of frequency converter，as shown in Figure 3.1.2. It is noted that the power interface and motor interface can't be interchanged，otherwise the frequency converter would be burned out.

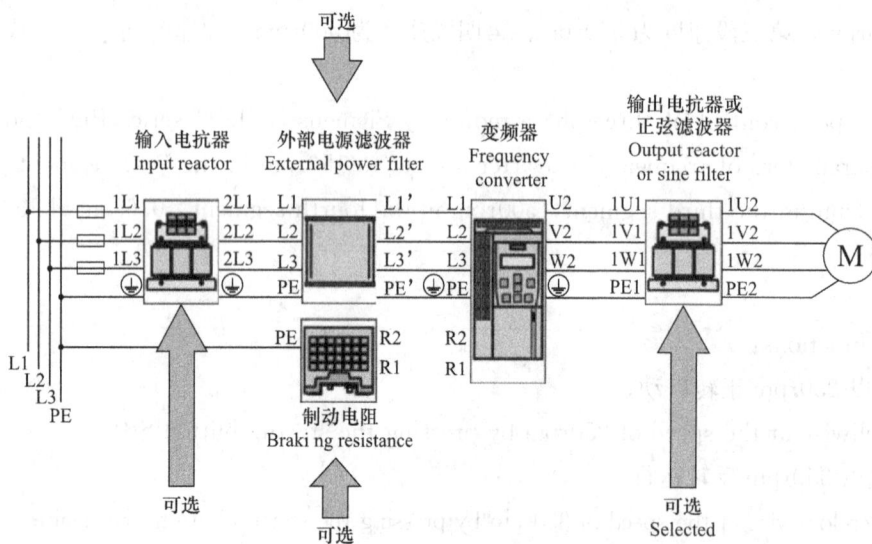

图 3.1.1　G120C 变频器及其组件的连接
Figure 3.1.1　G120C frequency converter，and connections among its components

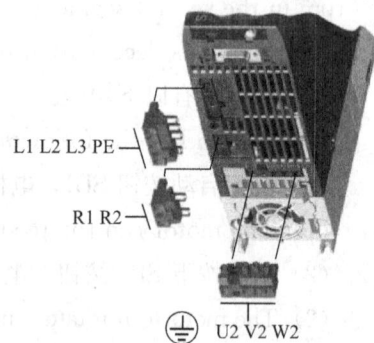

图 3.1.2　变频器底部的接口
Figure 3.1.2　Bottom interface of frequency converter

2. 连接用于变频器控制的接口

2. To connect the interfaces for controlling the frequency converter

拆下操作面板（如果有）并打开前盖才可以操作控制单元正面的接口，各端子排和接口的名称如图 3.1.3 所示。

The interfaces on the front of control units can be operated only after removing the operator panel（if any）and opening the front cover，and the names of terminal boards and interfaces as shown in Figure 3.1.3.

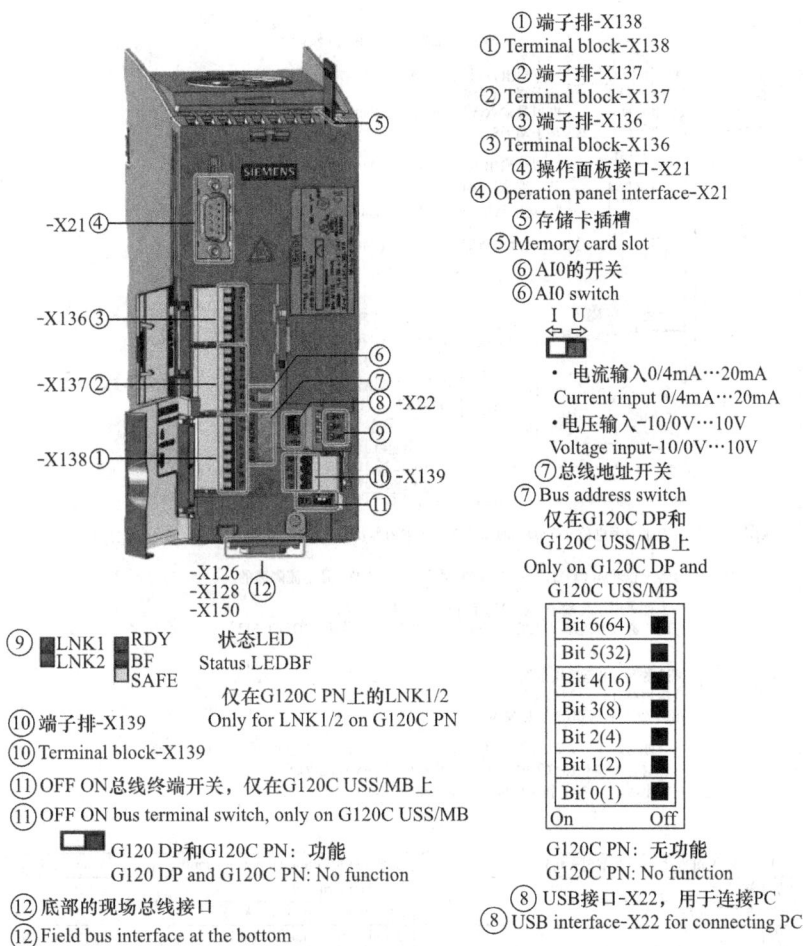

①端子排-X138
① Terminal block-X138
②端子排-X137
② Terminal block-X137
③端子排-X136
③ Terminal block-X136
④操作面板接口-X21
④ Operation panel interface-X21
⑤存储卡插槽
⑤ Memory card slot
⑥AI0的开关
⑥ AI0 switch

I U

• 电流输入0/4mA…20mA
Current input 0/4mA…20mA
• 电压输入-10/0V…10V
Voltage input-10/0V…10V
⑦总线地址开关
⑦ Bus address switch
仅在G120C DP和
G120C USS/MB上
Only on G120C DP and
G120C USS/MB

Bit 6(64)	
Bit 5(32)	
Bit 4(16)	
Bit 3(8)	
Bit 2(4)	
Bit 1(2)	
Bit 0(1)	
On	Off

G120C PN：无功能
G120C PN: No function
⑧USB接口-X22，用于连接PC
⑧ USB interface-X22 for connecting PC

⑨ LNK1 RDY　状态LED
LNK2 BF　Status LEDBF
SAFE
仅在G120C PN上的LNK1/2
Only for LNK1/2 on G120C PN
⑩端子排-X139
⑩ Terminal block-X139
⑪OFF ON总线终端开关，仅在G120C USS/MB上
⑪ OFF ON bus terminal switch, only on G120C USS/MB
G120 DP和G120C PN：功能
G120 DP and G120C PN: No function
⑫底部的现场总线接口
⑫ Field bus interface at the bottom

图 3.1.3　变频器的正面接口
Figure 3.1.3　Front interfaces of frequency converter

各端子排的功能如图 3.1.4 所示。其中，DI 指的是数字量输入端子，DO 指的是数字量输出端子，AI 指的是模拟量输入端子，AO 指的是模拟量输出端子。它们的接口电路与能接受的电压都在图 3.1.4 中有详细的说明，它们的功能需要通过设定变频器的参数来定义。

The functions of all terminal boards are shown in Figure 3.1.4. Of which DI is the digital input terminal，DO is the digital output terminal，AI is the analog input terminal，and AO is the analog output terminal. Their interface circuits and acceptable voltages are detailed in Figure 3.1.4，and their functions should be defined by setting the parameters of frequency converter.

3. 变频器的接通电机和停止电机的指令

3. Motor startup and stop instructions of frequency converter

接通电源电压后，变频器通常都会进入"接通就绪"状态，在该状态下，变频器会一直等待接通电机的命令。在收到 ON 指令后，变频器会接通电机，进入运行状态。在发出 OFF1 指令后，变频器对电机进行制动，直至静止。在电机静止后，变频器会将其关闭。变频器又回到"接通就绪"状态。除了 OFF1 指令让电机停止之外，变频器还有 OFF2 和 OFF3 指令能让电机停止。OFF2 指令发出后，变频器立即关闭电机，不对其进行制动，电机依靠惯性停止。OFF3 指令发出后，变频器从"运行"状态进入"快速停止"状态。"快速停止"时变频器以 OFF3 减速时间使电机制动。在电机静止后，变频器会将其关闭。OFF2 和 OFF3 都不是让电机正常停止的指令，一般在变频器出现故障时或非正常运行状态下使用。

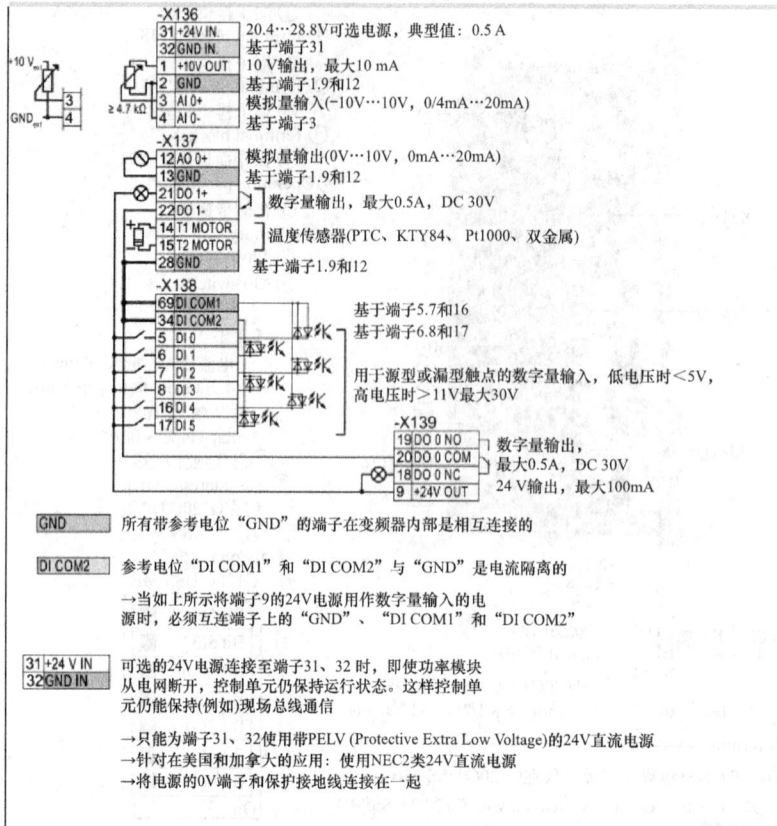

GND	所有带参考电位"GND"的端子在变频器内部是相互连接的

DI COM2	参考电位"DI COM1"和"DI COM2"与"GND"是电流隔离的

→当如上所示将端子9的24V电源用作数字量输入的电源时,必须互连端子上的"GND"、"DI COM1"和"DI COM2"

31 +24 V IN 32 GND IN	可选的24V电源连接至端子31、32时,即使功率模块从电网断开,控制单元仍保持运行状态。这样控制单元仍能保持(例如)现场总线通信

→只能为端子31、32使用带PELV (Protective Extra Low Voltage)的24V直流电源
→针对在美国和加拿大的应用:使用NEC2类24V直流电源
→将电源的0V端子和保护接地线连接在一起

20.4…28.8V可选电源,典型值: 0.5A	20.4…28.8V optional power supply, typical value: 0.5A
基于端子31	Based on terminal 31
10V输出,最大10mA	10V output, up to 10mA
基于端子1.9和12	Based on terminal 1.9 and 12
模拟量输入(-10V…10V, 0/4mA…20mA)	Analog quantity input (-10V…10V, 0/4mA…20mA)
基于端子3	Based on terminal 3
模拟量输出(0V…10V, 0mA…20mA)	Analog quantity output (0V…10V, 0mA…20mA)
基于端子1.9和12	Based on terminals 1.9 and 12
数字量输出,最大0.5A, DC 30V	Digital output, up to 0.5 A, DC 30V
温度传感器(PTC、KTY84、Pt1000、双金属)	Temperature sensors (PTC, KTY84, Pt1000, bimetal)
基于端子1.9和12	Based on terminals 1.9 and 12
基于端子5.7和16	Based on terminals 5.7 and 16
基于端子6.8和17	Based on terminals 6.8 and 17
用于源型或漏型触点的数字量输入,低电压时<5V,高电压时>11V最大30V	Digital input of source or sink pattern contacts, <5V at low voltage and>11V at high voltage but maximum 30V
数字量输出,最大0.5A, DC 30V	Digital output, Max. 0.5A, DC 30V
24V输出,最大100mA	24V output, up to 100mA
所有带参考电位"GND"的端子在变频器内部是相互连接的	All terminals with reference potential "GND" are interconnected inside the frequency converter
参考电位"DI COM1"和"DI COM2"与"GND"是电流隔离的	The reference potentials "DI COM1" and "DI COM2" are galvanically isolated from "GND"
→当如上所示将端子9的24V电源用作数字量输入的电源时,必须互连端子上的"GND"、"DI COM1"和"DI COM2"	When using the 24V power supply of terminal 9 as shown above to power the digital inputs, it is required to interconnect "GND", "DI COM1" and "DI COM2" on the terminals
可选的24V电源连接至端子31、32时,即使功率模块从电网断开,控制单元仍保持运行状态。这样控制单元仍能保持(例如)现场总线通信	When the optional 24V is connected to terminals 31 and 32, the control unit remains operational even if the power module gets away from the grid. In this way, the control unit can still maintain the fieldbus communication, for example
→只能为端子31、32使用带PELV (Protective Extra Low Voltage)的24V直流电源	→ Use only 24V DC power supply with PELV (Protective Extra Low Voltage) for terminals 31 and 32
→针对在美国和加拿大的应用:使用NEC2类24V直流电源	→ For applications in the US and Canada: Use NEC Class2 power supply(DC 24V)
→将电源的0V端子和保护接地线连接在一起	→Connect 0V terminal of the power supply to the protective ground wire

图 3.1.4 变频器端子排的功能

Figure 3.1.4 Functions of frequency converter's terminal boards

After switching on the power，the frequency converter generally enters into the "Ready" state under which it will always wait for the motor startup instruction. After receiving ON instruction，the frequency converter will turn on the motor to enter the running state. After issuing OFF1 instruction，the frequency converter will restrict the motor rotation until to stop. After stopping the rotation，the frequency converter will turn off the motor. The frequency converter returns to the "Ready" state. Except that OFF1 instruction can stop the motor，its OFF2 and OFF3 instructions can also stop the motor. After issuing OFF2 instruction，the frequency converter immediately turns off the motor without braking，and the motor will be stopped depending on its inertia. After issuing OFF3 instruction，the frequency converter enters into the "Quick Stop" state from "Run" state. At the time of "Quick Stop"，the frequency converter stops the motor within the deceleration time for OFF3. After stopping the rotation，the frequency converter will turn off the motor. OFF2 and OFF3 are the instructions to abnormally stop the motor，and often applied in the event that the frequency converter goes wrong or under the abnormal run state.

变频器通过两个或三个指令控制电机的正反转，不同的控制方法如图 3.1.5 所示。

The frequency converter controls the positive and negative rotation of motor by two or three instructions，and the different control methods as shown in Figure 3.1.5.

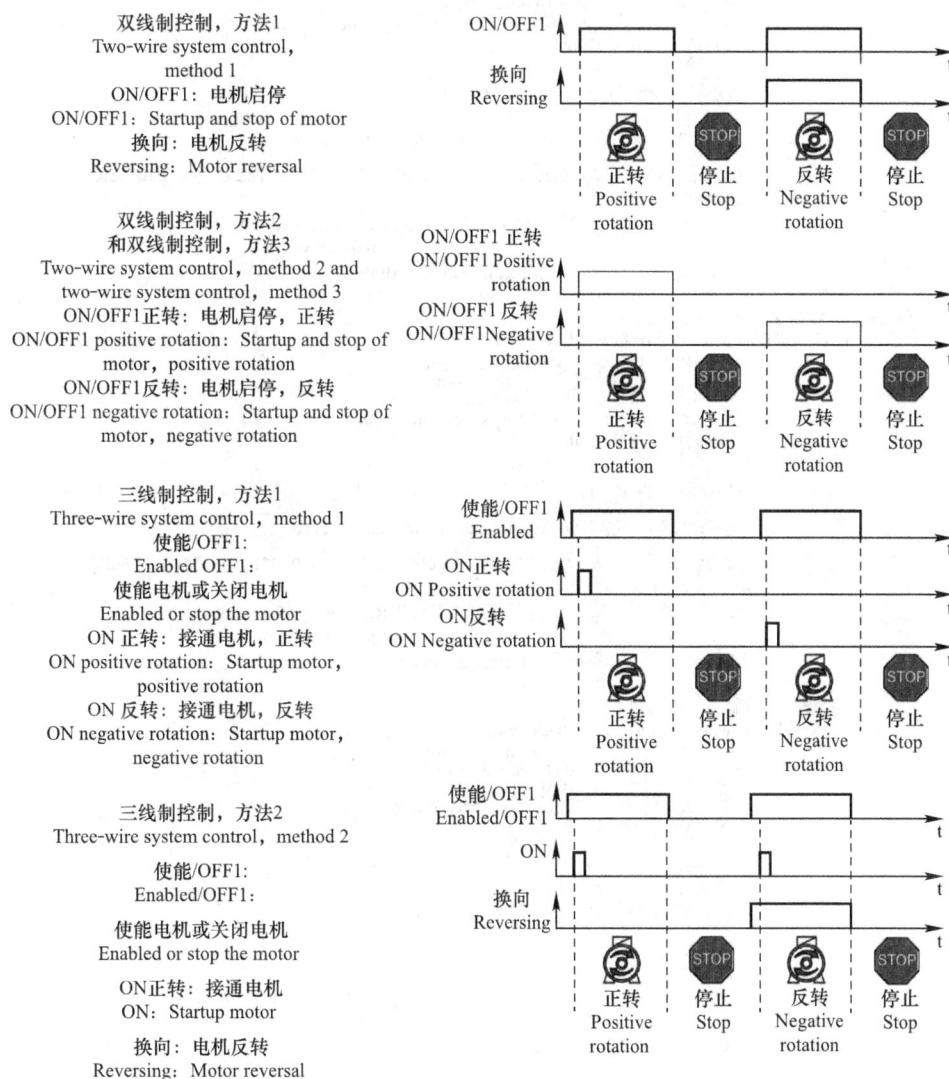

图 3.1.5 电机正反转的控制方法

Figure 3.1.5 Methods for controlling the positive and negative rotation of motor

4. 变频器的预设值

4. Preset values of frequency converter

变频器提供了若干种预设值，它事先设定好了端子排中某些端子的功能。在使用变频器前，只需选择某个预设值，而不需要对这些端子的功能一一进行设置了，这样就可以快速地设置好变频器的各项参数。变频器的各种预设值如图 3.1.6 所示，在快速调试时通过设置 P0015 的值进行选择。

The frequency converter is assigned with several preset values，and it sets the functions of some terminals in the terminal boards in advance. Before using the frequency converter，certain preset value must be selected without the need for setting the functions of these terminals one by one such that various parameters of frequency converter can be quickly set. All preset values of frequency converter are shown in Figure 3. 1. 6，and will be selected by setting the value of P0015 at the time of quick debugging.

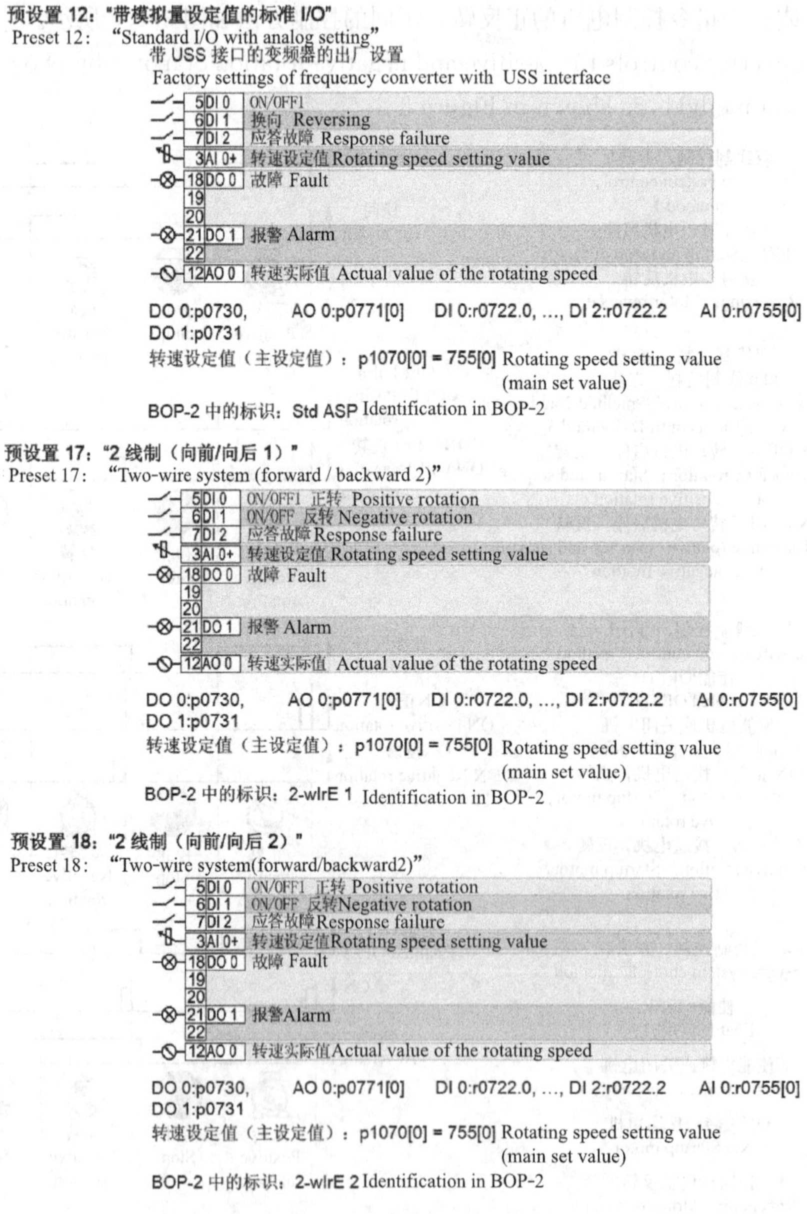

预设置 12："带模拟量设定值的标准 I/O"
Preset 12: "Standard I/O with analog setting"
带 USS 接口的变频器的出厂设置
Factory settings of frequency converter with USS interface

- 5 DI 0 ON/OFF1
- 6 DI 1 换向 Reversing
- 7 DI 2 应答故障 Response failure
- 3 AI 0+ 转速设定值 Rotating speed setting value
- 18 DO 0 故障 Fault
- 19
- 20
- 21 DO 1 报警 Alarm
- 22
- 12 AO 0 转速实际值 Actual value of the rotating speed

DO 0:p0730,　　AO 0:p0771[0]　　DI 0:r0722.0, ..., DI 2:r0722.2　　AI 0:r0755[0]
DO 1:p0731
转速设定值（主设定值）：p1070[0] = 755[0] Rotating speed setting value
(main set value)
BOP-2 中的标识：Std ASP Identification in BOP-2

预设置 17："2 线制（向前/向后 1）"
Preset 17: "Two-wire system (forward / backward 2)"

- 5 DI 0 ON/OFF1 正转 Positive rotation
- 6 DI 1 ON/OFF 反转 Negative rotation
- 7 DI 2 应答故障 Response failure
- 3 AI 0+ 转速设定值 Rotating speed setting value
- 18 DO 0 故障 Fault
- 19
- 20
- 21 DO 1 报警 Alarm
- 22
- 12 AO 0 转速实际值 Actual value of the rotating speed

DO 0:p0730,　　AO 0:p0771[0]　　DI 0:r0722.0, ..., DI 2:r0722.2　　AI 0:r0755[0]
DO 1:p0731
转速设定值（主设定值）：p1070[0] = 755[0] Rotating speed setting value
(main set value)
BOP-2 中的标识：2-wIrE 1 Identification in BOP-2

预设置 18："2 线制（向前/向后 2）"
Preset 18: "Two-wire system(forward/backward2)"

- 5 DI 0 ON/OFF1 正转 Positive rotation
- 6 DI 1 ON/OFF 反转 Negative rotation
- 7 DI 2 应答故障 Response failure
- 3 AI 0+ 转速设定值 Rotating speed setting value
- 18 DO 0 故障 Fault
- 19
- 20
- 21 DO 1 报警 Alarm
- 22
- 12 AO 0 转速实际值 Actual value of the rotating speed

DO 0:p0730,　　AO 0:p0771[0]　　DI 0:r0722.0, ..., DI 2:r0722.2　　AI 0:r0755[0]
DO 1:p0731
转速设定值（主设定值）：p1070[0] = 755[0] Rotating speed setting value
(main set value)
BOP-2 中的标识：2-wIrE 2 Identification in BOP-2

图 3.1.6　变频器的预设值（一）

Figure 3. 1. 6　Preset values of frequency converter（一）

预设置 19："3 线制（使能/向前/向后）"
Preset 19: "Three-wire system(enabled/forward/backward)"

DO 0:p0730,   AO 0:p0771[0]   DI 0:r0722.0, …, DI 4:r0722.4   AI 0:r0755[0]
DO 1:p0731

转速设定值（主设定值）：p1070[0] = 755[0] Rotating speed setting value
(main set value)

BOP-2 中的标识：3-wlrE 1 Identification in BOP-2

预设置 20："3 线制（使能/正转/反转）"
Preset 20: "Three-wire system(enabled/positive rotation/negative rotation)"

DO 0:p0730,   AO 0:p0771[0]   DI 0:r0722.0, …, DI 4:r0722.4   AI 0:r0755[0]
DO 1:p0731

转速设定值（主设定值）：p1070[0] = 755[0] Rotating speed setting value
(main set value)

BOP-2 中的标识：3-wlrE 2 Identification in BOP-2

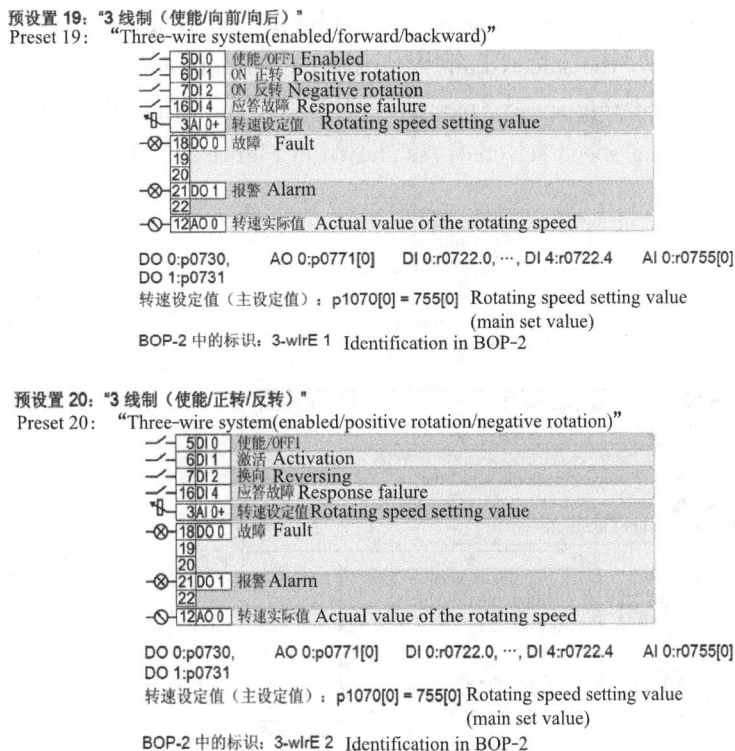

图 3.1.6　变频器的预设值 （二）

Figure 3.1.6　Preset values of frequency converter（二）

5. 使用 BOP-2 操作面板进行快速调试

5. To quickly debug with BOP-2 operator panel

在使用变频器之前，需要对其进行快速调试，其中的一种方法是使用 BOP-2 操作面板。在接通变频器的电源后，按下面板上的 ESC 键，再通过按下向上/向下的箭头键，直至显示 SETUP 菜单，最后按下 OK 键进入快速调试。根据电机的参数和使用需求按步骤对变频器进行调试，如图 3.1.7 所示。

Before using the frequency converter，it must be quickly debugged，and one of methods is to use BOP-2 operator panel. After switching on the frequency converter，operator will successively press the key ESC on the operator panel，the up/down arrow key until to display the SETUP menu，and the key OK to enter into the quick debugging. The frequency converter is debugged by steps based on the motor parameters and usage requirements，as shown in Figure 3.1.7.

（二）将转速固定设定值设为设定值源

（Ⅱ）To set the fixed rotating speed set values to the sources of set values

在很多应用中，只需要电机在通电后以固定转速运转，或在不同的固定转速之间来回切换，例如，输送带在接通后只使用两个不同的速度运行。

In many applications，the motor only requires rotating at the fixed rotating speed or at different rotating speeds which are changed over after being energized，for example，the conveyer belt only operates at different speeds after being energized.

1. 转速固定设定值与主设定值互联

1. Interconnection between fixed rotating speed set value and main set value

在通过外部端子或者通信指令选择了某个固定转速设定值后，参数 r1024 显示选中并生效的转速固定设定值。将此值与转速的主设定值（参数 P1070）关联，意味着将当前电机的转速设置为选中的转速固定值，如图 3.1.8 所示。

After certain fixed rotating speed set value is selected by the external terminal or communication instruction, the parameter r1024 shows the selected and effective fixed rotating speed set value. The value is correlated to the main set value of rotating speed (parameter P1070), indicating that the rotating speed of current motor is set to the fixed rotating speed selected, as shown in Figure 3.1.8.

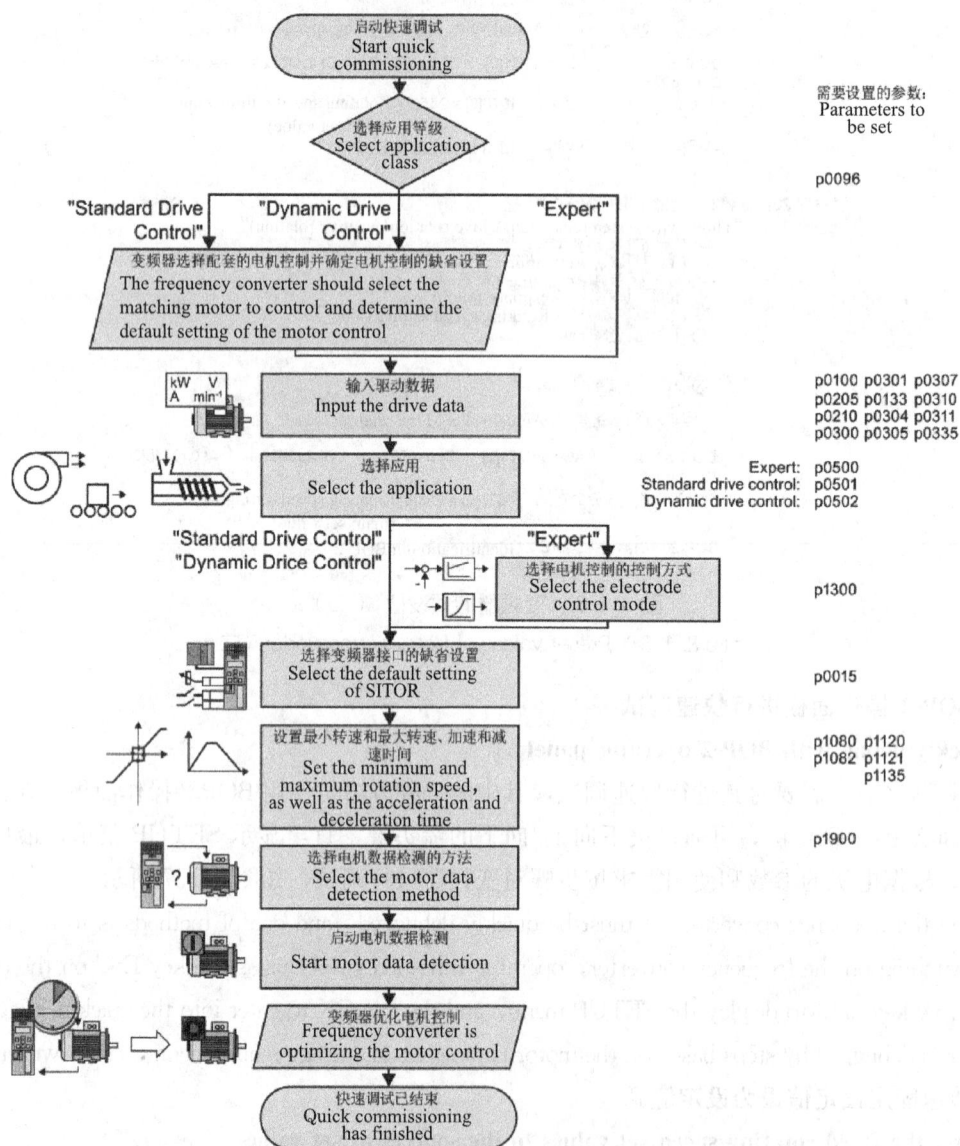

图 3.1.7 快速调试的步骤

Figure 3.1.7 Steps of quick debugging

2. 转速固定设定值的直接或二进制选择

2. Direct or binary selection of fixed rotating speed set values

变频器提供了两种选择转速固定设定值的方法：转速固定设定值的直接选择和转速固定设定值的二进制选择。直接选择的方法是添加 1～4 个转速固定设定值，通过这 4 个值之间的叠加关系，可得到最多 16 个不同的转速固定设定值，如图 3.1.9 所示。二进制选择的方法是设置 16 个不同的转速固定设定值，通过四个选择位的不同组合，从这 16 个中选择一个转速固定设定值，如图 3.1.10 所示。

The frequency converter has two methods for selecting the fixed rotating speed set values; Direct selection and binary selection of fixed rotating speed set values. The direct selection method is to add 1-4 fixed rotating speed set values，and obtain at most 16 different fixed rotating speed set values depending their superposition relations，as shown in Figure 3.1.9. The binary selection method is to set 16 different fixed rotating speed set values，and select one value from the different combinations of four selection bits，as shown in Figure 3.1.10.

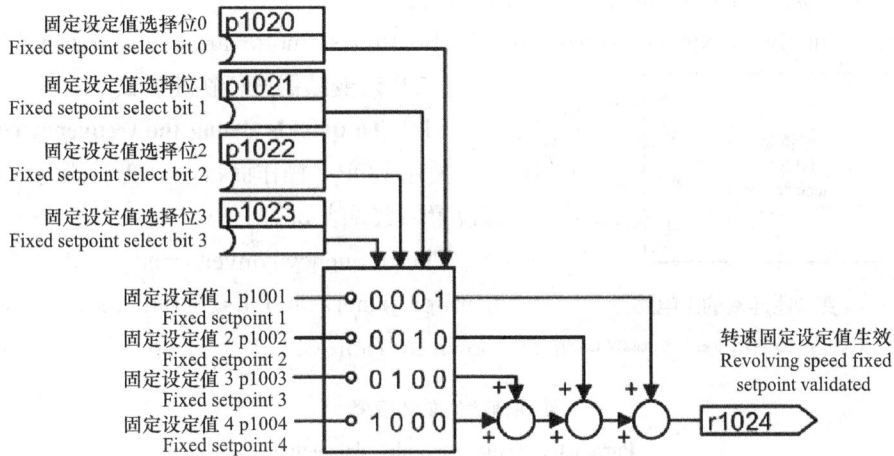

图 3.1.8 直接选择转速固定设定值的简易功能图

Figure 3.1.8 Functional diagram for direct selection of fixed rotating speed set values

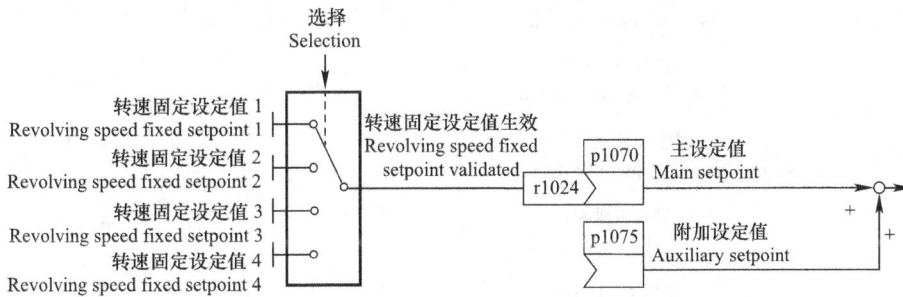

图 3.1.9 转速固定设定值设为设定值源

Figure 3.1.9 Set the fixed rotating speed set values to the set value sources

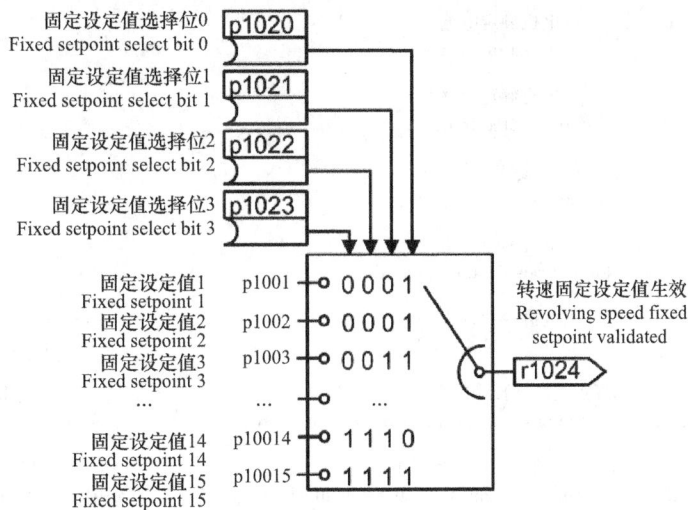

图 3.1.10 二进制选择转速固定设定值的简易功能图

Figure 3.1.10 Functional diagram for binary selection of fixed rotating speed set values

三、任务实施

Ⅲ. Task Implementation

（一）连接变频器与电机的主电路

（Ⅰ）To connect the main circuit between frequency converter and motor

将三相电源、变频器、三相电机如图 3.1.11 所示连接。

The three-phase supply, frequency converter and three-phase motor are connected by Figure 3.1.11.

图 3.1.11 变频器连接的主电路

Figure 3.1.11 Main circuit to connect the frequency converter

（二）对变频器进行快速调试

（Ⅱ）To quickly debug the frequency converter

使用 BOP-2 操作面板，对变频器进行快速调试，按步骤设置参数见表 3.1.1。

The frequency converter is quickly debugged by BOP-2 operator panel, and its parameters are set by steps, as shown in Table 3.1.1

快速调试的参数设置　　　　　　　　　　　　　　　　　表 3.1.1

Parameter setting for quick debugging　　　　　　　　　**Figure 3.1.1**

参数名称 Parameter name	参数功能 Parameter function	设置值 Set value
DRV APPL（P96）	设置应用等级 Set the application class	Standard Drive Control
EUR/USA（P100）	设置电机标准 Set the motor standard	KW 50HZ
INV VOLT（P210）	设置变频器的输入电压 Set the input voltage of frequency converter	380
MOT TYPE（P300）	设置电机类型 Set the motor type	INDUCT
87HZ	电机 87Hz 运行 Motor running at 87Hz	No
MOT VOLT（P304）	电机额定电压 Motor's rated voltage	380
MOT CURR（P305）	电机额定电流 Motor's rated current	0.23
MOT POW（P307）	电机额定功率 Motor's rated frequency	0.04
MOT FREQ（P310）	电机额定频率 Motor's rated frequency	50
MOT RPM（P311）	电机额定转速 Motor's rated rotating speed	1400
MOT COOL（P335）	电机冷却方式 Motor's cooling mode	SELF（自然冷却） SELF（natural cooling）
TEC APPL（P501）	选择点击闭环控制的基础设置 Select and click on the basic settings of closed-loop control	VEC STD（恒定负载） VEC STD（constant load）
MAc PAr（P15）	选择变频器接口的预设值 Select the preset value of frequency converter interface	18［2线制（向前/向后 2）］ 18［two-wire system（forward/backward 2）］
MIN RPM（P1080）	电机最小转速 Motor's minimum rotating speed	0

续表

参数名称 Parameter name	参数功能 Parameter function	设置值 Set value
MAX RPM（P1082）	电机最大转速 Motor's maximum rotating speed	1500
RAMP UP（P1120）	电机的斜坡加速时间 Motor's slope acceleration time	0.6
RAMP DWN（P1121）	电机的斜坡减速时间 Motor's slope deceleration time	0.4
OFF3 RP（P1135）	符合 OFF3 指令的斜坡减速时间 Slope deceleration time corresponding to OFF3 instruction	0
MOT ID（P1900）	选择电机数据检测方式 Select the detection mode of motor data	STILL（测量静止状态下的电机数据） STILL（measure the motor data under the static state)
FINISH	结束快速调试 Complete the quick debugging	Yes

如果在快速调试时选择了一种电机数据检测方式，例如，在静止时测量电机数据，那么在快速调试后，变频器输出报警 A07991。如果此时电机已冷却到环境温度，就可以使用 BOP-2 操作面板检测电机数据。首先按下"HAND/AUTO"键，操作面板中显示手动运行图标 🖐，接着按下绿色的启动按钮 🔲 开始电机数据检测。在进行电机数据检测期间，BOP-2 上的"MOT-ID"会闪烁。如果变频器再次输出报警 A07991，变频器会等待新的 ON 指令用于启动旋转测量。如果变频器不输出报警 A07991，变频器会关闭电机并将变频器控制由 HAND 切换为 AUTO。

If one detection mode of motor data is selected at the time of quick debugging，for example the measurement of motor data under the static state，the frequency converter outputs the alarm signal A07991. If the motor has been cooled to the ambient temperature，BOP-2 operator panel can be used for detecting the motor data. The operator firstly presses the key "HAND/AUTO"，and then presses the green startup button 🔲 after the manual running icon 🖐 appearing on the operator panel to begin the detection of motor data. The "MOT-ID" indicator on the BOP-2 flashes during the detection period of motor data. If the frequency converter outputs the alarm A07991，it will wait the new ON instruction for measuring the startup rotation. If the frequency converter doesn't output the alarm A07991，it will turn off the motor and change its mode to AUTO from HAND.

（三）设置变频器参数

（Ⅲ）To set the parameters of frequency converter

按表 3.1.2 设置变频器参数。

The parameters of frequency converter are set based on Table 3.1.2.

变频器参数设置   表 3.1.2

Parameter setting of frequency converter   **Table 3.1.2**

参数 Parameter	值 Value	说明 Description
P3330	r722.0	设置双线制/三线制控制指令 1（ON/OFF1 正转）的信号源为 DI 0 Set the signal source of two-wire/three-wire system control instruction 1（ON/OFF1 positive rotation）to DI 0
P3331	r722.1	设置双线制/三线制控制指令 2（ON/OFF1 反转）的信号源为 DI 1 Set the signal source of two-wire/three-wire system control instruction 2（ON/OFF1 negative rotation）to DI 1

参数 Parameter	值 Value	说明 Description
P1000	3	设置转速设定值来源为转速固定设定值 Set the source of rotating speed set values to the fixed rotating speed set values
P1001	150	设置转速固定设定值 1 为 150rpm Set the fixed rotating speed set value 1 to 150rpm
P1002	250	设置转速固定设定值 2 为 250rpm Set the fixed rotating speed set value 2 to 250rpm
P1003	350	设置转速固定设定值 3 为 350rpm Set the fixed rotating speed set value 3 to 350rpm
P1004	500	设置转速固定设定值 4 为 500rpm Set the fixed rotating speed set value 4 to 500rpm
P1016	2	设置选择转速固定设定值的模式为二进制 Set the selection method of fixed rotating speed set values to the binary selection
P1020	r722.2	选择转速固定设定值的信号源为 DI 2 Select the signal source of fixed rotating speed set value to DI 2
P1021	r722.3	选择转速固定设定值的信号源为 DI 3 Select the signal source of fixed rotating speed set value to DI 3
P1022	r722.4	选择转速固定设定值的信号源为 DI 4 Select the signal source of fixed rotating speed set value to DI 4
P1070	r1024	设置主设定值的信号源为转速固定设定值 Set the signal source of main set values to the fixed rotating speed set values

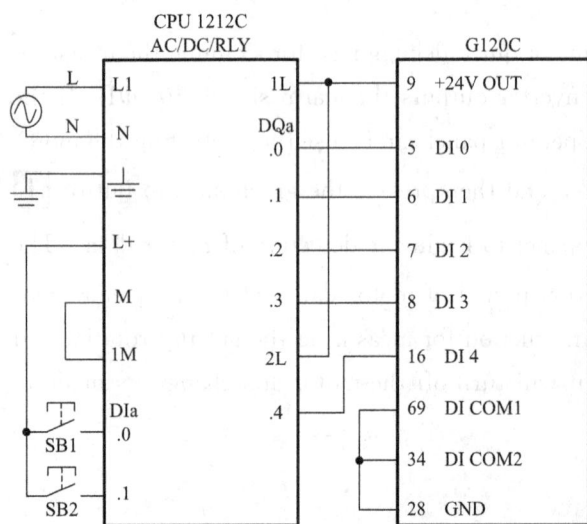

图 3.1.12 I/O 接线图

Figure 3.1.12 I/O wiring diagram

（四）连接 I/O 设备

（Ⅳ）To connect I/O device

按照控制要求，I/O 接线图如图 3.1.12 所示。

Based on the control requirements, the I/O wiring diagram is shown in Figure 3.1.12.

（五）在 TIA 博途中新建项目和设备组态

（Ⅴ）To create a new item and device configuration in TIA Portal

由于本任务无需连接触摸屏，因此不需要在设备组态时在 CPU 模块的"保护"选项中勾选"允许远程伙伴（PLC、HMI、OPC）使用 PUT/GET 通信访问"。也不需要在"系统和时钟存储器"中勾选"启用时钟存储器字节"。

Because the connection to touch screen isn't required in executing this task, user will not check "Allow the remote partners (PLC, HMI and OPC) to use PUT/GET communication access" in the "Protection" option of CPU module at the time of device configuration. There is also no need for checking "Enable the bytes of clock timer" in "System and clock memory".

（六）建立变量表

（Ⅵ）To create the variable table

按照控制要求，定义变量表如图 3.1.13 所示。

Define the variable table is shown in Figure 3.1.13 based on the control requirements.

PLC变量								
	名称	变量表	数据类型	地址	保持	在 H..	可从..	注释
1	SB1启动按钮	默认变量表	Bool	%I0.0		☑	☑	
2	SB2停止按钮	默认变量表	Bool	%I0.1		☑	☑	
3	电机正转	默认变量表	Bool	%Q0.0		☑	☑	
4	电机反转	默认变量表	Bool	%Q0.1		☑	☑	
5	电机速度选择位1	默认变量表	Bool	%Q0.2		☑	☑	
6	电机速度选择位2	默认变量表	Bool	%Q0.3		☑	☑	
7	电机速度选择位3	默认变量表	Bool	%Q0.4		☑	☑	

图 3.1.13 变量表

Figure 3.1.13 Variable table

(七) 编写梯形图程序

(Ⅶ) To write the ladder logic program

对 SB1 按下的次数进行计数。当达到要求的次数时，相应的输出得电，控制电机的正反转和速度。当按下 SB2 时，复位所有输出，电机停止，同时复位计数器。

To count the number of times SB1 is pressed. When the required number of times is reached, the corresponding output is energized to control the positive and negative rotation, and the speed of motor. All outputs are reset, the motor is stopped, and the counter is also reset when SB2 is pressed.

梯形图程序如图 3.1.14 所示。

The ladder diagram program is shown in Figure 3.1.14.

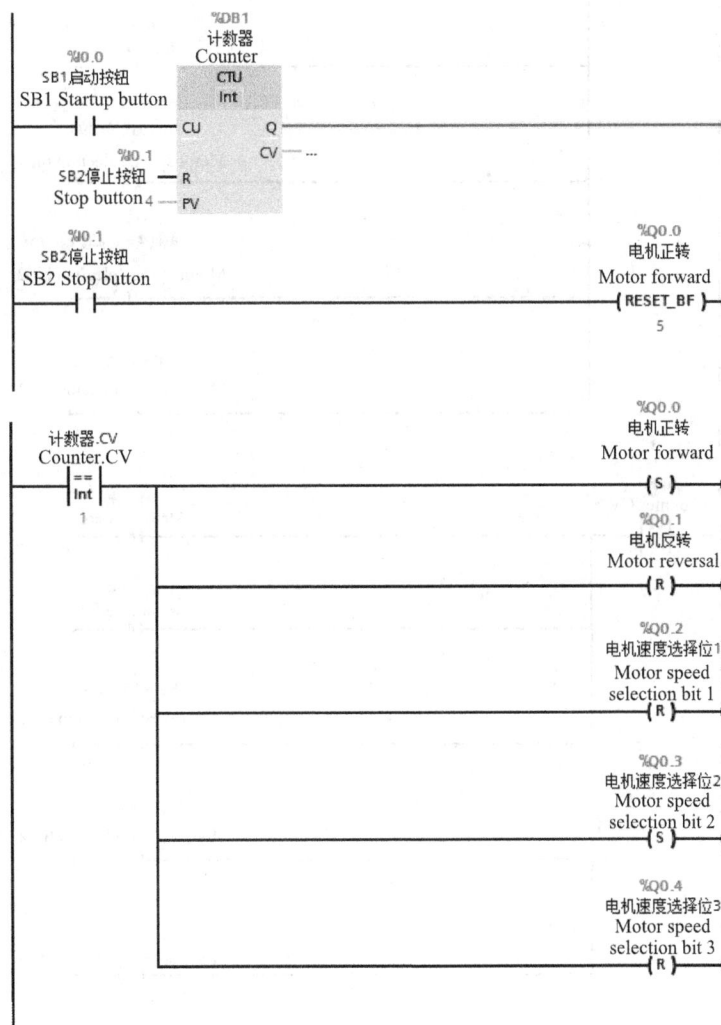

图 3.1.14 梯形图程序 （一）

Figure 3.1.14 Ladder logic program （一）

91

计数器 .CV
Counter.CV
==
Int
2

%Q0.0
电机正转
Motor forward
(R)

%Q0.1
电机反转
Motor reversal
(S)

%Q0.2
电机速度选择位1
Motor speed selection bit 1
(S)

%Q0.3
电机速度选择位2
Motor speed selection bit 2
(S)

%Q0.4
电机速度选择位3
Motor speed selection bit 3
(R)

计数器 .CV
Counter.CV
==
Int
3

%Q0.0
电机正转
Motor forward
(S)

%Q0.1
电机反转
Motor reversal
(R)

%Q0.2
电机速度选择位1
Motor speed selection bit 1
(S)

%Q0.3
电机速度选择位2
Motor speed selection bit 2
(R)

%Q0.4
电机速度选择位3
Motor speed selection bit 3
(R)

计数器 .CV
Counter.CV
==
Int
4

%Q0.0
电机正转
Motor forward
(S)

%Q0.1
电机反转
Motor reversal
(R)

%Q0.2
电机速度选择位1
Motor speed selection bit 1
(R)

%Q0.3
电机速度选择位2
Motor speed selection bit 2
(R)

%Q0.4
电机速度选择位3
Motor speed selection bit 3
(S)

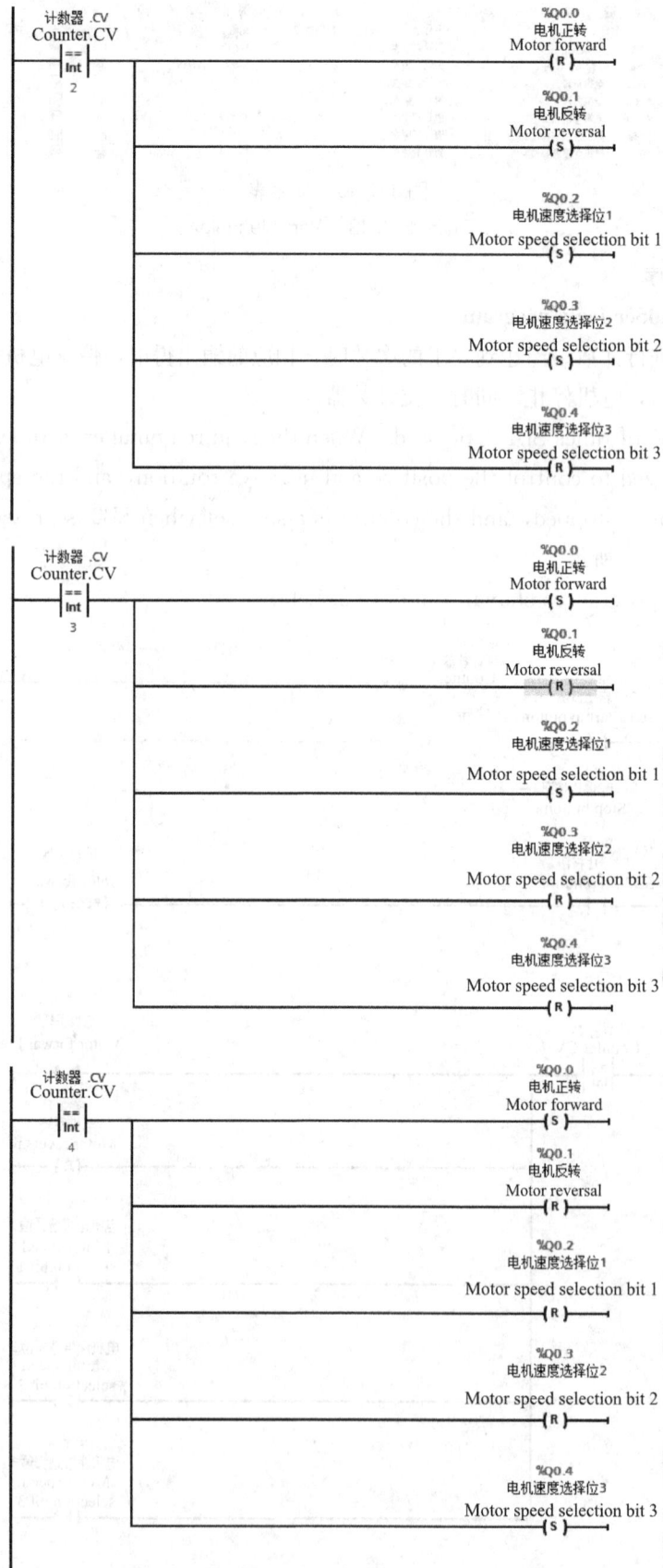

图 3.1.14 梯形图程序 （二）

Figure 3.1.14 Ladder logic program （二）

（八）调试

（Ⅷ）Debugging

完成以上步骤后，将编辑好的 TIA 博途项目下载至 PLC，按照控制要求逐项对系统进行调试，直到所有功能都实现为止。

After the above steps，the edited TIA Portal items are downloaded to the PLC，and then the system will be debugged item by item based on the control requirements until that all functions are realized.

任务二 基于模拟量控制的三相电机调速

Task Ⅱ Speed Control of Three-phase Motor Based on the Analog Control

一、任务分析

Ⅰ. Task Analysis

用西门子 S7-1200 系列 PLC 与 G120C 变频器组成一个三相电机速度控制系统，要求根据输入的模拟量电压对电机的速度进行调节。控制要求如下：

To assemble Siemens S7-1200 SERIES PLC and G120C frequency converter into one speed control system for three-phase motor to control the motor speed based on the analog input voltage. The control requirements are shown as follows：

（1）当输入电压在 0～3V 之间时，按下按钮 SB1，电机以 1000rpm 正转运行；

（1）When the input voltage is between 0V and 3V，the motor can rotate clockwise at the speed of 1000rpm by pressing the button SB1；

（2）当输入电压在 3～5V 之间时，按下按钮 SB1，电机以 700rpm 反转运行；

（2）When the input voltage is between 3V and 5V，the motor can rotate counterclockwise at the speed of 700rpm by pressing the button SB1；

（3）当输入电压在 5～8V 之间时，按下按钮 SB1，电机以 500rpm 正转运行；

（3）When the input voltage is between 5V and 8V，the motor can rotate clockwise at the speed of 500rpm by pressing the button SB1；

（4）当输入电压在 8～10V 之间时，按下按钮 SB1，电机以 200rpm 反转运行；

（4）When the input voltage is between 8V and 10V，the motor can rotate counterclockwise at the speed of 200rpm by pressing the button SB1；

（5）按下停止按钮 SB2，电机停止；

（5）The motor can stop the rotation by pressing the stop button SB2；

（6）变频器参数设置为加速时间 0.5s，减速时间 0.2s。

（6）The parameters of frequency converter are set to the acceleration time 0.5s and the deceleration 0.2s.

在上一任务中，转速的设定是通过固定转速设定值实现的，但是这样的方法占用 PLC 输出口较多，很不经济。为了更好地完成本任务，需要学习用 S7-1200 输入/输出模拟量的方法、S7-1200 的移动值指令和西门子 G120C 变频器的模拟量调速方法。

In the last task，the rotating speed can be set with the fixed rotating speed set values，but many PLC output ports will be occupied，therefore the method is not very cost-effective. In order to better complete this task，it is necessary to learn how to apply the input/output analog of S7-1200，the movement value instruction of S7-1200，and the analog speed control of Siemens G120C frequency converter.

二、相关知识技能

Ⅱ. Relevant Knowledge & Skills

（一）S7-1200 输入/输出模拟量的方法

（Ⅰ）Input/Output analog of S7-1200

1. 西门子 S7-1200 信号模块与信号板的地址分配

1. Address assignment of Siemens S7-1200 signal module and signal board

在设备组态过程中，CPU、信号板和信号模块的 I、Q 地址是自动分配的。在前面的学习中已经简单介绍过设备视图（图 3.2.1）和设备概览视图（图 3.2.2）。在设备概览视图中，可以看到 CPU 集成的 I/O 点和信号模块的字节地址。

The I and Q addresses of CPU，signal board and signal module are automatically assigned in the process of device configuration. The device view （Figure 3.2.1）and device overview （Figure 3.2.2）have been simply introduced in the previous learning. We can see the byte address of CPU integrated I/O points and signal module from the device overview.

CPU 1212C 集成的 8 点数字量输入字节地址自动分配为 0（I0.0～I0.7），6 点数字量输出字节地址自动分配为 0（Q0.0～Q0.5）。CPU 集成的模拟量输入点的地址为 IW64 和 IW66，每个通道占一个或两个字节。

0（I0.0 to I0.7）is automatically assigned to the byte addresses of the CPU 1212C integrated 8-point digital inputs，and 0（Q0.0 to Q0.5）is automatically assigned to the byte addresses of 6-point digital output. The addresses of CPU integrated analog input points are IW64 and IW66，and each channel occupies one word or two bytes.

DI、DQ 的地址以字节为单位分配，如果没有用完分配给它的某个字节中所有的位，剩余的位也不能再作他用，如前所述的例子中，Q0.6 和 Q0.7 就不能再使用。模拟量输入，模拟量输出的地址以组为单位分配，每一组有两个输入/输出点，在前述的例子中，IW64 和 IW66 就是一组模拟量输入。

The addresses of DI and DQ are assigned in the unit of byte，if all bits in a certain byte assigned can't be fully occupied，the remaining bits can't be used any longer，as shown in the previous example，Q0.6 and Q0.7 can't be used any longer. The addresses of analog input and output are assigned in the unit of group，each group has two input/output points，and as shown in the previous example，IW64 and IW66 belong to one group of analog input.

从设备概览视图还可以看到分配给各插槽的信号模块的输入、输出字节地址。例如，插槽 2 的 AQ 模块的 2 点模拟量输出地址为 QW96 和 QW98。

The input and output byte addresses assigned to the signal module in each slot can also be seen from the device overview. For example，the addresses of 2-point analog output of AQ module in No. 2 slot are QW96 and QW98.

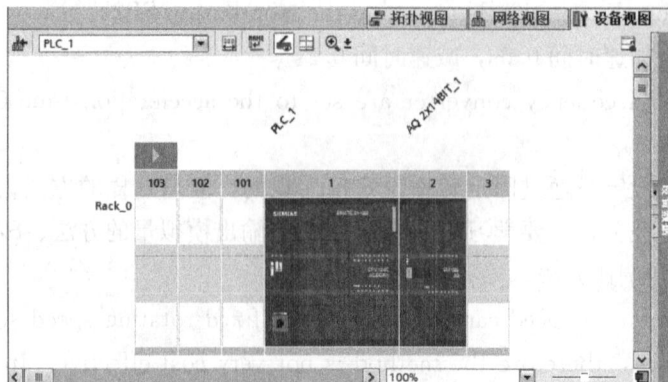

图 3.2.1　设备视图

Figure 3.2.1　Device view

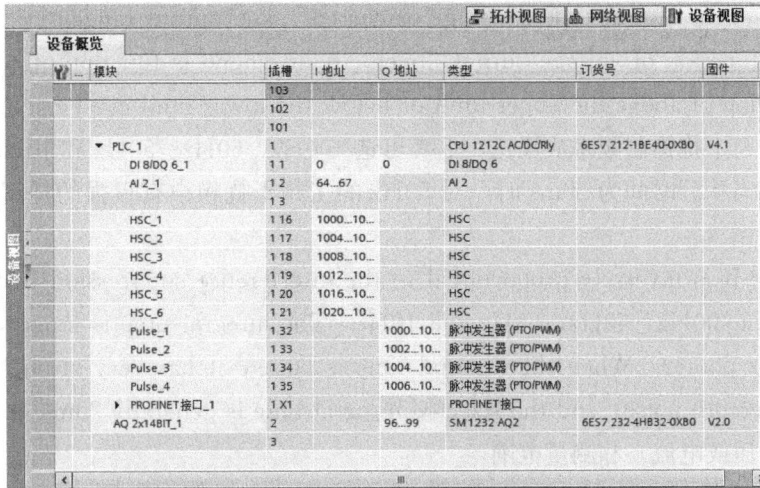

图 3.2.2　设备概览视图

Figure 3.2.2　Device Overview

选择设备概览中某个插槽的模块，可以修改自动分配的 I、Q 地址。建议采用自动分配的地址，不要修改它。但是在编程时必须使用分配给各 I/O 点的地址。

The automatically-assigned I and Q addresses can be modified by selecting the module in each slot in the device overview. It is suggested to use the automatically-assigned addresses and not modify them. However, the addresses assigned to each I/O point must be used in the programming process.

2. 模拟量输入的参数设置

2. Parameter setting of analog input

在设备视图中选中 CPU 模块，在巡视窗口中的"属性＞常规"选项卡中，选择"AI 2＞模拟量输入"，可以设置模拟量输入的各项参数，如图 3.2.3 所示。

The various parameters of analog input can be set by selecting the CPU module in the device overview, and selecting "AI 2＞Analog output" from the "Attributes＞General" tabs in the inspector window, as shown in Figure 3.2.3.

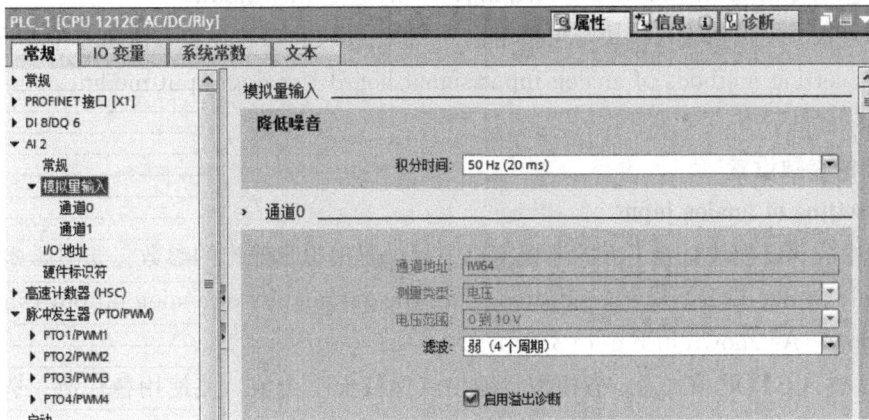

图 3.2.3　模拟量输入的参数设置

Figure 3.2.3　Parameter setting of analog input

需要设置各输出点的输出类型（电压或电流）和输出范围。可以激活电压输出的短路诊断功能，电流输出的断路诊断功能，以及超出上限值或低于下限值的溢出诊断功能。

It is necessary to set the output type (voltage or current) and output range of each output point. The short-circuit diagnosis function of voltage output, open-circuit diagnosis function of current output, and o-verflow diagnosis function of above the upper limit or below the lower limit can be actuated.

（1）积分时间。它与干扰抑制频率成反比，后者可选 400Hz、60Hz、50Hz 和 10Hz。积分时间越长，精度越高，快速性越差。当积分时间为 20ms 时，对 50Hz 的工频干扰噪声有很强的抑制作用，所以一般选择积分时间为 20ms。

（1）Integral time. It is inversely proportional to the interference suppression frequency, and the later one can be selected from 400Hz, 60Hz, 50Hz and 10Hz. The integral time is longer, the accuracy is longer, and the rapidity is poorer. When the integral time is 20ms, it has very strong suppression on 50Hz power-frequency interference noise, and the selected integral time is generally 20ms.

（2）测量类型（电压或电流）和测量范围。

（2）Measurement type (voltage or current) and measurement range.

（3）A-D 转换得到模拟值的滤波等级。模拟值的滤波处理可以减轻干扰的影响，这对缓慢变化的模拟量信号（例如温度测量信号）是很有意义的。滤波处理根据系统规定的转换次数来计算转换后的模拟值的平均值，有"无、弱、中、强"这 4 个等级，它们对应的计算平均值的模拟量采样值的周期数分别为 1、4、16、32。所选的滤波等级越高，滤波后的模拟值越稳定，但是测量的快速性越差。

（3）Filtering grade of analog value obtained by A-D conversion. The filtering processing of analog value can relieve the effect of interference and it is very significant to the analog signals which change slowly (for example the temperature measurement signal). The filtering processing will calculate the average of converted analog values based on the times of conversions specified for the system, there are four grades "none, poor, moderate and strong", and their corresponding number of analog sampling cycles for calculating the average is respectively 1, 4, 16 and 32. The selected filtering grade is higher, the analog value after the filtering processing is more stable, but the rapidity of measurement is poorer.

（4）设置诊断功能，可以选择是否启用断路和溢出诊断功能。只有 4～20mA 输入才能检测是否有断路故障。

（4）The open-circuit and overflow diagnosis functions can be enabled or disabled by setting the diagnosis functions. The open-circuit fault can be detected only with 4—20mA input.

模拟量输入信号板、模拟量输入模块与模拟量输入/输出模块的参数设置方法基本上相同。

The parameter setting methods of analog input signal board, analog input module, analog input/output module are basically same.

3. 模拟量输出的参数设置

3. Parameter setting of analog input

选中设备视图中的 AQ 模块，在下方的巡视窗口中可设置模拟量输出的参数。如图 3.2.4 所示。

The AQ module in the device view is selected, and the parameters of analog output can be set in the below inspector window. As shown in Figure 3.2.4.

可以设置 CPU 进入 STOP 模式后，各模拟量输出点保持上一个值，或使用替代值。选中后者时，可以设置各点的替代值。

The operator can set that each analog output point remains the last value or uses the alternative value after CPU entering into STOP mode. If the later is selected, the alternative value of each point can be set.

需要设置各输出点的输出类型（电压或电流）和输出范围。可以激活电压输出的短路诊断功能，电流输出的断路诊断功能，以及超出上限值或低于下限值的溢出诊断功能。

It is necessary to set the output type (voltage or current) and output range of each output point. The short-circuit diagnosis function of voltage output, open-circuit diagnosis function of current output, and overflow diagnosis function of above the upper limit or below the lower limit can be actuated.

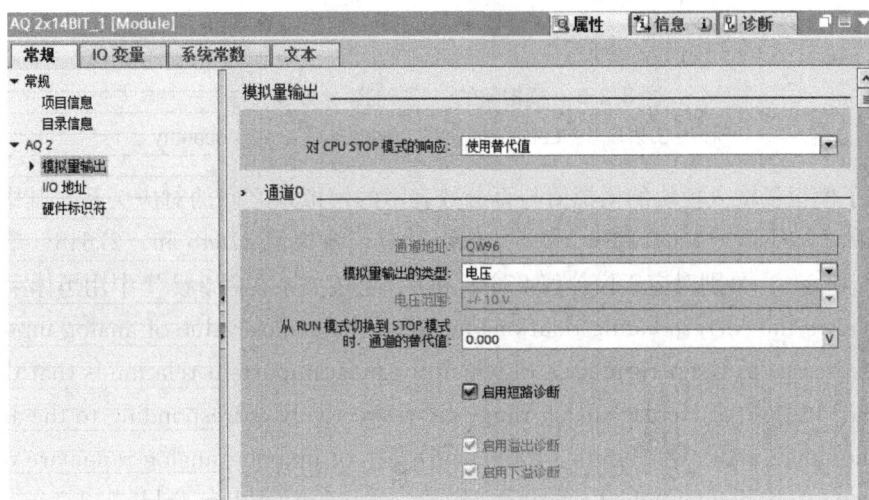

图 3.2.4 模拟量输出的参数设置

Figure 3.2.4 Parameter setting of analog input

CPU 集成的模拟量输出点、模拟量输出信号板与模拟量输入/输出模块的参数设置方法基本上相同。

The parameter setting methods of CPU integrated analog input point, analog output signal board, analog input/output module are basically same.

4. 模拟量转换的模拟值

4. Analog value of analog conversion

模拟量输入/输出点中模拟量对应的数字称为模拟值。模拟值用 16 位二进制补码（整数）来表示。最高位（第 15 位）为符号位，正数的符号位为 0，负数的符号位为 1。

The number corresponding to the analog quantity in the analog input/output point is called as the analog value. The analog value is expressed with 16-bit binary complement number (integer). The highest digit (15th digit) is the sign bit, and the sign bits of positive number and negative number are respectively 0 and 1.

当模拟量输入或输出点的转换精度小于 16 位时，模拟值被自动左移，使其最高的符号位在 16 位字的最高位。模拟值左移后未使用的低位则填入"0"，这种处理方法称为"左对齐"。假设模拟值的精度为 14 位，左移 2 位后未使用的低位（第 0～1 位）为 0，相当于实际的模拟值被乘以 4，如图 3.2.5 所示。

When the number of bits for conversion accuracy of analog input/output point is less than 16, the analog value automatically shifts leftward, thus the highest sign bit is the highest bit of 16-bit word. The unused low bit is filled with "0" after the analog value shifting rightward, and such method is called as "left justifying". The number of bits for the accuracy of analog value is assumed as 14, and the unused lowest bit (0 to 1st bit) after shifting 2 bits rightward is 0, which is equivalent to four times of actual analog value, as shown in Figure 3.2.5.

这种处理方法的优点在于模拟量的量程与移位处理后的数字的关系是固定的，与左对齐之前的转换值位数（即模拟量模块的分辨率）无关。

The advantages of such processing method is that the relation between the range of analog quantity and the figure after the shifting is fixed and unrelated to the number of bits of conversion value (i. e. the resolution of analog quantity module) before the left justifying.

分辨率 Resolution	模拟值 Analog value															
位 Bit	15	14	13	12	11	10	9	8	7	6	5	4	3	2	1	0
16位	0	1	0	0	0	1	1	0	0	1	0	1	1	1	1	1
14位	0	1	0	0	0	1	1	0	0	1	0	1	1	1	0	0

图 3.2.5 模拟量的 "左对齐" 处理方法

Figure 3.2.5 "Left justifying" method of analog quantity

表 3.2.1 给出了模拟量输入模块的模拟值与百分数表示的模拟量之间的对应关系，其中最重要的关系是双极性模拟量量程的上、下限（100％和－100％）分别对应于模拟值 27648 和－27648，单极性模拟量量程的上、下限（100％和 0％）分别对应于模拟值 27648 和 0。上述关系在表 3.2.1 中用黑体字表示。

Table 3.2.1 shows the corresponding relations between the analog value of analog input module and the analog quantities expressed as the percentage，of which the most important relation is that the upper and lower limits（100％ and -100％）of bipolar analog range are respectively corresponding to the analog value 27648 and -27648，and the upper and lower limits（100％ and 0％）of unipolar analog range are respectively corresponding to 27648 and 0. The mentioned relations are expressed in bold in Table 3.2.1.

模拟值与模拟量的对应关系　　　　　　　　　　　　　　表 3.2.1

Corresponding relations between analog values and analog quantities　　　Table 3.2.1

范围 Range	双极性 Bipolarity				单极性 Unipolarity			
	百分比 Percentage	十进制 Decimal	十六进制 Hexadecimal	±10V	百分比 Percentage	十进制 Decimal	十六进制 Hexadecimal	0-20mA
上溢出，断电 Upper overflow, power off	118.515％	32767	7FFFH	11.851V	118.515％	32767	7FFFH	23.70mA
超出范围 Out of range	117.589％	32511	7EFFH	11.759V	117.589％	32511	7EFFH	23.52mA
正常范围 Normal range	**100.000％**	**27648**	**6C00H**	**10V**	**100.000％**	**27648**	**6C00H**	**20mA**
	0％	**0**	**0H**	**0V**	**0％**	**0**	**0H**	**0mA**
	－100.000％	**－27648**	**9400H**	**－10V**				
低于范围 Below the range	－117.593％	－32512	810HH	－11.759V				
下溢出，断电 Lower overflow, power off	－118.519％	－32768	8000H	－11.851V				

（二）移动值指令

（Ⅱ） Movement value instruction

移动值指令 MOVE（图 3.2.6）用于将 IN 输入端的源数据传送给 OUT1 输出的目的地址，并且转换为 OUT1 允许的数据类型（与是否进行 ICE 检查有关），源数据保持不变。IN 和 OUT1 的数据类型可以是位字符串、整数、浮点数、定时器、日期时间、CHAR、WCHAR、STRUCT、ARRAY、ICE 定时器/计数器数据类型、PLC 数据类型，IN 还可以是常数。

The movement value instruction MOVE（Figure 3.2.6）is used for transmitting the source data on the IN input terminal to the designated address of OUT1 output，and conversing it into the allowable data type of

OUT1 (related to execution of IEC check), and the source data is unchanged. The data type of IN and OUT1 can be the bit character string, integer, floating number, timer, date and time, CHAR, WCHAR, STRUCT, ARRAY, ICE timer/counter data type and PLC data type, and IN can be a constant.

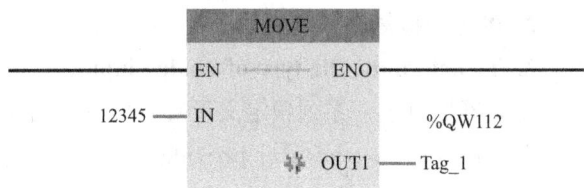

图 3.2.6 移动值指令

Figure 3.2.6 Movement value instruction

可用于 S7-1200 CPU 的不同数据类型之间的数据传送请参见 MOVE 指令的在线帮助。如果输入 IN 数据类型的位长度超出 OUT1 数据类型的位长度，则源值的高位会丢失。如果输入 IN 数据类型的位长度小于输出 OUT1 数据类型的位长度，则目标值的高位会被改写为 0。

The data transmissions available among different data types of S7-1200 CPU are shown in the online help of MOVE instruction. If the bit length of input IN data type is longer than that of OUT1 data type, the high bits of source value may be lost. If the bit length of input IN data type is shorter than that of OUT1 data type, the high bits of target value may be rewritten to 0.

MOVE 指令允许有多个输出，单击 OUT1 前面的 ![icon]，将会增加一个输出，增加的输出的名称为 OUT2，以后增加的输出的编号按顺序排列。用鼠标右键单击某个输出的短线，执行快捷菜单中的"删除"命令，将会删除该输出参数。删除后自动调整剩下输出的编号。

MOVE instruction can allow many outputs, one output will be added by clicking on the front ![icon] of OUT1, the name of output added is OUT2, and the serial number of outputs added subsequently will be arranged in sequence. When the "Delete" command in the shortcut menu is executed by right clicking on the short line of a certain output, the output parameter can be deleted. Then, the serial number of remaining outputs will be automatically modified.

(三) 西门子 G120C 变频器的模拟量调速

(Ⅲ) Analog speed control of Siemens G120C frequency converter

1. 互联模拟量输入

1. Interconnected analog input

当您选择不带模拟量输入功能的标准设置时，必须将主设定值的参数和一个模拟量输入互联在一起。在通过外部端子 AI 0 接入了某个模拟量电压或电流时，参数 r0755 [0] 是模拟量输入的当前值。将此值与转速的主设定值（参数 P1070）关联，意味着将当前电机的转速设置为模拟量输入值，如图 3.2.7 所示。

When you select the standard setting without the analog output function, the parameter of main set value must be interconnected to one analog input. When a certain analog voltage or current is connected through the external terminal AI 0, the parameter r0755 [0] is the current value of analog input. The value is correlated to the main set value of rotating speed (parameter P1070), indicating that the rotating speed of current motor is set to the analog input value, as shown in Figure 3.2.7.

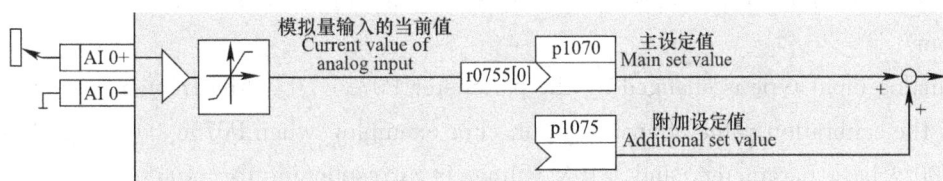

图 3.2.7 将 AI 0 设为设定值源

Figure 3.2.7 Set AI 0 to the source of set values

2. 确定模拟量输入的类型

2. To determine the type of analog input.

变频器提供了一系列预定义设置，可以使用参数 P0756［0］进行选择，见表 3.2.2。

The frequency converter provides a series of predefined settings，and they can be selected by using the parameter P0756［0］，as shown in Table 3.2.2.

参数 P0756［0］的预定义值　　　　　　　　　　　　　　　　　　　　表 3.2.2

Predefined values of parameter P0756［0］　　　　　　　　　　　　Table 3.2.2

AI 0 的输入方式 Input mode of AI 0	输入值的范围 Range of input value	P0756［0］的值 Value of P0756［0］
单极性电压输入 Unipolar voltage input	0～+10 V 0～+10V	0
单极性电压输入，受监控 Unipolar voltage input，monitored	+2～+10 V +2～+10V	1
单极性电流输入 Unipolar current input	0～+20 mA 0～+20mA	2
单极性电流输入，受监控 Unipolar current input，monitored	+4～+20 mA +4～+20mA	3
双极性电压输入 Bipolar voltage input	-10～+10 V -10～+10V	4
没有连接传感器 No connection to sensor	无 None	8

另外，还必须设置模拟量输入类型的选择开关，如图 3.2.8 所示。该开关位于控制单元正面保护盖的后面。

Additionally，the selective switch for the analog input types also must be set，as shown in Figure 3.2.8. The switch is located at the back of the front cover of control unit.

- 电压输入：开关位置 U(出厂设置)
Voltage input：Switch position U(factory setting)
- 电流输入：开关位置 I
Current input：Switch position I

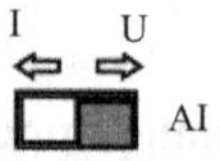

图 3.2.8　模拟量输入类型的选择开关

Figure 3.2.8　Selective switch for analog input types

3. 确定转速与模拟值之间的关系

3. To determine the relation between the rotating speed and analog value

用 P0756［0］修改模拟量输入的类型后，变频器会自动调整模拟量输入的定标。例如，当 P0756[0]=4 时，10V 的电压对应 100% 的基准参数，-10V 的电压对应 -100% 的基准参数。如果将一个模拟输入信号（例如 r755［0］）连接到转速设定值（例如 P1070［0］），则当前百分比形式的输入值通过参考转速（P2000）周期性的被转换成绝对的转速设定值。转换的公式为：模拟量输入对应的转速值（rpm）＝输入电压（V）/10 ×参考转速（rpm）。

After the analog input type is changed by the parameter P0756［0］，the frequency converter will automatically adjust the calibration value of analog input. For example，when P0756[0]=4，10V voltage is corresponding to 100% basic parameter，and -10V voltage is corresponding to -100% basic parameter. If one analog input signal （for example；r755［0］）is connected to the rotating speed set value （for example；P1070［0］），the current percentile input value can be periodically converted into the absolute rotating speed set value by the rotating speed reference value (P2000). The conversion formula is shown as follows: The rotating speed (rpm) corresponding to the analog input＝input voltage (V)/10×rotating speed reference value (rpm).

三、任务实施

Ⅲ. Task Implementation

(一) 连接变频器与电机的主电路

(Ⅰ) To connect the main circuit between frequency converter and motor

参照前图3.1.11将三相电源、变频器、三相电机连接起来。

The three-phase supply, frequency converter and three-phase motor are connected by reference to Figure 3.1.11.

图 3.2.9　模拟量与基准参数的对应关系

Figure 3.2.9　Corresponding relation between analog quantity and standard parameter

(二) 对变频器进行快速调试

(Ⅱ) To quickly debug the frequency converter

按照前一任务介绍的方法对变频器进行快速调试，并进行电机数据检测。

The quick debugging of frequency converter and the detection of motor data are carried by the methods prescribed in the last task.

(三) 设置变频器参数

(Ⅲ) To set the parameters of frequency converter

按表3.2.3设置变频器参数。

The parameters of frequency converter are set based on Table 3.2.3.

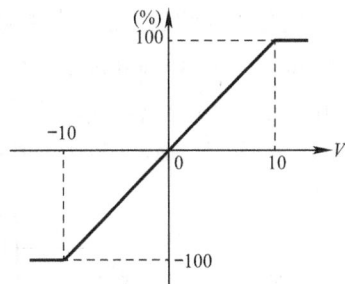

变频器参数设置　　　　　　　　　　　　　　表 3.2.3

Parameter setting of frequency converter　　　　　　Table 3.2.3

参数 Parameter	值 Value	说明 Description
P3330	r722.0	设置双线制/三线制控制指令 1 (ON/OFF1 正转) 的信号源为 DI 0 Set the signal source of two-wire/three-wire system control instruction 1 (ON/OFF1 positive rotation) to DI 0
P3331	r722.1	设置双线制/三线制控制指令 2 (ON/OFF1 反转) 的信号源为 DI 1 Set the signal source of two-wire/three-wire system control instruction 2 (ON/OFF1 negative rotation) to DI 1
P0756	4	设置模拟量输入的类型为双极电压输入 Set the analog input type to the bipolar voltage input
P1000	2	设置转速设定值来源为模拟量设定值 Set the source of rotating speed set values to the analog set values
P1070	r755 [0]	设置主设定值的信号源为模拟输入的当前输入值 Set the signal source of main set value to the current input value of analog input

(四) 连接 I/O 设备

(Ⅳ) To connect I/O device.

按照控制要求，I/O接线图如图3.2.10所示。

Based on the control requirements, the I/O wiring diagram is shown in Figure 3.2.10.

(五) 在 TIA 博途中新建项目和设备组态

(Ⅴ) To create a new item and device configuration in TIA Portal

在本任务的实施中，需要在设备组态时添加 AQ 模块，订货号为 6ES7 232-4HB32-0XB0。再按照本任务相关知识技能中介绍的方法对模拟量输入和输出点进行参数设置。

It is necessary to add AQ module at the time of device configuration in the process of implementing this task，the order number；6ES7 232-4HB32-0XB0. Then the parameters of analog input and output points are set by the methods prescribed in the relevant knowledge and skills of this task.

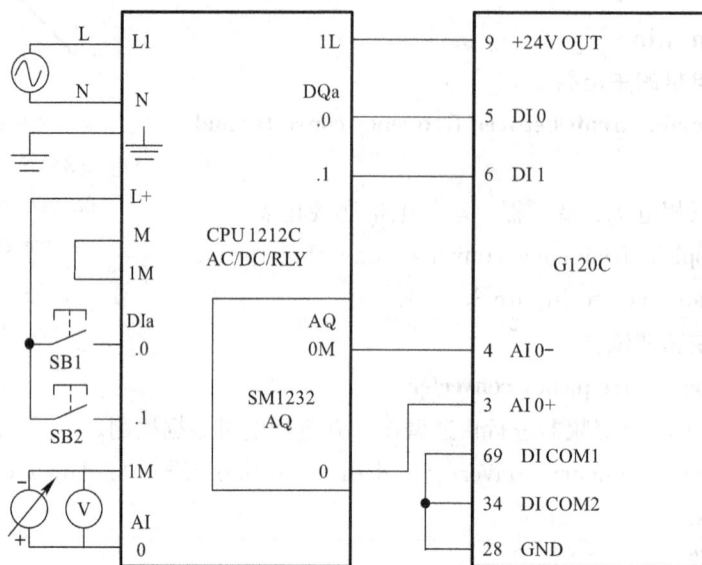

图 3. 2. 10 I/O 接线图

Figure 3. 2. 10 I/O wiring diagram

（六）建立变量表

（Ⅵ）**To create the variable table**

按照控制要求，定义变量表如图 3.2.11 所示。

Define the variable table as shown in Figure 3. 2. 11 based on the control requirements.

		名称	变量表	数据类型	地址	保持	在 H...	可从 ...	注释
		PLC 变量							
1		SB1启动按钮	默认变量表	Bool	%I0.0		✓	✓	
2		SB2停止按钮	默认变量表	Bool	%I0.1		✓	✓	
3		正转启动运行	默认变量表	Bool	%Q0.0		✓	✓	
4		反转启动运行	默认变量表	Bool	%Q0.1		✓	✓	
5		电机转速模拟量输出值	默认变量表	Int	%QW96		✓	✓	
6		电压输入模拟值	默认变量表	Int	%IW64		✓	✓	

图 3. 2. 11 变量表

Figure 3. 2. 11 Variable table

（七）编写梯形图程序

（Ⅶ）**To write the ladder logic program**

先计算出电压输入值 0V、3V、5V、8V、10V 的对应模拟值为 0、8294、13824、22118、27648。再计算出各转速对应的模拟值，分别为 1000rpm 对应 18432，700rpm 对应 12902，500rpm 对应 9216，200rpm 对应 3686。编程时，先用比较指令对输入电压进行比较。在确定电压的范围后，按下启动按钮，设置或重置相应的电机正反转控制位置，同时将转速对应的模拟值传送给模拟量输出点。变频器在接收到模拟值后，将其换算成转速，并以该转速启动运行。

Firstly，the analog values corresponding to the voltage input values 0V，3V，5V，8V and 10V are respectively calculated as 0，8294，13824，22118 and 27648. Then the analog values corresponding to all rotating speed values 1000rpm，700rpm，500rpm and 200rpm are calculated as 18432，12902，9216 and 3686. The input voltage values are compared by the comparison instruction in the programming process. After determining the voltage range，the startup button is pressed to set or reset the bits corresponding to the positive and negative rotation of motor，and meanwhile transmit the analog value corresponding to the rotating speed to the analog output point. The frequency converter will convert the received analog value into the rotating speed，and start up the motor at the rotating speed.

梯形图程序如图 3.2.12 所示。

The ladder diagram program is shown in Figure 3.2.12.

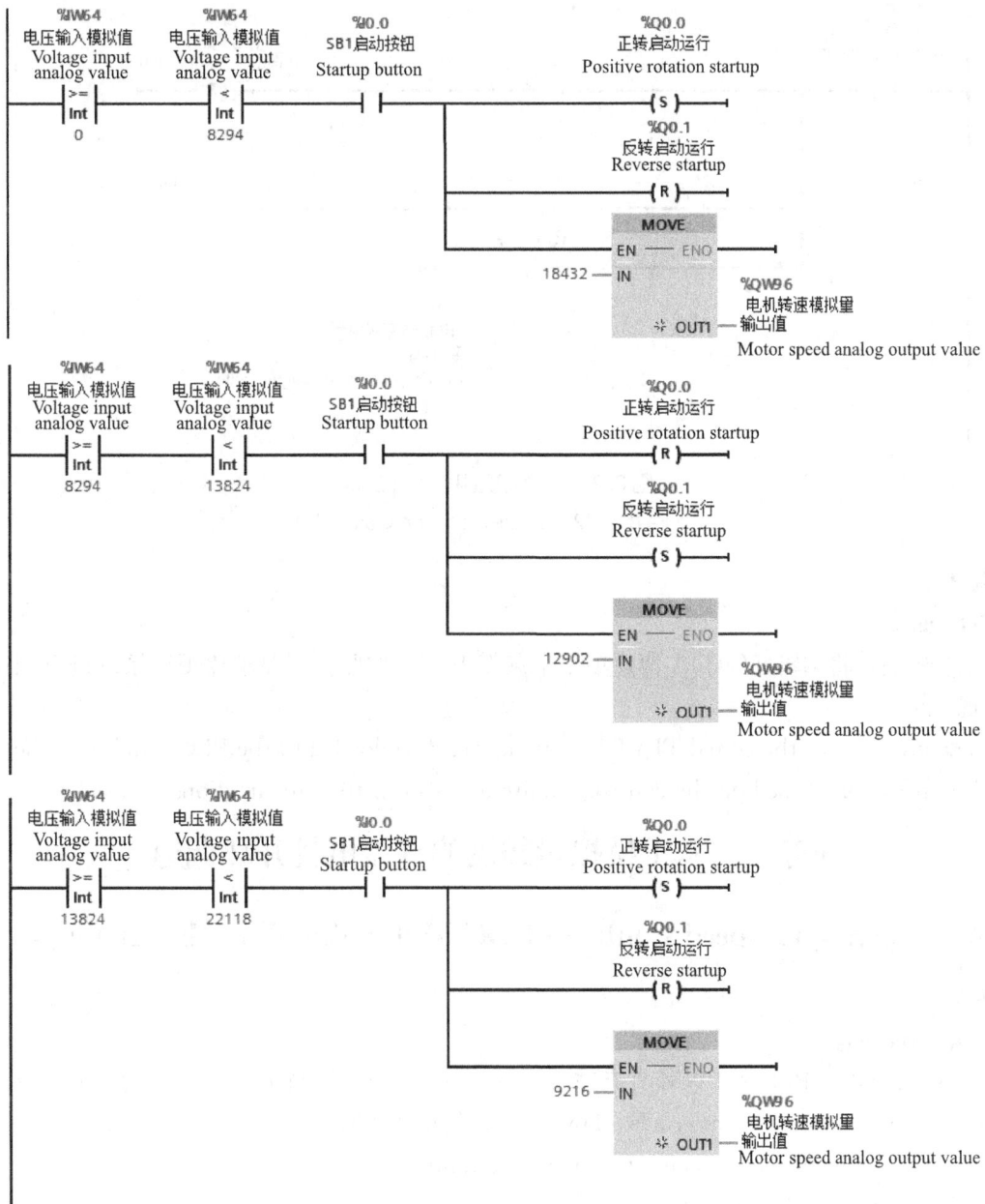

图 3.2.12　梯形图程序 （一）

Figure 3.2.12　Ladder logic program （Ⅰ）

103

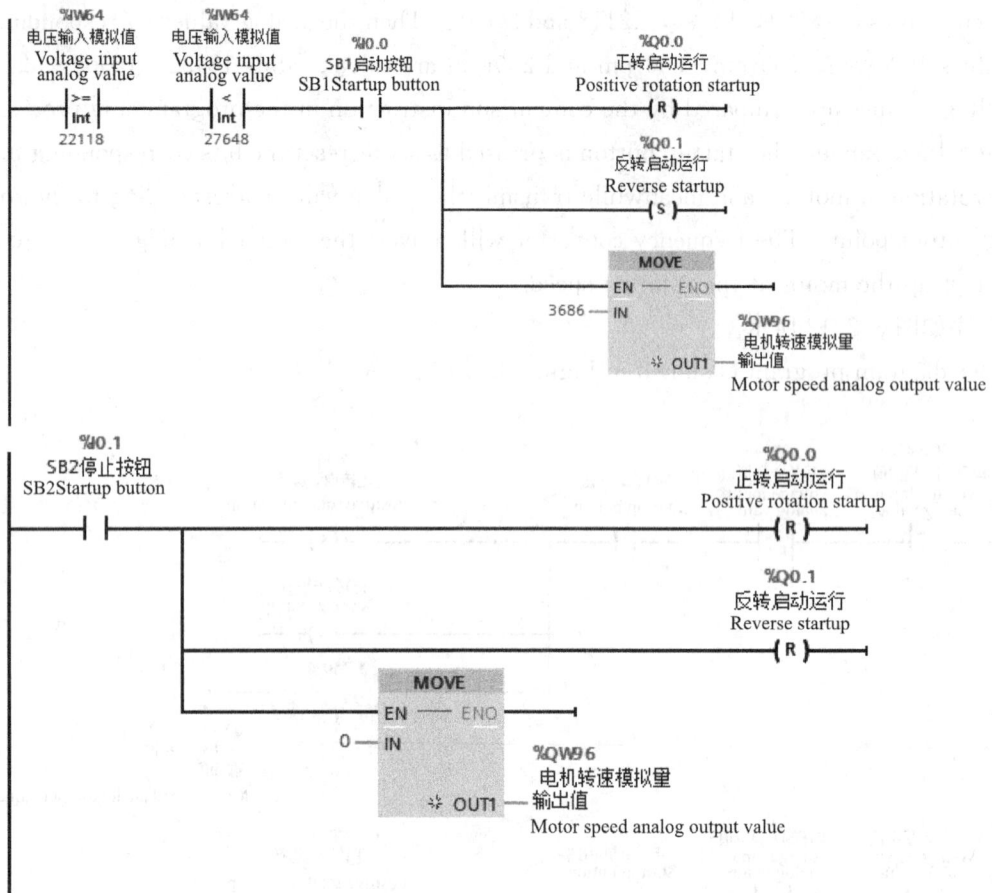

图 3.2.12　梯形图程序（二）

Figure 3.2.12　Ladder logic program（Ⅱ）

（八）调试

（Ⅷ）Debugging

完成以上步骤后，将编辑好的 TIA 博途项目下载至 PLC，按照控制要求逐项对系统进行调试，直到所有功能都实现为止。

After the above steps，the edited TIA Portal items are downloaded to the PLC，and then the system will be debugged item by item based on the control requirements until that all functions are realized.

任务三　基于模拟量控制的直流电机开环调速

TaskⅢ　Open-loop Speed Control of DC Motor Based on the Analog Control

一、任务分析

Ⅰ. Task Analysis

用人机界面（HMI）、PLC 和 PWM 直流调速器创建一个直流电机调速控制系统。控制要求如下：

Assemble the human-machine interface（HMI），PLC and PWM DC governor into a speed control system for DC system. The control requirements are shown as follows：

（1）监控画面如图 3.3.1 所示；

（1）The monitoring interface is shown in Figure 3.3.1；

（2）点击正转启动按钮和反转启动按钮可以让直流电机正转或反转启动运行。点击停止按钮可以让直流电机停止；

（2）The DC motor can be started up clockwise or counterclockwise by clicking on the positive rotation startup button and negative rotation startup button. The DC motor can be stopped by clicking on the stop button；

（3）每点击一次加速和减速按钮，电机调速端子上所加的调速电压分别增加和减少1V；

（3）The voltage applied on the speed control terminal of motor is respectively increased and decreased by 1V whenever the acceleration and deceleration buttons are clicked；

（4）调速电压的输入范围为0～5V，可在人机界面上显示该电压值。

（4）The input range of speed control voltage is between 0V and 5V, which can be displayed on the HMI.

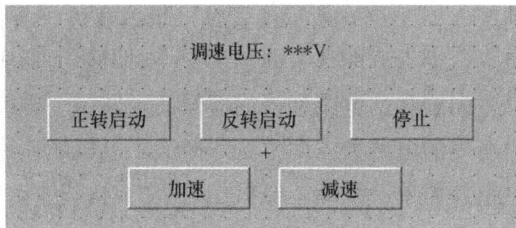

图 3.3.1　监控画面

Figure 3.3.1　Monitoring interface

为了完成本任务，首先需要对涉及的设备进行选型，根据现有条件，人机界面选择昆仑通态的 MCGS 触摸屏，PLC 选择西门子 S7-1200 系列 PLC。通过完成本任务，可以学习到直流电机的工作原理、西门子 S7-1200 的四则运算指令和 MCGS 嵌入版的设备通道处理这三个方面的知识。

For the purpose of this task, it is firstly necessary to select the required types of devices involved, of which MCGS touch screen（HMI）, and Siemens S7-1200 PLC will be selected under the current conditions. By completing this task, you can learn about the knowledge in three aspects, i. e. the working principles of DC motor, four operational orders of Siemens S7-1200, and the device channel processing of embedded MCGS.

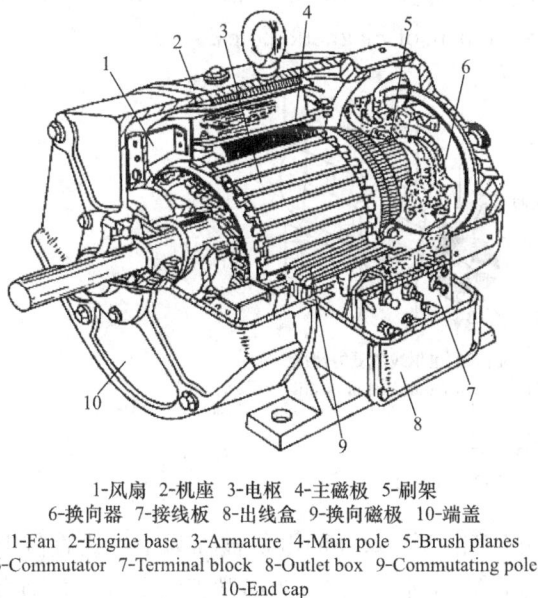

1-风扇 2-机座 3-电枢 4-主磁极 5-刷架
6-换向器 7-接线板 8-出线盒 9-换向磁极 10-端盖
1-Fan 2-Engine base 3-Armature 4-Main pole 5-Brush planes
6-Commutator 7-Terminal block 8-Outlet box 9-Commutating pole
10-End cap

图 3.3.2　直流电机的结构

Figure 3.3.2　Structure of DC motor

二、相关知识技能

II. Relevant Knowledge & Skills

（一）直流电机的工作原理

（I）Working principle of DC motor

1. 直流电机的调速原理

1. Working principle of DC motor

直流电机是将直流电能转化为机械能的电磁装置。直流电机分为定子和转子两部分。定子主要由主磁极、换向极、电刷和机座构成。转子主要由电枢铁心、电枢线圈和换向器构成。直流电机的结构如图 3.3.2 所示。

The DC motor is an electromagnetic device converting the DC electrical energy into the mechanical energy. The DC motor consists of stator and rotor. The stator is mainly composed of main magnetic pole, commutating pole, brush and base. The rotor is mainly composed of armature core, armature coil and commutator. The structure of DC motor is shown in Figure 3.3.2.

直流电机的工作原理是直流电压通过电刷和换向器在电枢线圈中产生的恒定电流在主磁极构成的磁场中受到电磁力的作用而旋转。如图 3.3.3（a）所示，电刷 A、B 分别接到直流电源的正负极上，此时电枢线圈中有电流流过。在磁场作用下，N 极性下导体 ab 受力方向从右向左，S 极下导体 cd 受力方向从左向右。这一对电磁力形成逆时针方向的电磁转矩。当电磁转矩大于阻转矩时，电机转子逆时针方向旋转。当电机旋转到图 3.3.3 (b) 所示位置时，原 N 极性下导体 ab 转到 S 极下，受力方向从左向右，原 S 极下导体 cd 转到 N 极下，受力方向从右向左。这一对电磁力形成逆时针方向的电磁转矩。电机在该电磁转矩的作用下继续逆时针方向旋转。电刷和换向器保证了同一个磁极下线圈边中的电流始终是一个方向，继而保证了该磁极下线圈边所受的电磁力方向不变，保证了电机能连续地旋转，以实现将电能转换成机械能。当然，实际的直流电机电枢并非单一线圈，磁极也并非一对。

The working principles of DC motor are that the constant current is generated in the armature coil while the DC voltage is flowing through the brush and commutator，and the constant current is affected by the electromagnetic force in the magnetic field created by main magnetic pole，thus driving the DC motor to rotate. As shown in Figure 3.3.3 (a)，the brushes A and B are respectively connected to the positive and negative poles of DC power supply，and the current flows through the armature coil at this moment. Under the action of magnetic field，the force direction of conductor ab under N pole is from right to left，and the force direction of cd under S pole is from left to right. This pair of magnetic forces produces the counterclockwise electromagnetic torque. When the electromagnetic torque is larger than the resistance torque，the motor rotor rotates counterclockwise. When the motor rotates to the position shown in Figure 3.3.3 (b)，the conductor ab originally under N pole rotates to the position under S pole，its force direction is from left to right，and the conductor cd originally under S rotates to the position under N pole，its force direction is from right to left. This pair of magnetic forces produces the counterclockwise electromagnetic torque. The motor can continue rotating counterclockwise under the action of this electromagnetic torque. The brushes and commutator can ensure that the current in the coil edge under the same pole still faces one direction，thus ensuring that the direction of magnetic force on the coil edge under this pole is unchanged and the motor can continuously rotate to realize the conversion from electrical energy to mechanical energy. Certainly，the armature of actual DC motor really isn't the single coil，and also there isn't only one pair of magnetic poles.

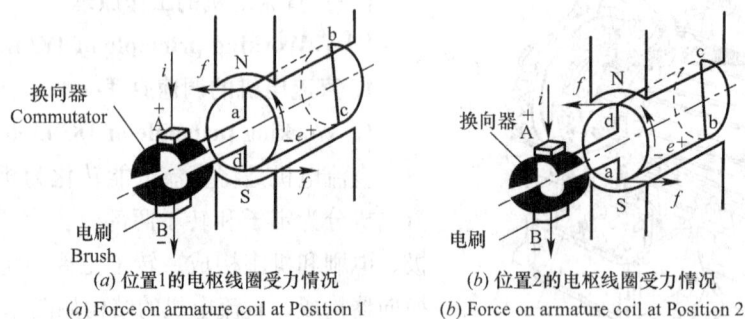

(a) 位置1的电枢线圈受力情况
(a) Force on armature coil at Position 1

(b) 位置2的电枢线圈受力情况
(b) Force on armature coil at Position 2

图 3.3.3 直流电机的工作原理
Figure 3.3.3 Working principle of DC motor

直流电机的主磁极可由永磁材料提供，如稀土、铁氧体、铝镍钴等，这样的电机被称为永磁式直流电机。主磁极也可以通过励磁线圈产生的电磁场提供，根据励磁方式的不同，可以分为串励直流电机、并励直流电机、他励直流电机和复励直流电机，如图 3.3.4 所示。

The main pole of DC motor can be provided by the permanent magnet materials such as tombarthite，ferrite and AlNiCo，such motors are called as the permanent magnet DC motors or provided by the electromagnetic field generated by the magnet exciting coils，such motors can further classified into the series excited DC motor，parallel-excited DC motor，separately-excited DC motor and compound excited DC motor by the exciting mode，as shown Figure 3.3.4.

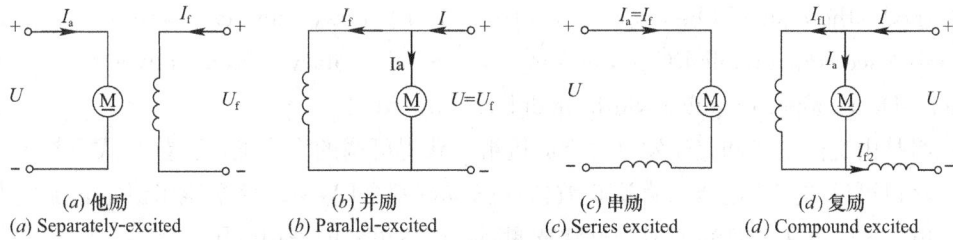

(a) 他励
(a) Separately-excited

(b) 并励
(b) Parallel-excited

(c) 串励
(c) Series excited

(d) 复励
(d) Compound excited

图 3.3.4　直流电机的励磁方式

Figure 3.3.4　Exciting modes of DC motors

2. 直流电机的调速方法

2. Speed control method of DC motor

直流电机的转速公式：

Rotating speed formula of DC motor：

$$n = \frac{U - IR}{K_e \Phi}$$

式中　n——转速（r/min）；

　　U——电枢电压（V）；

　　I——电枢电流（A）；

　　R——电枢回路总电阻（Ω）；

　　Φ——励磁磁通（Wb）；

　　K_e——由电机结构决定的电动势常数。

转速	Speed
电枢电压	Armature voltage
电枢电流	Armature current
电枢回路总电阻	Total armature loop resistance
励磁磁通	Exciting flux
由电机结构决定的电动势常数	The electromotive force constant determined by the structure of the motor

由公式可知，调节转速有三种方法，即改变电枢供电电压 U，调节电枢回路电阻 R 或减弱励磁磁通 Φ。这三种调速方法中，改变电阻 R 只能有级调速且效率低下，只适用于小型直流电机的调速和大功率直流电机的启动控制；减弱励磁磁通 Φ 虽然能够平滑调速且效率高，但只能在负载较轻时在额定速度以上做小范围的弱磁升速；调节电枢电压 U 能在较大范围内无级平滑调速。因此，调压调速是直流电机调速的主要方法。

Acording to the formula，there are three speed control methods，i. e. changing the supply voltage U of armature，regulating the circuit resistance R of armature or weakening the exciting flux Φ. In the three speed control methods，changing the resistance R can only realize the step speed control with the lower efficiency and is only applicable to the speed control of small-sized DC motor and the startup control of high-power DC motor；although weakening the exciting flux Φ can smoothly control the speed with high efficiency，it can only weaken the flux and increase the speed within small range at the speed of exceeding the rated rotating speed under the lower load；regulating the armature voltage U can smoothly realize the stepless speed control with a larger range. Therefore，the speed control by voltage regulation is the main speed control method of DC motor.

调压调速需要有专门向电机供电的可控直流电源。常用的可控直流电源有三种，分别是：旋转变流机组、静止式可控整流器、直流斩波器或脉宽调制变换器。

For this method, there should be a special controllable DC power supply to the motor. There are three types of commonly-used controllable DC power supply, i. e. the rotary current converter set, the static controllable rectifier, DC chopper or pulse width modulated inverter.

旋转变流机组是由交流电机和直流发电机组成机组，获得可调的直流电压。静止式可控整流器是用静止式的可控整流器获得可调的直流电压。直流斩波器或脉宽调制（PWM）变换器用恒定直流电源或不控整流电源供电，利用电力电子开关器件斩波或进行脉宽调制，产生可变的平均电压。

The rotary current converter set is composed of AC motor and DC generator to obtain the adjustable DC voltage. The static controllable rectifier is used for obtaining the adjustable DC voltage. The DC chopper or pulse width modulated （PWM） inverter is powered by a constant DC power or uncontrollable rectifier power, and uses the electrical switching element to chop the wave or modulate the pulse width to generate the variable average voltage.

（二）S7-1200 的四则运算指令

（Ⅱ） Four operational orders of S7-1200

数学函数指令中的 ADD、SUB、MUL 和 DIV 分别是加、减、乘、除指令，它们执行的操作见表 3.3.1。操作数的数据类型可选整数（SInt、Int、DInt、USInt、UInt、UDInt）和浮点数 Real，IN1 和 IN2 可以是常数。IN1、IN2 和 OUT 的数据类型应相同。

The mathematical function instructions ADD, SUB, MUL and DIV respectively indicate the addition, deletion, multiplication and division, and their operations are shown in Table 3.3.1. The data types of operand can be integer （SInt, Int, Dint, USInt, UInt and UDInt） and floating number Real, and IN1 and IN2 can be the constant. The data types of IN1, IN2 and OUT should be same.

整数除法指令将得到的商截尾取整后，作为整数格式的输出 OUT。

The integer division instruction will truncate and round off the quotient, and output the integer value.

ADD 和 MUL 指令允许有多个输入，单击方框中参数 IN2 后面的 ⁂，将会增加输入 IN3，以后增加的输入的编号依次递增。

ADD and MUL instructions allow many inputs, the operator can add the input IN3 by clicking on the back ⁂ of parameter IN2 in the box, and the serial number of inputs added subsequently will be added in sequence.

数学函数指令　　　　　　　　　　　　　表 3.3.1

Mathematical function instructions　　　　　Table 3.3.1

指令 Instructions	描述 Description	指令 Instructions	描述 Description	表达式 Expression
ADD	IN1+IN2=OUT	SQR	计算平方 Calculate the square	IN2=OUT
SUB	IN1-IN2=OUT	SQRT	计算平方根 Calculate the quadratic sum	$\sqrt{IN}$=OUT
MUL	IN1 * IN2=OUT	LN	计算自然对数 Calculate the natural logarithm	ln （IN）=OUT
DIV	IN1/IN2=OUT	EXP	计算指数值 Calculate the exponential value	e^{IN}=OUT
MOD	返回除法的余数 Return the remainder	SIN	计算正弦值 Calculate the Sine value	sin （IN）=OUT

指令 Instructions	描述 Description	指令 Instructions	描述 Description	表达式 Expression
NEG	将输入值的符号取反（求二进制的补码） Negate the symbol of input value (obtain the binary complement number)	COS	计算余弦值 Calculate the Cosine value	cos（IN）＝OUT
INC	将参数 IN/OUT 的值加 1 Add 1 to the value of parameter IN/OUT	TAN	计算正切值 Calculate the tangent value	tan（IN）＝OUT
DEC	将参数 IN/OUT 的值减 1 Delete 1 from the value of IN/OUT	ASIN	计算反正弦值 Calculate the Arcsine value	arc sin（IN）＝OUT
ABS	求有符号整数和实数的绝对值 Obtain the absolute value of signed integer and real number	ACOS	计算反余弦值 Calculate the Arccosine value	arc cos（IN）＝OUT
MIN	获取最小值 Obtain the minimum	ATAN	计算反正切值 Calculate the Arctan value	arc tan（IN）＝OUT
MAX	获取最大值 Obtain the maximum	EXPT	取幂 Exponentiation	$IN1^{IN2}$＝OUT
LIMIT	将输入值限制在制定的范围内 Limit the input value within the specified scope	FRAC	提取小数 Extract the decimal part	—

（三）MCGS 嵌入版的设备通道处理

（Ⅲ）Device channel processing of embedded MCGS

1. 通道处理设置

1. Channel processing settings

在实际应用中，经常需要对从设备中采集到的数据或输出到设备的数据进行预处理，以得到实际需要的工程物理量，如从 A-D 转换通道采集进来的数据一般都为电压 mV 值，需要进行量程转换或查表计算等处理才能得到所需的物理量。

In the actual application, it is often needed to preprocess the data collected from the device or the data output to the device to obtain the actual engineering physical quantity, if the data collected from A-D conversion channel is generally the voltage value mV, the required physical quantity can be obtained only after the range conversion, lookup table, calculation and other processing.

MCGS 嵌入版系统对设备采集通道的数据可以进行八种形式的数据处理，包括：多项式计算、倒数计算、开方计算、滤波处理、工程转换计算、函数调用、标准查表计算、自定义查表计算。在 MCGS 嵌入版设备编辑窗口下，选中某个设备通道，再点击右边的"设备通道处理设置"按钮，如图 3.3.5 所示，即可在弹出的通道处理设置窗口中对设备通道处理进行组态，如图 3.3.6 所示。MCGS 嵌入版从上到下顺序进行计算处理，每行计算结果作为下一行计算输入值，通道值等于最后计算结果值。单击每种处理方法前的数字按钮，即可把对应的处理内容增加到右边的处理内容列表中，"上移"和"下移"按钮改变处理顺序，"删除"按钮删除选定的处理项，单击"设置"按钮，弹出处理参数设置对话框，可以对编辑过的某个处理方法进行设置，其中，倒数、开方、滤波处理不需设置参数，故没有对应的对话框弹出。

The embedded MCGS has eight types of data processing with respect to the data of device collection channel, including: Polynomial calculation, reciprocal calculation, square root extraction calculation, filtering processing, project converter calculation, function call, standard look-up table calculation, and custom look-up table calculation. In the device edit window of embedded MCGS, you can select a certain device channel, click on the right "device channel processing setting" button, as shown in Figure 3.3.5, and configure the device channel processing in the channel processing setting window popped up (as shown in Figure 3.3.6). The embedded MCGS performs the computing process from top to bottom, each line of computed result is used as the next line of calculated input value, and the channel value is equal to the final calculated result. You can click on the number button before each processing method to add the corresponding processing content to the right list of processing content, the "Up" and "Down" buttons to change the sequence of processing, the "Delete" button to delete the selected processing item, and the "Setting" button to pop up the parameter setting box and set the edited certain processing method, of which the parameter setting isn't required for the reciprocal calculation, square root extraction and filtering processing, therefore there is no corresponding box to be popped up.

图 3.3.5　打开通道处理设置窗口

Figure 3.3.5　Open the channel processing setting window

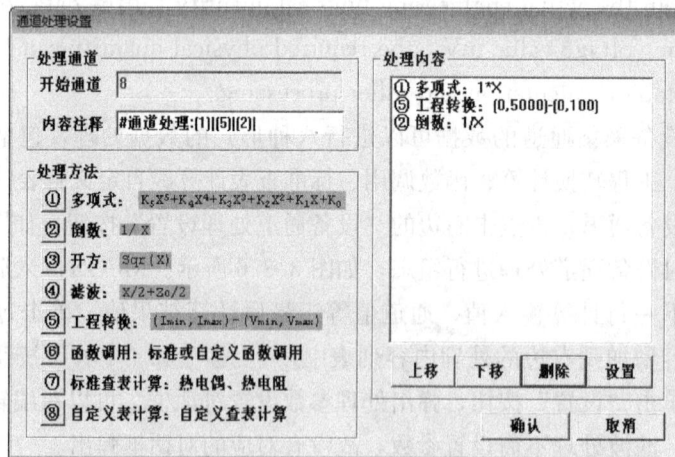

图 3.3.6　通道处理设置窗口

Figure 3.3.6　Channel processing setting window

对输入通道的处理顺序是：

The processing sequence of input channel is shown as follows：

（1）通过设备构件从外部设备读取数据。

（1）Read the data from the external devices by the device components.

（2）按处理内容列表设置的处理内容，从上到下顺序计算处理，第一行使用通道从外部设备读取数据作为计算输入值，其他行使用上一行的计算结果作为输入值。

（2）Perform the calculation processing from top to bottom based on the list of processing contents，of which the first line uses the data read from the external device by the channel as the calculated input value，and other lines use the calculated result of previous line as the input value.

（3）最后一行计算结果作为通道的值。

（3）Use the calculated result of final line as the channel value.

（4）根据所建立的设备通道和实时数据库的连接关系，把通道的值送入实时数据库中的指定数据对象。

（4）Assign the channel value to the designated data object in the real-time database based on the connecting relation established between the device channel and real-time database.

对输出通道的处理顺序是：

The processing sequence of output channel is shown as follows：

（1）根据所建立的设备通道和实时数据库的连接关系，把实时数据库中的指定数据对象的值读入到通道。

（1）Read the value of specified data object in the real-time database into channel based on the connection between the established device channel and real-time database.

（2）按处理内容列表设置的处理内容，从上到下顺序计算处理，第一行使用通道从 MCGS 嵌入版中读取的数据作为计算输入值，其他行使用上一行的计算结果作为输入值。

（2）Perform the calculation processing from top to bottom based on the list of processing contents，of which the first line uses the data read from the embedded MCGS by the channel as the calculated input value，and other lines use the calculated result of previous line as the input value.

（3）最后一行计算结果作为通道的值。

（3）Use the calculated result of final line as the channel value.

（4）通过设备构件把通道的数据输出到外部设备。

（4）Output the channel data to the external device through the device components.

2. 多项式计算处理

2. Polynomial calculation processing

在八种设备通道处理方式中，最常见的就是多项式处理方式。多项式可设置的处理参数有 K0～K5，可以将其设置为常数，也可以设置成指定通道的值（通道号前面加 "!"），另外，还应选择参数和计算输入值 X 的乘除关系，如图 3.3.7 所示。

In the eight device channel processing methods，the most common is the polynomial calculation processing. The available processing parameters K0 to K5 can be set as the constant or the value of designated channel（"!" added before the channel number），and additionally，the multiplication-division relation between the parameter and the calculated input value X should also be selected，as shown in Figure 3.3.7.

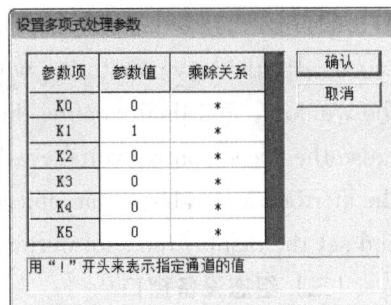

参数项	参数值	乘除关系
K0	0	*
K1	1	*
K2	0	*
K3	0	*
K4	0	*
K5	0	*

用 "!" 开头来表示指定通道的值

图 3.3.7　设置多项式处理参数
Figure 3.3.7　Set the polynomial processing parameters

三、任务实施

Ⅲ. Task Implementation

（一）构建实时数据库

（Ⅰ）To create one real-time database

分析控制要求，可以知道需要新建三个开关型变量控制直流电机的正反转和停止，同时还需要两个开关型变量对直流电机的转速进行控制。最后，需要新建一个数值型变量显示调速电压。实时数据库变量见表3.3.2。

It can be known from the analysis of control requirements that it is necessary to create three switch-type variables to control the positive and negative rotation of DC motor，and meanwhile create two switch-type variables to control the rotating speed of DC motor. Finally，it is necessary to create one numeric type variable to display the speed control voltage. The variables of real-time database are shown in Table 3. 3. 2.

实时数据库变量表　　　　　　　　　　　　　　　表 3. 3. 2
Variables of real-time database　　　　　　　　　Table 3. 3. 2

变量名 Variable name	类型 Type	报警 Alarm
正转 Positive rotation	开关型 Switch type	
反转 Negative rotation	开关型 Switch type	
停止 Stop	开关型 Switch type	
加速 Acceleration	开关型 Switch type	
减速 Deceleration	开关型 Switch type	
调速电压 Speed control voltage	数值型 Numeric type	

按照项目一中介绍的操作步骤，根据表3.2.5构建实时数据库。

The real-time database is created by the steps introduced in Item I based on Table 3. 2. 5.

（二）组态用户窗口

（Ⅱ）To configure the user window

按照项目一中介绍的操作步骤，新建监控窗口，并为窗口创建合适的图形对象，最后定义各对象的动画连接。在这里需要注意的是，因为任务要求需要显示调速电压，因此需要在标签图形对象的属性中勾选输出显示选项，如图3.3.8所示，并在"显示输出"选项卡中关联输出显示的变量和设置输出数据的显示方式，如图3.3.9所示。

Create the new monitoring window by the steps introduced in Item I，create the proper graphic objects to the window，and finally define the animation connection with each object. Here，it should be noted that because the speed control voltage will be displayed in this task，you should check the output display option in the attributes of tab graphic object as shown in Figure 3. 3. 8，and correlate the variables of output display and set the display mode of output data in the "Display Output" tab control as shown in Figure 3. 3. 9.

（三）组态设备窗口

（Ⅲ）To configure the device window

按照项目一中介绍的操作步骤，组态设备窗口，按表3.3.3创建PLC和实时数据库变量的通道连接。由于"调速电压"是数值型变量，需要关联PLC中的模拟量输出值。根据表3.2.1所示，该值是一个16位的有符号变量，因此需要在新建设备通道时选择相应的数据类型，如图3.3.10所示。

Configure the device window by the steps introduced in Item I，and create the channel connection between PLC and real-time database variable according to Table 3. 2. 6. Because the "speed control voltage" is the numeric type variable，it should be correlated to the analog output value in PLC. Table 3. 2. 1 shows that the value is a 16-bit signed variable，thus the corresponding data type should be selected while creating a new device channel，as shown in Figure 3. 3. 10.

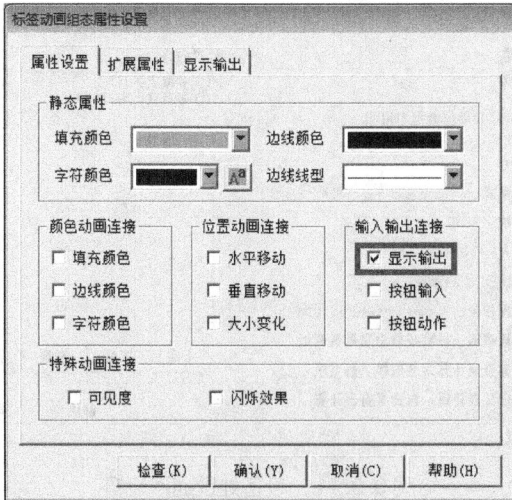

图 3.3.8 勾选标签图形对象的显示输出功能

Figure 3.3.8　Check the display output function of tab graphic object

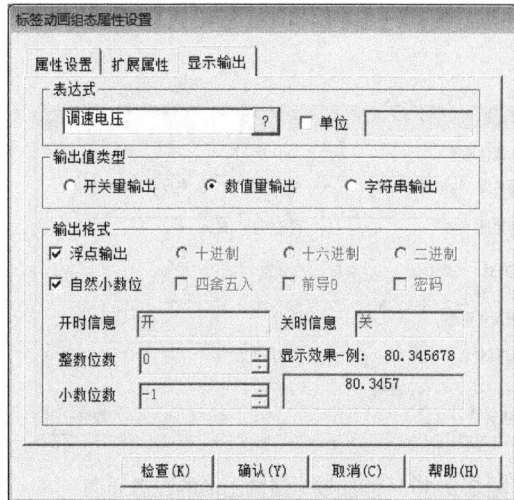

图 3.3.9 设置标签图形对象的显示输出属性

Figure 3.3.9　Set the display output attributes of tab graphic object

设备通道和连接变量　　　　　　表 3.3.3

Device channels and connection variables　　　　Table 3.3.3

连接变量 Connection variables	通道名称 Channel name
正转 Positive rotation	只写 M0.0 Write M0.0 only
反转 Negative rotation	只写 M0.1 Write M0.1 only
停止 Stop	只写 M0.2 Write M0.2 only
加速 Acceleration	只写 M0.3 Write M0.3 only
减速 Deceleration	只写 M0.4 Write M0.4 only
调速电压 Speed control voltage	只读 QWB96 Read QWB96 only

此外，还要对读入的电压值进行通道处理。根据表 3.2.1，可得出调速电压的计算公式为 10×模拟量输出值/27648，设置设备处理通道如图 3.3.11 所示。

Additionally，the voltage value read should be carried out the channel processing. Table 3.2.1 shows that the calculation formula of speed control voltage is 10×analog output value/27648，thus the device processing channel is set as shown in Figure 3.3.11.

（四）连接 I/O 设备

（Ⅳ）To connect I/O device

按照控制要求，I/O 接线图如图 3.3.12 所示。

Based on the control requirements，the I/O wiring diagram is shown in Figure 3.3.12.

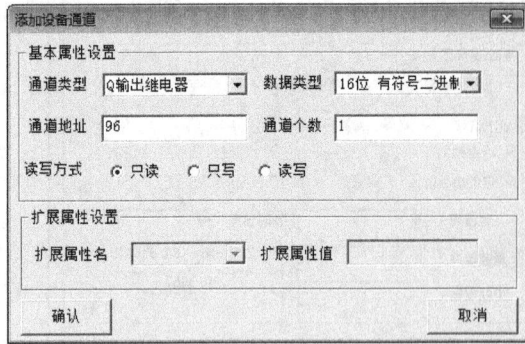

图 3.3.10 添加模拟量输出值的设备通道
Figure 3.3.10 Add the device channel of analog output value

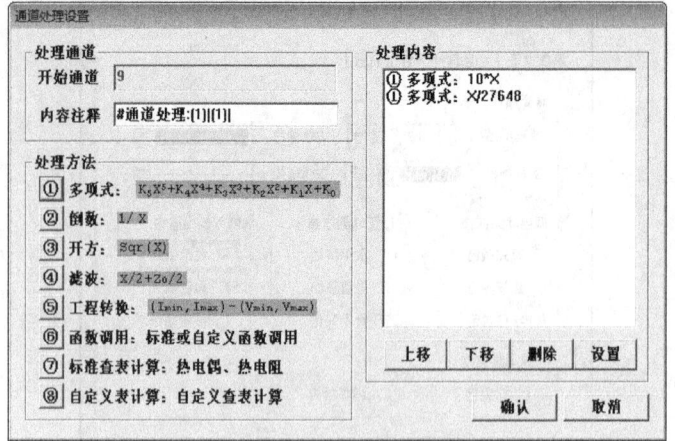

图 3.3.11 设备处理通道设置
Figure 3.3.11 Set the device processing channel

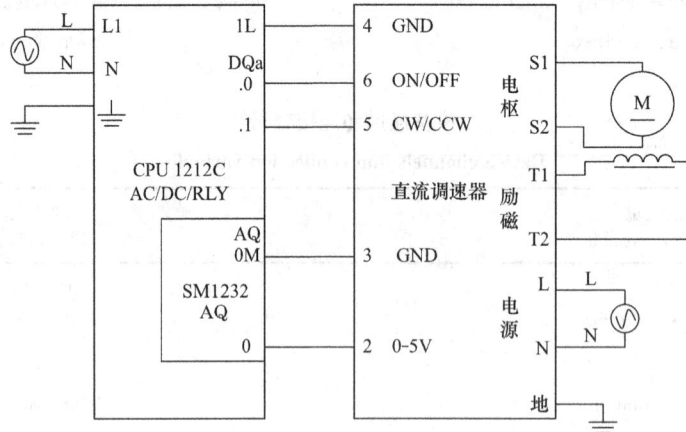

图 3.3.12 I/O 接线图
Figure 3.3.12 I/O wiring diagram

（五）在 TIA 博途中新建项目和设备组态

（Ⅴ）To create a new item and device configuration in TIA Portal

在本任务的实施中，需要在设备组态时添加 AQ 模块，订货号为 6ES7 232-4HB32-0XB0。再按照本模块任务二中相关知识技能中介绍的方法对模拟量输出点进行参数设置。

It is necessary to add AQ module at the time of device configuration in the process of implementing this task，the order number：6ES7 232-4HB32-0XB0. Then the parameters of analog output point are set by the methods prescribed in the relevant knowledge and skills of task II in this module.

（六）建立变量表

（Ⅵ）To create the variable table

按照控制要求，定义变量表如图 3.3.13 所示。

Define the variable table as shown in Figure 3.3.13 based on the control requirements.

（七）编写梯形图程序

（Ⅶ）To write the ladder logic program

按下正转按钮时，置位电机启动/停止位和正转/反转位。按下反转按钮时，置位电机启动/停止位，复位电机正转/反转位。按下停止按钮时，复位电机启动/停止位和正转/反转位，并将调速电压置零。加速按钮的每个上升沿到来，调速电压增加 1V，输出的模拟值相应增加 2765，当电压上升到 5V 时，也就是模拟值为 13824 时，不再增加；减速按钮的每个上升沿到来，调速电压减少 1V，输出的模拟值相应减少 2765，当电压下降到 0V 时，也就是模拟值为 0 时，不再减少。

Set the startup/stop bit and positive/negative rotation bit of motor by pressing the positive rotation button. Set the startup/stop bit and reset the positive/negative rotation bit of motor by pressing the negative rotation button. Reset the startup/stop bit and positive/negative rotation bit of motor, and reset the speed control voltage by pressing the stop button. The speed control voltage is increased by 1V, the analog output value is correspondingly increased by 2765 whenever the each rising edge of acceleration arrives, and when the voltage is increased to 5V, i. e. the analog value is 13824, it will not be increased any longer; The speed control voltage is decreased by 1V, the analog output value is correspondingly decreased by 2765 whenever the each rising edge of deceleration arrives, and when the voltage is decreased to 0V, i. e. the analog value is 0, it will not be decreased any longer.

PLC 变量

	名称	变量表	数据类型	地址	保持	在 H...	可从 ...	注释
1	正转	默认变量表	Bool	%M0.0	☐	☑	☑	
2	反转	默认变量表	Bool	%M0.1	☐	☑	☑	
3	停止	默认变量表	Bool	%M0.2	☐	☑	☑	
4	加速	默认变量表	Bool	%M0.3	☐	☑	☑	
5	减速	默认变量表	Bool	%M0.4	☐	☑	☑	
6	调速电压	默认变量表	Int	%QW96	☐	☑	☑	
7	电动机启动/停止	默认变量表	Bool	%Q0.0	☐	☑	☑	
8	电动机正转/反转	默认变量表	Bool	%Q0.1	☐	☑	☑	

图 3.3.13 变量表

Figure 3.3.13 Variable table

梯形图程序如图 3.3.14 所示。

The ladder diagram program is shown in Figure 3.3.14.

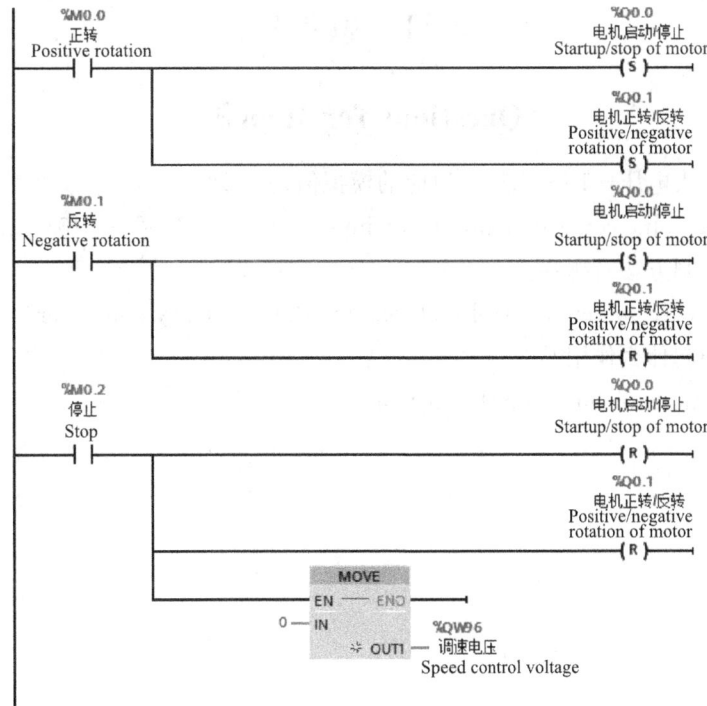

图 3.3.14 梯形图程序（一）

Figure 3.3.14 Ladder logic program（一）

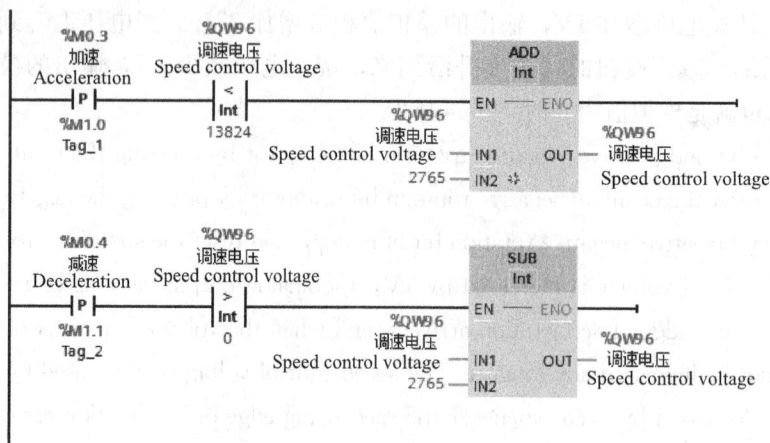

图 3.3.14　梯形图程序（二）

Figure 3.3.14　Ladder logic program（二）

（八）调试

（Ⅷ）Debugging

完成以上步骤后，将编辑好的 MCGS 嵌入版工程和 TIA 博途项目分别下载至触摸屏和 PLC，再用编程线将二者连接起来，这样一个直流电机的调试控制系统就建立起来了。按照控制要求逐项对系统进行调试，直到所有功能都实现为止。

After the above steps, the embedded MCGS engineering and TIA Portal items have been edited are respectively downloaded to the touch screen and PLC, the both are connected with the programming line such that one debugging control system of DC motor is finally created. The system is debugged item by item according to the control requirements until that all functions are realized.

项目三思考题

Questions for Item Ⅲ

1. 模拟量输入点的输入电压－10～＋10V 对应的模拟值是多少？

1. What is the analog value corresponding to the input voltage －10V—＋10V on the analog input point?

2. 变频器的双线制控制方法有哪些？

2. What are the two-wire system control methods for the frequency converter?

3. 直流电机中的电刷起什么作用？

3. What are the functions of bushes in DC motor?

项目四 步进电机和伺服电机的运动控制
Item Ⅳ Motion Control of Stepping Motor and Servo Motor

任务一 步进电机的开环运动控制
Task Ⅰ Open-loop Motion Control of Stepping Motor

一、任务分析
Ⅰ. Task Analysis

用人机界面（HMI）、PLC、步进电机驱动器和一个自由运动的步进电机创建一个步进电机开环运动控制系统。

Assemble the human-machine interface (HMI), PLC, stepping motor driver and free-moving stepping motor into a open-loop motion control system of stepping motor.

（1）监控画面如图 4.1.1 所示。

（1）The monitoring interface is shown in Figure 4.1.1.

（2）按下正转点动按钮和反转点动按钮可以让步进电机正向点动或反向点动运行。

（2）The stepping motor can be joggled clockwise or counter-clockwise by pressing the positive rotation inching button and negative rotation inching button.

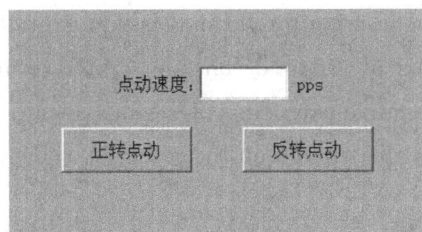

图 4.1.1 监控画面
Figure 4.1.1 Monitoring interface

（3）步进电机点动速度由组态画面中的输入框设定，单位为脉冲每秒（pps）。

（3）The inching speed of stepping motor is set in the input box of the configuration interface, and the unit is pulse per second (pps).

（4）设定的点动速度范围为 1000～5000pps。

（4）The range of inching speed set is between 1000 and 5000pps.

为了完成本任务，首先需要对涉及的设备进行选型，根据现有条件，人机界面选择昆仑通态的 MCGS 触摸屏，PLC 选择西门子 S7-1200 系列 PLC。通过完成本任务，可以学习到步进电机的工作原理、西门子 S7-1200 的轴工艺对象、启动/禁用轴指令和"点动"模式下的移动轴指令这四个方面的知识。

For the purpose of this task, it is firstly necessary to select the required types of devices involved, of which MCGS touch screen, and Siemens S7-1200 PLC will be selected under the current conditions. By completing this task, you can learn about the knowledge in four aspects, i. e. the working principles of stepping motor, and the shaft process object, the enabled/disabled shaft instructions and the shaft movement instruction under the "Inching" mode of Siemens S7-1200.

二、相关知识技能

Ⅱ. Relevant Knowledge & Skills

（一）步进电机的工作原理

（Ⅰ）Working principle of stepping motor

1. 步进电机的结构组成

1. Structure of stepping motor

步进电机是一种把电脉冲信号变成机械角位移的开环控制电机。步进电机由定子和转子组成。定子由硅钢片叠成，有一定数量的磁极和绕组。转子是用硅钢片叠成或用软磁性材料做成的凸极结构。步进电机的结构如图4.1.2所示。

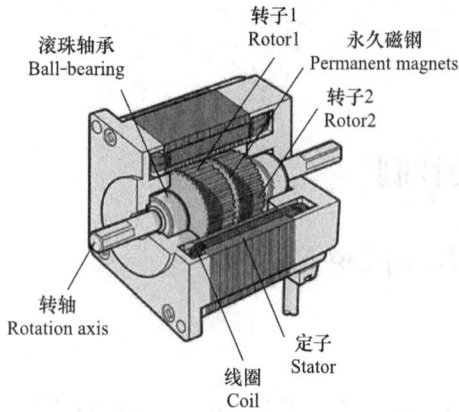

图4.1.2 步进电机的结构

Figure 4.1.2 Structure of stepping motor

The stepping motor is a kind of open-loop control motor converting the electric pulse signal into the mechanical angle displacement. The stepping motor consists of stator and rotor. The stator is composed of silicon steel sheets and a certain quantity of magnetic poles and windings. The rotor is the salient pole structure made of silicon steel sheet stacked or soft magnetic material. The structure of stepping motor is shown in Figure 4.1.2.

按照力矩产生原理，步进电机可分为励磁式、反应式和混合式。按照定子励磁相数，可分为三相、四相、五相和六相。以三相步进电机为例，其定子上有6个磁极，两个相对的磁极组成一相，每个磁极上有5个齿。而转子上均匀分布着40个齿。如图4.1.3所示。

The stepping motor can be classified into the exciting, reactive and compound types by the principles of generating torque and classified into the three-phase, four-phase, five-phase and six-phase types by the number of excitation phases. For an example of three-phase stepping motor, its stator has 6 magnetic poles, two relative magnetic poles compose into one phase, and each magnetic pole has 5 teeth. There are 40 teeth evenly distributed on the rotor. As shown in Figure 4.1.3.

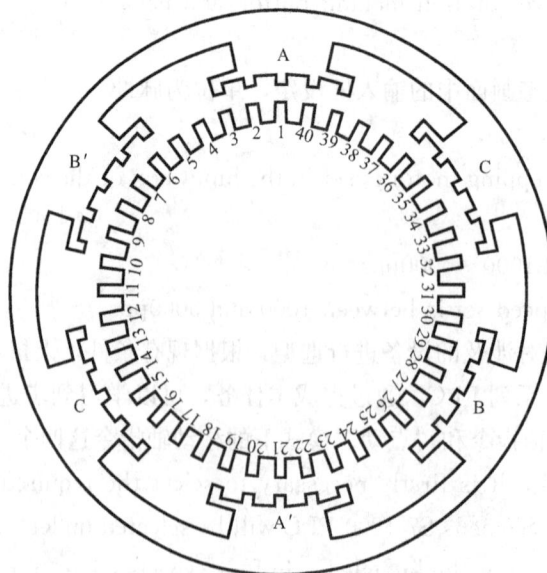

图4.1.3 步进电机的定子和转子原理图

Figure 4.1.3 Schematic diagram of stepping motor's stator and rotor

2. 步进电动机旋转的原因

2. Rotation reason of stepping motor

错齿是步进电动机旋转的根本原因。由于 6 个磁极之间的角度是 $60°$，而每个转子齿的角度是 $9°$，因此，当 A 相通电时，转子齿和 A 相定子齿对齐，但不能与 C 相定子齿对齐，转子齿与距其最近的 C 相定子齿相差 $3°$，如图 4.1.4 (a) 所示。A 相断电、C 相通电后，转子转过 $3°$ 使转子齿与 B 相定子齿对齐，如图 4.1.4 (b) 所示。同理，C 相断电，B 相通电后，转子再转 $3°$，如图 4.1.4 (c) 所示。

The staggered teeth is the basic reason for the rotation of stepping motor. The angle among 6 magnetic poles is $60°$ and the angle of each rotor tooth is $9°$, thus when A phase is energized, the rotor tooth is aligned with A phase stator tooth, but it can't be aligned with C phase rotor tooth, the rotor tooth is $3°$ distant from its closest C phase stator tooth, as shown in Figure 4.1.4 (a). After A phase is de-energized and C phase is energized, the rotor rotates $3°$ to make the rotor tooth align with B phase stator tooth, as shown in Figure 4.1.4 (b). Similarly, after C phase is de-energized and B phase is energized, the rotor rotates $3°$ again as shown in Figure 4.1.4 (c).

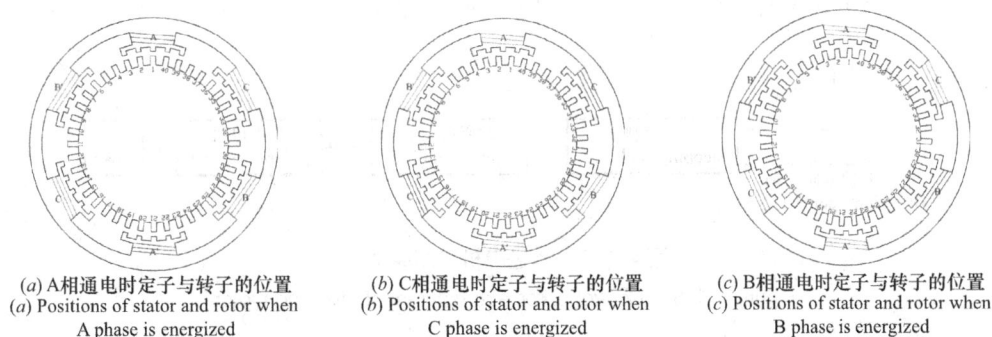

(a) A相通电时定子与转子的位置
(a) Positions of stator and rotor when A phase is energized

(b) C相通电时定子与转子的位置
(b) Positions of stator and rotor when C phase is energized

(c) B相通电时定子与转子的位置
(c) Positions of stator and rotor when B phase is energized

图 4.1.4 步进电机的旋转原因

Figure 4.1.4 Rotation reason of stepping motor

绕组每通电一次（即运行一拍），转子转过的角度叫作步距角，显然，本例的步距角是 $3°$。转子的旋转方向取决于三相绕组的通电顺序，改变通电顺序即可改变转向。若通电顺序为 A→B→C→A，则步进电动机反转。

The rotating angle of rotor every time the winding is energized (i. e. run for one step) is called as the stepping angle, and obviously, the stepping angle in the example is $3°$. The rotational direction of rotor depends on the sequence of energizing three-phase windings, and can be changed only by changing the sequence. If the sequence is A→B→C→A, the stepping motor rotates counterclockwise.

在上述步进电机的工作过程中，三相绕组每次只有一相通电，而且一个循环周期共包括三个脉冲，这被称为三相单三拍的工作方式。

In the working process of above stepping motor, every time only one of three phases is energized, and one cycle period includes three pulses, thus such working mode is called as three-phase single-three stepping.

除了这种工作方式，还有三相单双六拍和三相双三拍两种工作方式。三相单双六拍的绕组通电顺序为 A→AB→B→BC→C→CA→A。此种工作方式的步距角为三相单三拍的二分之一。三相双三拍的绕组通电顺序为 AB→BC→CA→AB。

Except for the working mode, there are three-phase double-six stepping and three-phase double-three stepping. Under the three-phase double-six stepping mode, the sequence of energizing windings is A→AB→B→BC→C→CA→A. In such working way, the stepping angle is half of three-phase single-three stepping.

Under the three-phase double-three stepping mode，the sequence of energizing windings is AB→BC→CA→AB.

3. 步进电机的控制原理

3. Control principles of stepping motor

当定子绕组按一定顺序轮流通电时，转子将沿一定方向一步步转动。因此步进电机绕组是按一定通电方式工作的，为实现该种轮流通电，要将控制脉冲按规定的通电方式分配到电机的每相绕组。步进控制器的作用是把输入的脉冲转换成环型脉冲，以控制步进电机，并能进行正反转控制。功率放大器的作用是把步进电机输出的环型脉冲放大，以驱动步进电机转动。如图 4.1.5 所示。

When the stator windings are energized by turns in a certain sequence，the rotor will rotate step by step along a certain direction. Therefore，the windings of stepping motor is enabled by a certain electrification method，and the control pulses should be assigned to each phase of motor windings based on the specified electrification method in order to realize the circular electrification. The stepping controller is used for converting the input pulse into ring pulse to control the stepping motor and its positive and negative rotation. The power amplifier is used for amplifying the ring pulse output from the stepping motor to drive the rotation of stepping motor. As shown in Figure 4.1.5.

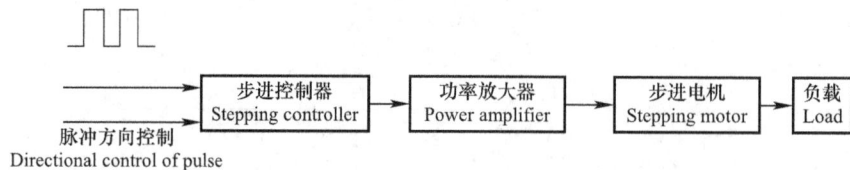

图 4.1.5　步进电机控制原理图

Figure 4.1.5　Control schematic diagram of stepping motor

4. Kinco 3M458 步进电机驱动器

4. Kinco 3M458 Stepping motor driver

Kinco 3M458 步进电机驱动器用于驱动三相步进电机。其供电电压为 DC 24～40V，可输出 3.0～5.8A 的相电流，有三路控制信号。一路是由 PLS＋和 PLS－端子输入的脉冲信号，用于控制步进电机的启停；一路是由 DIR＋和 DIR－端子输入的方向信号，用于控制步进电机的旋转方向；一路是 FREE＋和 FREE－端子输入的脱机信号，可以决定在驱动器上电后是锁住电机还是让电机处于自由状态。步进电机驱动器、步进电机和控制器的接线图如图 4.1.6 所示。

Kinco 3M458 Stepping motor driver is used for driving the three-phase motor. Its power voltage is DC 24 to 40V，it can output 3.0 to 5.8A phase current，and has three control signal circuits. One circuit is the pulse signal input by the PLS＋and PLS－terminals，which is used for controlling the startup and stop of stepping motor；One is the direction signal input from DIR＋and DIR－terminals，which is used to control the direction of rotation of the stepping motor；One way is an offline signal input from the FREE＋and FREE-terminals which determines whether the motor is locked or left free when the driver is powered on. The wiring diagram of the stepping motor driver，stepping motor and controller is shown in Figure 4.1.6.

在驱动器上有一个红色的 8 位 DIP 功能设定开关，如图 4.1.7 所示，可以用来设定驱动器的工作方式和工作参数。8 位 DIP 开关的功能见表 4.1.1。其中，细分设置开关的功能见表 4.1.2，输出相电流的设置见表 4.1.3。

There is a red 8-bit DIP function setting switch on the driver，as shown in Figure 4.1.7，which can be used to set the driver's operation modes and operating parameters. The functions of the 8-bit DIP switch is shown in Table 4.1.1. The function of the subdivision setup switch is shown in Table 4.1.2，and the setting of the output phase current is shown in Table 4.1.3.

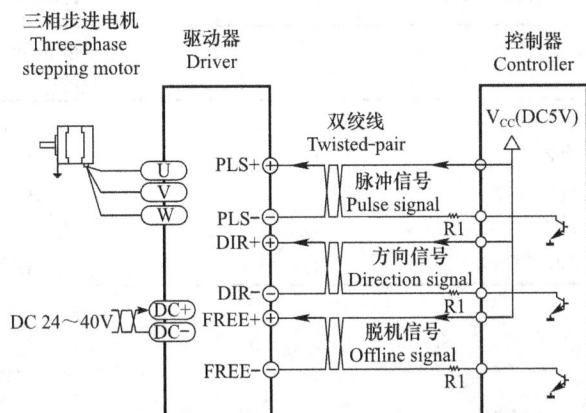

图 4.1.6 步进电机驱动器的接线图

Figure 4.1.6 Wiring diagram of stepping motor driver

图 4.1.7 快速调试的步骤

Figure 4.1.7 Steps of quick debugging

8 位 DIP 开关的功能

表 4.1.1

Function of 8-bit DIP switch

Table 4.1.1

开关序号 Switch serial number	ON 功能 ON Function	OFF 功能 OFF Function
DIP1~DIP3	细分设置用 Subdivision settings	细分设置用 Subdivision settings
DIP4	静态电流全流 Quiescent current full current	静态电流半流 Quiescent current half current
DIP5~DIP8	输出相电流设置用 Output phase current setting	输出相电流设置用 Output phase current setting

细分设置开关的功能

表 4.1.2

Functions of subdivision setup switches

Table 4.1.2

DIP1	DIP2	DIP3	细分 Subdivision
ON	ON	ON	400 步/转 400steps/rotation
ON	ON	OFF	500 步/转 500steps/rotation
ON	OFF	ON	600 步/转 600steps/rotation
ON	OFF	OFF	1000 步/转 1000steps/rotation
OFF	ON	ON	2000 步/转 2000steps/rotation
OFF	ON	OFF	4000 步/转 4000steps/rotation
OFF	OFF	ON	5000 步/转 5000steps/rotation
OFF	OFF	OFF	10000 步/转 10000steps/rotation

输出相电流设置

Output phase current setting

表 4.1.3
Table 4.1.3

DIP5	DIP6	DIP7	DIP8	输出电流 Output current
OFF	OFF	OFF	OFF	3.0A
OFF	OFF	OFF	ON	4.0A
OFF	OFF	ON	ON	4.6A
OFF	ON	ON	ON	5.2A
ON	ON	ON	ON	5.8A

（二）S7-1200 的轴工艺对象

（Ⅱ）S7-1200 shaft process object

1. PTO 的信号类型

1. Signal type of PTO

西门子 S7-1200 的 CPU 通过 PTO（脉冲串输出）为步进电机和伺服电机的运行提供运动控制功能。根据步进电机的设置，每个脉冲会使步进电机移动特定角度。例如，如果将步进电机设置为每转 1000 个脉冲，则每个脉冲电机移动 0.36°。步进电机的速度则通过每单位时间的脉冲数来确定。

The CPU of the Siemens S7-1200 provides motion control function for the operation of the stepping motor and servo motor via PTO (pulse strain output). Depending on the stepping motor setting, each pulse causes the stepping motor to move by a specific angle. For example, if the stepping motor is set to 1000 pulses per rotation, each pulse motor is moved by 0.36°. The speed of the stepping motor is determined by the number of pulses per unit time.

每一个具有轴工艺的 CPU 版本最多有 4 个 PTO，输出最大频率为 100kHz。PTO 有四种信号类型，分别是"PTO（脉冲 A 和方向 B）""PTO（正数 A 和倒数 B）""PTO（A/B 相移）""PTO（A/B 相移-四倍频）"。

Each CPU version with the shaft process has up to 4 PTOs and maximum output frequency is 100kHz PTO has four signal types: "PTO (pulse A and direction B)", "PTO (positive A and reciprocal B)", "PTO (A/B phase shift)" and "PTO (A/B phase shift-quadruple frequency)".

（1）PTO（脉冲 A 和方向 B）的一个输出控制脉冲，另一输出控制方向。若方向输出上输出 5V/24V，则驱动步进电机正向旋转；若方向输出上输出 0V，则驱动步进电机反向旋转，如图 4.1.8 所示。

（1）One output of the PTO (pulse A and direction B) controls pulse and the other output controls direction. If 5V/24V is output value on the directional output, the stepping motor is driven to rotate in the forward direction; If 0V is output value on the directional output, the stepping motor is driven to rotate in the reverse direction, as shown in Figure 4.1.8.

（2）PTO（正数 A 和倒数 B）的一个输出脉冲控制步进电机正方向旋转，另一个输出脉冲控制步进电机负方向旋转，如图 4.1.9 所示。

（2）One output pulse of the PTO (positive A and reciprocal B) controls the positive rotation of the stepping motor and the other output pulse controls the negative rotation of the stepping motor, as shown in Figure 4.1.9.

（3）"PTO（A/B 相移）"的两个输出均以指定速度产生脉冲，但相位相差 90°。如果信号 A 超前信号 B 90°，则驱动步进电机正向旋转；如果信号 B 超前信号 A 90°，则驱动步进电机反向旋转，如图 4.1.10 所示。

（3）Both outputs of "PTO (A/B phase shift)" generate pulses at the specified speed, but with a phase difference of 90°. If the signal A is ahead of the signal B by 90°, it will drive the stepping motor to rotate in the forward direction；If signal B is ahead of signal A by 90°，it will drive the stepping motor to rotate in reverse rotation，as shown in Figure 4.1.10.

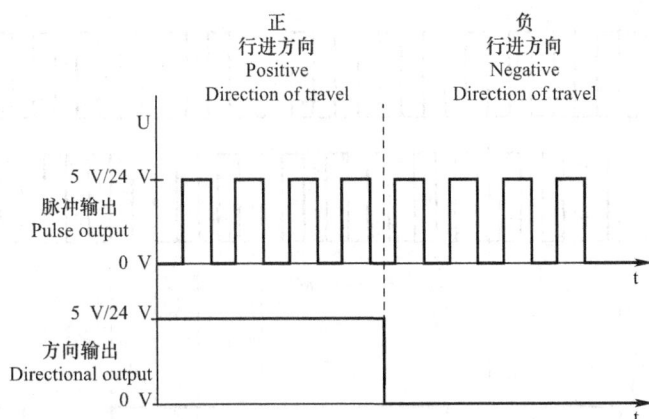

图4.1.8 PTO（脉冲A和方向B）的控制示意图

Figure 4.1.8 Control schematic diagram for pTO（Pulse A and Direction B）

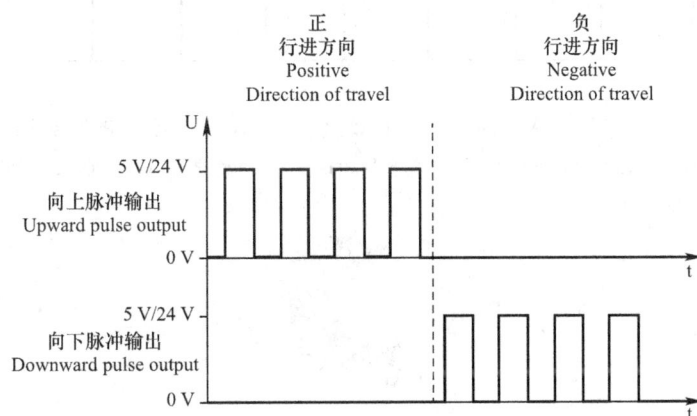

图4.1.9 PTO（正数A和倒数B）的控制示意图

Figure 4.1.9 Control schematic diagram for PTO（positive A and reciprocal B）

（4）"PTO（A/B相移-四倍频）"的两个输出均以指定速度产生脉冲，但相位相差90°，输出脉冲的频率减小到PTO（A/B相移）的1/4，如图4.1.10所示。

（4）Both outputs of the "PTO（A/B phase shift-quadruple frequency）" generate pulses at the specified speed，but with a 90-degree phase difference，the frequency of the output pulses is reduced to one quarter of the PTO（A/B phase shift），as shown in Figure 4.1.10.

2. 轴工艺对象

2. Shaft process object

西门子S7-1200的轴工艺对象用于组态机械驱动器的数据、驱动器的接口、动态参数以及其他驱动器属性，测试轴的功能、跟踪轴和驱动器的当前状态以及错误信息，因此轴工艺对象有"组态"（Configuration）、"调试"（Commissioning）和"诊断"（Diagnostics）三种工具，如图4.1.11所示。

The shaft process object of Siemens S7-1200 is used to configure data of mechanical drivers，driver interfaces，dynamic parameters，and other driver properties，test shaft functions，track current status of shafts and drivers，and error information，so the shaft process object has three tools，"Configuration"，"Commissioning"，and "Diagnostics"，as shown in Figure 4.1.11.

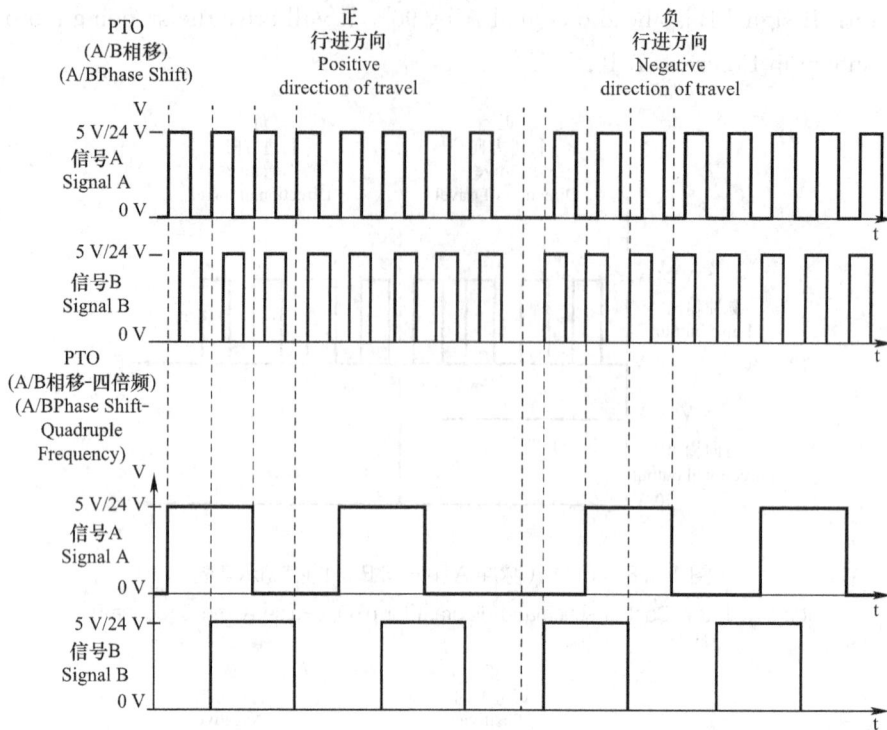

图 4.1.10 PTO（A/B 相移）和 PTO（A/B 相移-四倍频）的控制示意图

Figure 4.1.10 Control schematic diagram of PTO（A/B Phase Shift）and PTO（A/B phase shift-quadruple frequency）

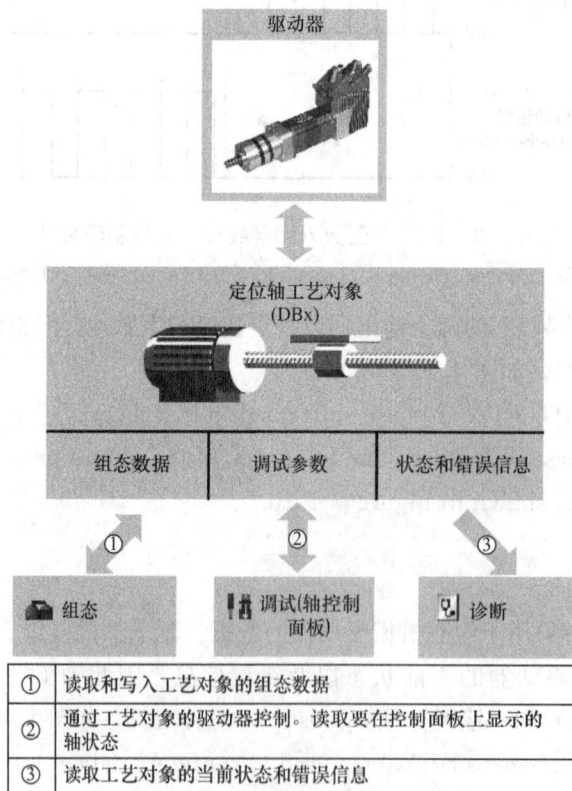

图 4.1.11 轴工艺对象功能示意图

Figure 4.1.11 Functional schematic diagram of shaft process object

使用"组态"工具可以组态定位轴工艺对象的以下属性：

Use the "Configuration" tool to configure the following attributes for locating axis process objects：

（1）要使用的 PTO（脉冲串输出）/PROFIdrive 驱动器/模拟量输出的选择选项和驱动器接口的组态。

（1）PTO（Pulse Train Output）/PROFIdrive drive/analog output selection options and configuration of drive interfaces.

（2）机械装置的属性和驱动器（机器/设备）的传动比参数。

（2）Mechanical devices'properties and drive ratio parameters of driver（machine/equipment）.

（3）位置限制和定位监控的属性。

（3）Position restriction and location monitoring properties.

（4）动态和归位的属性。

（4）Dynamic and returned properties.

（5）控制回路的参数。

（5）Parameters of the control loop.

使用"调试"工具即可测试轴的功能，无需创建用户程序。启动该工具时，将显示轴控制面板。轴控制面板提供了下列命令：

Use the "Commissioning" tool to test shaft function without creating user programs. When you start the tool，the axis control panel is displayed. The axis control panel provides the following commands：

（1）启用和禁用轴。

（1）Enable and disable axis.

（2）在点动模式下移动轴。

（2）Move shaft in inching mode.

（3）以绝对和相对方式定位轴。

（3）Position axis in absolute and relative manner.

（4）使轴归位。

（4）Return axis to original position.

（5）确认错误信息。

（5）Confirm error message.

可以为运动命令相应地调整动态值。轴控制面板同时显示当前的轴状态。利用基于 PROFIdrive/模拟量输出的驱动器连接，通过自整定，可确定最佳的控制环增益。

Dynamic values can be adjusted accordingly for motion commands. The axis control panel also displays the current axis status. Using a driver connection based onPROFIdrive/analog output，the optimal control loop gain can be determined by self-tuning.

使用"诊断"工具，可以跟踪轴和驱动器的当前状态和错误信息。

The "Diagnostics" tool can be used to track the current status and error information for shafts and drives.

3. 添加一个轴工艺对象

3. Add an shaft process object

在已创建的具有 S7-1200 CPU 模块的项目中，按照如下步骤添加轴工艺对象。在项目树 PLC_1 中，展开节点"工艺对象"，然后选择"添加新对象"。弹出新增对象窗口如图 4.1.12 所示。

In the created project with the S7-1200 CPU module，add the shaft process object as follows. In the project tree PLC_1，expand the node "Process Object" and select "Add New Object". The new object window pops up as shown in Figure 4.1.12.

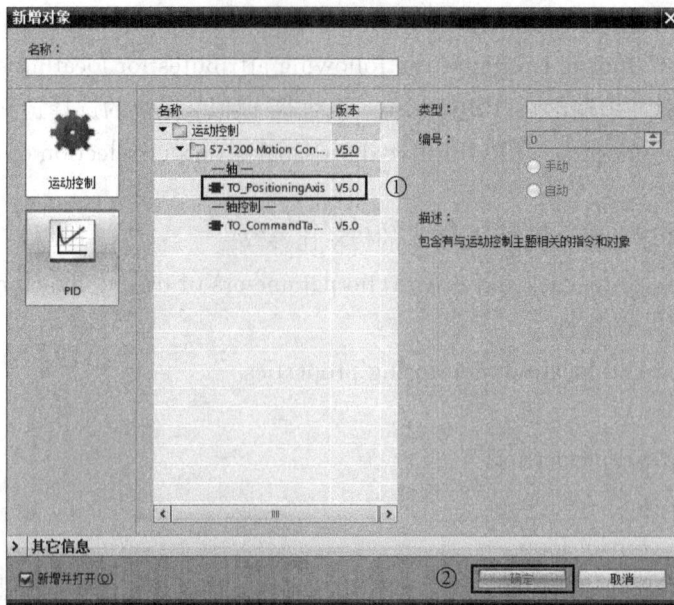

图 4.1.12 新增轴工艺对象

Figure 4.1.12 New shaft process object

选择"轴">"TO_PositioningAxis"(必要时可以重命名),然后单击确定。之后项目树中的工艺对象文件夹下就出现了刚才新建的轴工艺对象的名称。可以看到该轴工艺对象下有三个工具,"组态"、"调试"和"诊断"。组态工具已经自动打开,如图 4.1.13 所示。

Select "Shaft" > "TO_PositioningAxis" (you can rename it if necessary) and click OK. The name of the shaft process object you just created appears under the Process Objects folder in the project tree. You can see that there are three tools under the shaft process object, "Configuration", "Commissioning", and"Diagnosis". The configuration tool has been opened automatically, as shown in Figure 4.1.13.

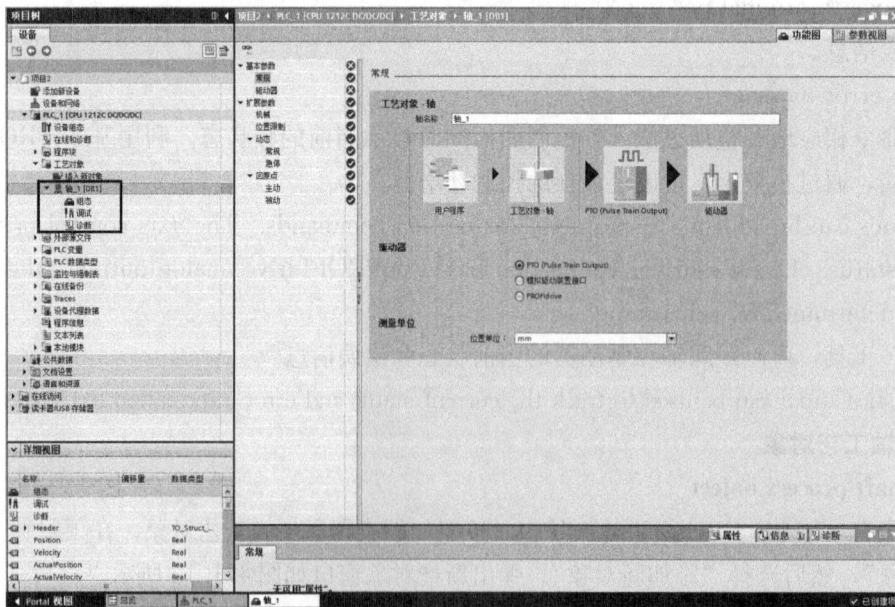

图 4.1.13 轴工艺对象组态窗口

Figure 4.1.13 Shaft process object configuration window

（三）启动/禁用轴指令

（Ⅲ）Enable/disable axis command

MC_Power 运动控制指令可启用或禁用轴。在启用或禁用轴之前，应确保以下条件：

The MC_Power motion control command enables or disables the shaft. Before enabling or disabling the shaft，ensure the following conditions are met：

（1）定位轴工艺对象已正确组态。

（1）The positioning shaft process object has been configured correctly.

（2）没有待决的启用/禁止错误。

（2）There are no pending enable/disable errors.

MC_Power 指令的参数见表 4.1.4。

The parameters of the MC_Power instruction is shown in Table 4.1.4.

MC_Power 指令的参数　　　　　　　　　　　　　　　表 4.1.4

Parameters of the MC_Power instruction　　　　　　　　Table 4.1.4

参数 Parameter	声明 Statement	数据类型 Data type	说明 Description
Axis	INPUT	TO_Axis	轴工艺对象 Shaft process object
Enable	INPUT	BOOL	FALSE（默认）：所有激活的任务都将按照参数化的"StopMode"而中止，并且轴也会停止 FALSE（Default）：All active tasks are stopped according to the parameterized "Stop Mode" and the axis is stopped TRUE：运动控制尝试启用轴 TRUE：Motion control attempts to enable the axis
StopMode	INPUT	INT	0：紧急停止：如果禁用轴的请求处于待决状态，则轴将以组态的急停减速度进行制动。轴在变为静止状态后被禁用 0：Emergency stop：If the request to disable the shaft is pending, the shaft will be braked at the configured sudden stop deceleration. The shaft is disabled when it becomes stationary 1：立即停止：如果禁用轴的请求处于待决状态，则轴将以组态的急停减速度进行制动。轴将根据驱动器中的组态进行制动，并转入停止状态 1：Stop immediately：If the request to disable the shaft is pending, the shaft will be braked at the configured sudden stop deceleration. The shaft will be braked according to the configuration in the drive and switched to the stopped state 2：带有加速度变化率控制的紧急停止：如果禁用轴的请求处于待决状态，则轴将以组态的急停减速度进行制动。如果激活了加速度变化率控制，会将已组态的加速度变化率考虑在内。轴在变为静止状态后被禁用 2：Emergency stop with acceleration change rate control：If the request to disable the shaft is pending, the shaft will be braked at the configured sudden stop deceleration. If acceleration change rate control is activated，the configured acceleration change rate will be taken into account. The shaft is disabled when it becomes stationary
Status	OUTPUT	BOOL	轴使能的状态 Axis enabled state FALSE：轴已禁用 FALSE：Axis disabled 1. 轴不会执行运动控制任务并且不接受任何新任务（例外：MC_Reset 任务） 1. The axis does not perform motion control tasks and does not accept any new tasks（exception：MC_Reset task） 2. 轴未回原点 2. Axis is not returned to original point 3. 禁用时，直到轴达到停止状态，状态才会更改为 FALSE 3. When disabled，the state does not change to FALSE until the shaft reaches a stop state

参数 Parameter	声明 Statement	数据类型 Data type	说明 Description
			TRUE：轴已启用 TRUE：Axis enabled 1. 轴已准备好执行运动控制任务 1. The axis is ready for motion control tasks 2. 轴启用时，直到信号"驱动器就绪"（Drive ready）进入未决，状态才会更改为 TRUE。如果在轴组态中未组态"驱动器就绪"（Drive ready）驱动器接口，状态会立即更改为 TRUE 2. When the shaft is enabled，the state does not change to TRUE until the signal "Drive Ready" is pending. If the Drive Ready driver interface is not configured in the shaft configuration，the status changes immediately to TRUE
Busy	OUTPUT	BOOL	FALSE：MC_Power 无效 FALSE：MC_Power void TRUE：MC_Power 已生效 TRUE：MC_Power activated
Error	OUTPUT	BOOL	FALSE：无错误 FALSE：No Errors TRUE：运动控制指令"MC_Power"或相关工艺发生错误出错。原因可在"ErrorID"和"ErrorInfo"参数中找到 TRUE：An error occurred in the motion control instruction "MC_Power" or related process. The cause of the error can be found in the "ErrorID" and "ErrorInfo" parameters
ErrorID	OUTPUT	WORD	参数"Error"的错误 ID Error ID for the "Error" parameters
ErrorInfo	OUTPUT	WORD	参数"ErrorID"的错误信息 ID Error information ID for the "ErrorID" parameters

（四）以"点动"模式移动轴指令
（Ⅳ）Move axis command in "inching" mode

通过运动控制指令"MC_MoveJog"，在点动模式下以指定的速度连续移动轴。例如，可以使用该运动控制指令进行测试和调试。为了使用 MC_MoveJog 指令，必须先启用轴。

Continuously move the shaft at the specified speed in the inching mode via the motion control command "MC_MoveJog". For example，this motion control instruction can be used for testing and commissioning. In order to use the MC_MoveJog instruction，you must first enable the shaft.

MC_MoveJog 指令的参数见表 4.1.5。

The parameters of the MC_MoveJog instruction are shown in Table 4.1.5.

MC_MoveJog 指令的参数　　　　　　　　　　　　　　　表 4.1.5
Parameters of the MC_MoveJog instruction　　　　　　　Table 4.1.5

参数 Parameter	声明 Statement	数据类型 Data type	说明 Description
Axis	INPUT	TO_SpeedAxis	轴工艺对象 Shaft process object
JogForward	INPUT	BOOL	如果参数值为 TRUE，则轴都将按参数"Velocity"中所指定的速度，正向移动 If the parameter value is TRUE，the shaft will move forward at the speed specified in the parameter "Velocity"

参数 Parameter	声明 Statement	数据类型 Data type	说明 Description
JogBackward	INPUT	BOOL	如果参数值为 TRUE，则轴都将按参数"Velocity"中指定的速度，反向移动 If the parameter value is TRUE, the shaft will move backward at the speed specified in the parameter "Velocity"

如果两个参数同时为 TRUE，轴将根据所组态的减速度直至停止。将通过参数"Error"、"ErrorID"和"ErrorInfo"指示错误

If both parameters are TRUE, the shaft stops according to the configured deceleration. Errors are indicated by the parameters "Error", "ErrorID", and "ErrorInfo".

Velocity	INPUT	REAL	点动模式的预设速度 Preset speed in inching mode 限值：启动/停止速度≤速度≤最大速度 Limit：Start/stop speed≤speed≤maximum speed
InVelocity	OUTPUT	BOOL	TRUE＝已达到参数"Velocity"中指定的速度 TRUE＝The speed specified in parameter "Velocity" has been reached
Busy	OUTPUT	BOOL	TRUE＝正在执行指令 TRUE＝Executing instruction
CommandAborted	OUTPUT	BOOL	TRUE＝任务在执行期间被另一任务中止 TRUE＝Task stopped by another task during execution
Error	OUTPUT	BOOL	TRUE＝任务执行期间出错。出错原因可在"ErrorID"和"ErrorInfo"参数中找到 TRUE＝Error during task execution. The cause of the error can be found in the "ErrorID" and "ErrorInfo" parameters
ErrorID	OUTPUT	WORD	参数"Error"的错误 ID Error ID for the "Error" parameters
ErrorInfo	OUTPUT	WORD	参数"ErrorID"的错误信息 ID Error information ID for the "ErrorID" parameters

三、任务实施
Ⅲ. Task Implementation

(一) 构建实时数据库
(Ⅰ) To create one real-time database

分析控制要求，可以知道需要新建两个开关型变量控制步进电机的正转点动和反转点动，还需要新建一个数值型变量显示点动速度。实时数据库变量见表 4.1.6。

By analyzing the control requirements，we can know that it is necessary to create two new switched variables to control the normal and reverse inching of the stepping motor，and to create a new numerical variable to display the pointing speed. The variables of real-time database is shown in Table 4.1.6.

实时数据库变量表 表 4.1.6
Variables of real-time database Table 4.1.6

变量名 Variable name	类型 Type	报警 Alarm
正转点动 Forward inch	开关型 Switch type	
反转点动 Reverse inch	开关型 Switch type	
点动速度 Inching speed	数值型 Numeric type	

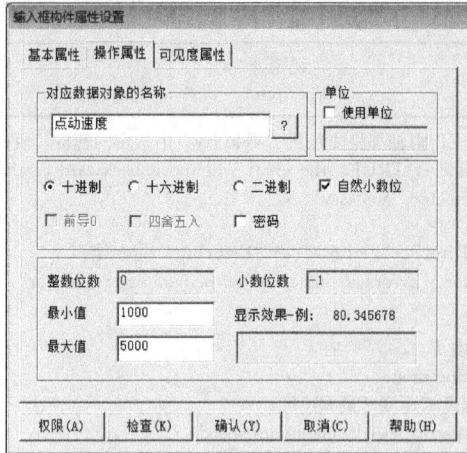

图 4.1.14 输入框构件属性设置
Figure 4.1.14 Input box component attribute settings

（二）组态用户窗口

（Ⅱ）To configure the user window

按照项目一中介绍的操作步骤，新建监控窗口和报警窗口，并为各窗口创建合适的图形对象，最后定义各对象的动画连接。在这里需要注意的是，因为任务要求需要由用户输入点动速度，且该速度需限定在一定范围内，因此需要新建输入框构件，并设置其属性如图 4.1.14 所示。

Create the new monitoring window and alarm window by the steps introduced in Item I, create the proper graphic objects to each window, and finally define the animation connection with each object. It is noted that since the task requires that the inching speed be entered by the user and that the speed be limited to a certain range, a new input box component needs to be created and its attributes are set as shown in Figure 4.1.14.

（三）组态设备窗口

（Ⅲ）To configure the device window

按照项目一中介绍的操作步骤，组态设备窗口，按表 4.1.7 创建 PLC 和实时数据库变量的通道连接。由于"点动速度"是数值型变量，需要关联 PLC 中的点动速度。根据表 4.1.5 所示，该值是一个 32 位的实数，因此需要在新建设备通道时选择 32 位浮点数的数据类型。另外，如果通道类型设置为 M 内部继电器，通道地址设置为 10，意味着"点动速度"这个变量与 S7-1200 中实数类型的存储器 MD10 连接在了一起。如图 4.1.15 所示。

Configure the device window by the steps introduced in Item I, and create the channel connection between PLC and real-time database variable according to Table 4.1.7. Because the "inching speed" is a numerical variable, the inching speed in the PLC needs to be associated. As Table 4.1.5 shows, this value is a 32-bit real number, so you need to select the data type of the 32-bit floating-point number when you create a new device channel. In addition, if the channel type is set to M internal relay, the channel address is set to 10, meaning that the variable "inching speed" is connected to the real-type memory MD10 in S7-1200. As shown in Figure 4.1.15.

设备通道和连接变量　　　　　　　　　　　　　　　　表 4.1.7
Device channels and connection variables　　　　　　Table 4.1.7

连接变量 Connection variables	通道名称 Channel name
正转 Positive rotation	只写 M2.0 Write M2.0 only
反转 Negative rotation	只写 M2.1 Write M2.1 only
点动速度 Inching speed	只写 MDF10 Write MDF10 only

（四）连接 I/O 设备

（Ⅳ）To connect I/O device

如图 4.1.6 所示，步进电机驱动器需要接收来自控制器的直流控制信号，因此 S7-1200 的 CPU 模块选择 DC/DC/DC 版本。按照控制要求，I/O 接线图如图 4.1.16 所示。

As shown in Figure 4.1.6, the stepping motor driver needs to receive DC control signals from the controller, so the CPU module of S7-1200 selects the DC/DC/DC version. Based on the control requirements, the I/O wiring diagram is shown in Figure 4.1.16.

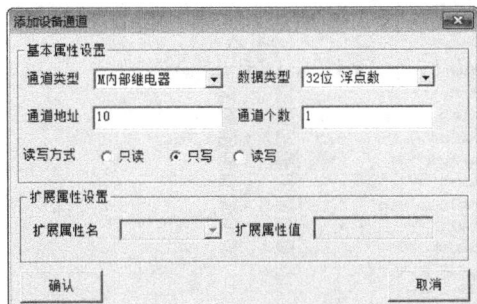

图 4.1.15　添加点动速度的设备通道

Figure 4.1.15　Device channel for adding inching speed

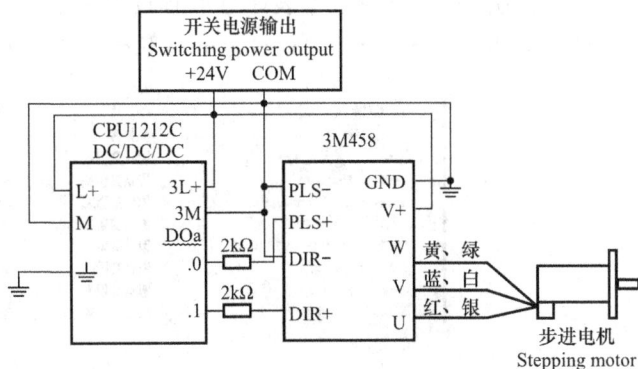

图 4.1.16　I/O 接线图

Figure 4.1.16　I/O wiring diagram

（五）调整步进电机驱动器

（Ⅴ）Adjust stepping motor driver

由于输入的脉冲信号范围在 1000~5000pps，因此调整步进驱动器的细分设置为 2000 步/转。其他设置为静态电流半流和相电流 5.8A。结合控制要求和表 4.1.1、4.1.2、4.1.3，设置步进电机驱动器的 8 位 DIP 见表 4.1.8。

Since the input pulse signal is between 1000 and 5000pps, adjust the stepper driver subdivision setting to 2000 steps/rotation. Other settings are: quiescent current half current and phase current 5.8A. According to the control requirements and Table 4.1.1, 4.1.2, 4.1.3, the 8-bit DIP setting for the stepping motor driver is shown in Table 4.1.8.

步进电机驱动器的 DIP 开关设置　　　　　　　　　　　表 4.1.8

DIP Switch Settings for Stepping Motor Drivers　　　　　**Table 4.1.8**

开关序号 Switch serial number	设置值 Set value
DIP1	OFF
DIP2	ON
DIP3	ON
DIP4	OFF
DIP5	ON
DIP6	ON
DIP7	ON
DIP8	ON

（六）在 TIA 博途中新建项目和设备组态

（Ⅵ）To create a new item and device configuration in TIA Portal

按照项目一中的步骤在 TIA 博途中新建项目和设备组态。在这里需要注意的是，本项目的 CPU 模块版本是 DC/DC/DC。由于编程中需要用到系统存储器，因此需要在 CPU 的属性中勾选启用系统存储器字节。

Create the new item and configure the device in TIA Portal by the steps in Item I. It is noted that the CPU module version of this project is DC/DC/DC. Because system memory is required for programming, the Enable System Memory byte needs to be checked in the properties of the CPU.

（七）建立变量表

（Ⅶ）To create the variable table

按照控制要求，定义变量表如图4.1.17所示。

Define the variable table as shown in Figure 4.1.17 based on the control requirements.

		名称	变量表	数据类型	地址	保持	在 H...	可从	注释
1		System_Byte	默认变量表	Byte	%MB1	☐	☑	☑	
2		FirstScan	默认变量表	Bool	%M1.0	☐	☑	☑	
3		DiagStatusUpdate	默认变量表	Bool	%M1.1	☐	☑	☑	
4		AlwaysTRUE	默认变量表	Bool	%M1.2	☐	☑	☑	
5		AlwaysFALSE	默认变量表	Bool	%M1.3	☐	☑	☑	
6		正转点动	默认变量表	Bool	%M2.0	☐	☑	☑	
7		反转点动	默认变量表	Bool	%M2.1	☐	☑	☑	
8		点动速度	默认变量表	Real	%MD10	☐	☑	☑	

图4.1.17 变量表

Figure 4.1.17 Variable table

（八）轴工艺对象组态及调试

（Ⅷ）Configuration and commissioning of axis process object

在TIA博途项目中新建轴工艺对象，在组态窗口中设置本任务所需的参数。由于步进电机是自由转动的，没有连接到任何传动机构上，因此，不需要设置位置限制及回原点参数。需要设置的内容有：

Create a new axis process object in the TIA Portal project and set the parameters required for this task in the configuration window. Since the stepping motor is free to rotate and is not connected to any drive mechanism, there is no need to set position limits and return parameters. The contents to be set are as follows:

1. 常规设置

1. General setting

在常规设置中，可以设置轴的名称。驱动器选择PTO，测量单位选择脉冲。如图4.1.18所示。

In general settings, you can set the name of the axis. The driver selects the PTO and the unit of measurement selects the pulse. As shown in Figure 4.1.18.

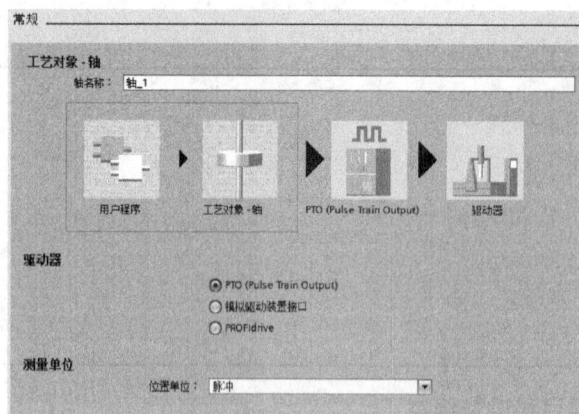

图4.1.18 轴工艺对象的常规设置

Figure 4.1.18 General setting of shaft process object

2. 驱动器设置

启用脉冲发生器，选择PTO的类型为脉冲A和方向B，脉冲信号由Q0.0提供，方向信号由Q0.1提供。如图4.1.19所示。

2. Drive settings

Enable the pulse generator and select the type of PTO as Pulse A and Direction B，the pulse signal is provided by Q0.0 and the direction signal is provided by Q0.1．As shown in Figure 4.1.19.

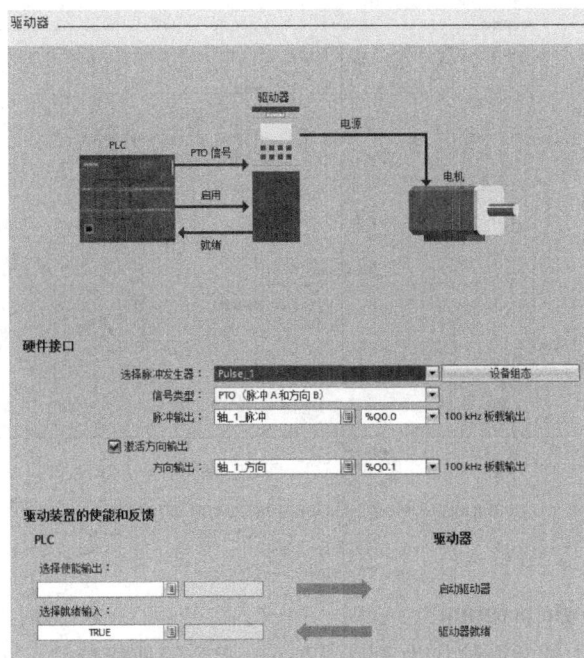

图 4.1.19　轴工艺对象的驱动器设置

Figure 4.1.19　Drive settings for shaft process object

3. 机械设置

根据步进电机驱动器的参数设置电机每转的脉冲数和位移，如图 4.1.20 所示。

3. Mechanical setting

Set the number of pulses per rotation and the displacement of the motor according to the parameters of the stepping motor driver，as shown in Figure 4.1.20.

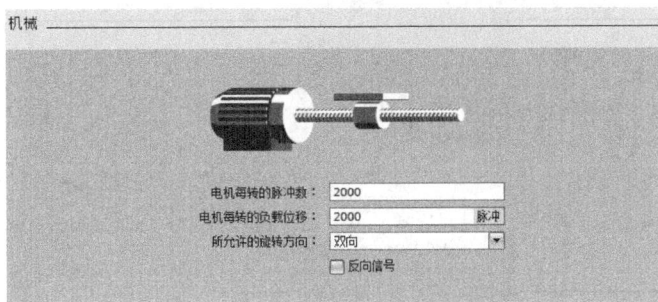

图 4.1.20　轴工艺对象的机械设置

Figure 4.1.20　Mechanical setting of shaft process object

4. 调试

选择调试工具，将 PLC 转入在线状态，可以打开轴控制面板，激活调试轴命令，对轴进行调试，如图 4.1.21 所示。注意，只有在程序块中启用/禁用轴指令没有激活的情况下才能调试轴。

4. Commissioning

Select the commissioning tool and put the PLC into online state．Open the shaft control panel，activate the commissioning shaft command and debug the shaft, as shown in Figure 4.1.21. Note that the shaft can only be debugged if the shaft enable/disable instruction is not active in the block.

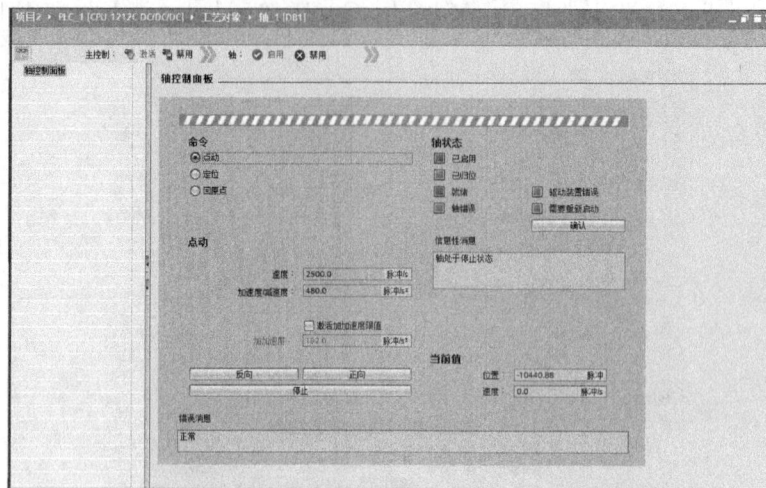

图 4.1.21 调试轴

Figure 4.1.21 Commissioning of shaft

（九）编写梯形图程序

（Ⅸ）To write the ladder logic program

先启用轴，用系统存储器中的常开位作为 MC_Power 指令的使能信号。再用 MC_MoveJog 指令对步进电机进行点动控制。

Enable the shaft first, and use the normally open bit in system memory as the enable signal for the MC_Power instruction. Then control the stepping motor in inching mode by MC_MoveJog instruction.

梯形图程序如图 4.1.22 所示。

The ladder diagram program is shown in Figure 4.1.22.

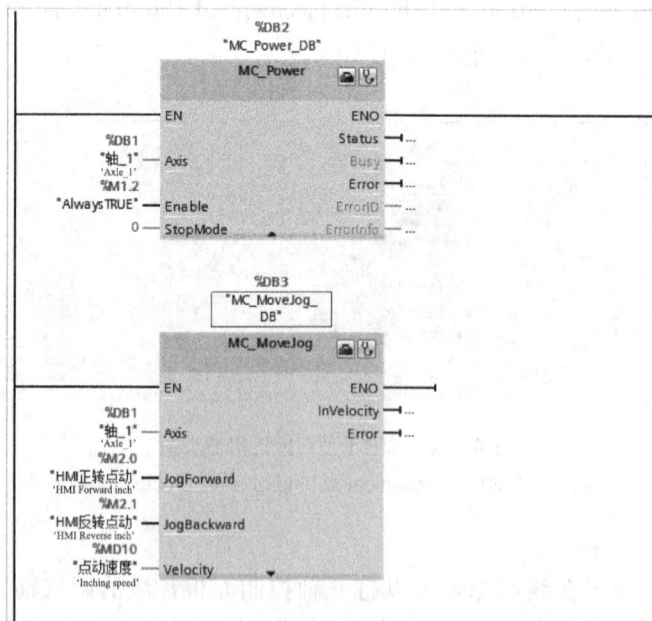

图 4.1.22 梯形图程序

Figure 4.1.22 Ladder logic program

（十）调试

（Ⅹ）Debugging

完成以上步骤后，将编辑好的 MCGS 嵌入版工程和 TIA 博途项目分别下载至触摸屏和 PLC，再用编程线将二者连接起来，这样一个步进电机点动控制系统就建立起来了。在输入框中输入点动的速度，再点击"正转点动"和"反转点动"按钮，观察步进电机的状态，直到所有功能都实现为止。

After the above steps, the embedded MCGS engineering and TIA Portal items have been edited are respectively downloaded to the touch screen and PLC, the both are connected with the programming line such that the stepping motor inching control system is finally created. Enter the inching speed in the input box, and then click the "Forward inching" and "Reverse inching" buttons to observe the state of the stepping motor until all functions are achieved.

任务二 基于光电编码器的步进电机运动控制

Task Ⅱ Motion Control of Stepping Motor Based on Photoelectric Encoder

一、任务分析

Ⅰ. Task Analysis

用人机界面（HMI）、PLC、步进电机驱动器、步进电机和小车运动单元创建一个基于光电编码器的步进电机运动系统。控制要求如下：

A stepping motor motion system based on photoelectric encoder is created by using HMI, PLC, stepping motor driver, stepping motor and trolley motion unit. The control requirements are shown as follows:

（1）小车通过丝杠由步进电机带动下在 A、B 之间做直线运动，A 点在丝杠的 10cm 刻线处，B 点在 20cm 刻线处，两点均安装有光电传感器 SQ1 和 SQ2 对小车进行限位。丝杠两端还安装有行程开关进行限位保护。步进电机每旋转一圈，小车行进 4mm。另有一光电编码器与步进电机同轴安装，记录小车行走的距离。如图 4.2.1 所示。

（1）A trolley moves linearly between A and B through a lead screw driven by a stepping motor, a point A is at a 10cm scribing line of the lead screw, and a point B is at a 20cm scribing line, and photoelectric sensors SQ1 and SQ2 are installed at both points to limit the position of the trolley. Stroke switches are also installed at both ends of the lead screw for limit protection. For each rotation of the stepping motor, the trolley travels 4mm. Another photoelectric encoder is installed coaxially with the stepping motor to record the distance traveled by the trolley. As shown in Figure 4.2.1.

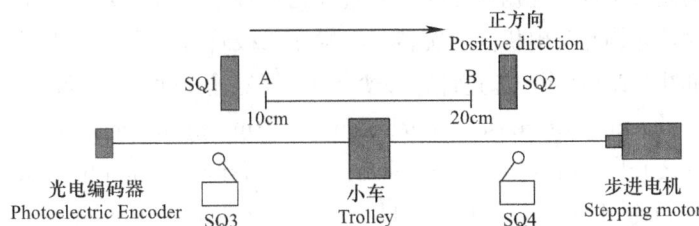

图 4.2.1 小车运动示意图

Figure 4.2.1 Schematic diagram of trolley movement

（2）监控画面如图 4.2.2 所示。

（2）The monitoring interface is shown in Figure 4.2.2.

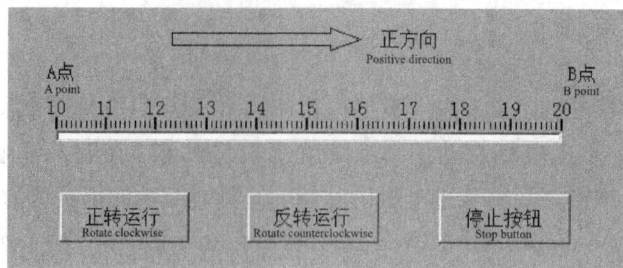

图 4.2.2　小车运动监控画面

Figure 4. 2. 2　Trolley movement monitoring interface

（3）定义从 A 点运动到 B 点为正方向，按下正转按钮，小车向正方向运行；按下反转按钮，小车向反方向运行；按下停止按钮，小车停止。当运动到 A、B 点的限位传感器 SQ1 或 SQ2 处，小车自动停止。小车运动的速度为 4mm/s。

（3）Define the movement from point A to point B as the positive direction，press the forward turning button，and the trolley runs in the positive direction；Press the reverse button and the trolley runs in the opposite direction. Press the stop button to stop the trolley. The trolley stops automatically when it moves to the limit sensors SQ1 or SQ2 at points A and B. The trolley moves at a speed of 4mm/s.

（4）监控画面可实时显示小车的当前位置。

（4）The monitoring interface can display the current position of the trolley in real time.

为了完成本任务，首先需要对涉及的设备进行选型，根据现有条件，人机界面选择昆仑通态的 MCGS 触摸屏，PLC 选择西门子 S7-1200 系列 PLC。通过完成本任务，可以学习到光电编码器的工作原理、西门子 S7-1200 的高速计数器、以预定义速度移动轴指令和暂停轴指令这四个方面的知识。

For the purpose of this task，it is firstly necessary to select the required types of devices involved，of which MCGS touch screen，and Siemens S7-1200 PLC will be selected under the current conditions. By completing this task，you can learn about the working principle of the photoelectric encoder，the high-speed counter of Siemens S7-1200，instruction to move axis at predefined speed，and the instruction to pause axis.

二、相关知识技能
Ⅱ. Relevant Knowledge & Skills
（一）光电编码器的工作原理
（Ⅰ）Working principle of photoelectric encoder

光电编码器，是一种通过光电转换，将输出轴上的机械几何位移量转换成脉冲或数字量的传感器。光电编码器是由光栅盘和光电检测装置组成。光栅盘是在一定直径的圆板上等分地开通若干个长方形孔。由于光电码盘与电机同轴，电机旋转时，光栅盘与电机同速旋转，经发光二极管等电子元件组成的检测装置检测输出若干脉冲信号，其原理示意图如图 4.2.3 所示。通过计算每秒光电编码器输出脉冲的个数，就能反映当前电机的转速。

Photoelectric encoder is a kind of sensor which converts the mechanical geometric displacement on the output shaft into pulse or digital quantity by photoelectric conversion. The photoelectric encoder is composed of grating disk and photoelectric detection device. The grating disk is a circular plate with a certain diameter，which is divided into several rectangular holes. As the photoelectric code disc is coaxial with the motor，when the motor rotates，the grating disc rotates at the same speed with the motor，and a detection device composed of electronic components such as light emitting diodes detects and outputs several pulse signals，the schematic diagram of which is shown in Figure 4. 2. 3. The current motor speed can be reflected by calculating the number of output pulses per second of the photoelectric encoder.

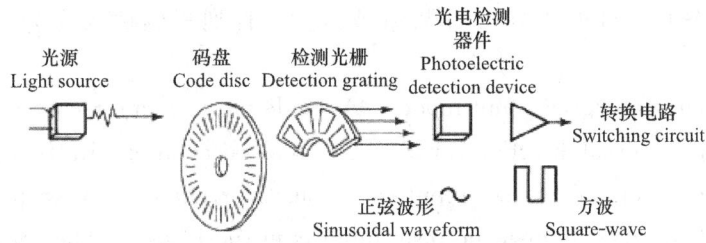

图 4.2.3 光电编码器原理示意图

Figure 4.2.3 Schematic diagram of photoelectric encoder

一般来说，根据旋转编码器产生脉冲的方式的不同，可以分为增量式、绝对式以及复合式三大类。增量式编码器是直接利用光电转换原理输出三组方波脉冲 A、B 和 Z 相；A、B 两组脉冲相位差 90，用于辨别方向：当 A 相脉冲超前 B 相时为正转方向，而当 B 相脉冲超前 A 相时则为反转方向，如图 4.2.4 所示。Z 相为每转一个脉冲，用于基准点定位。

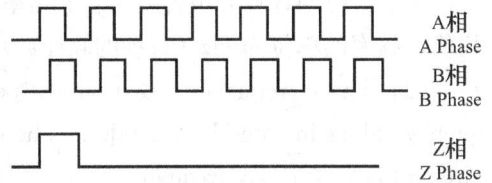

图 4.2.4 增量式编码器输出的三组方波脉冲

Figure 4.2.4 Three groups of square wave pulses output by incremental encoder

Generally speaking，according to the different ways in which the rotary encoder generates pulses，it can be divided into three categories：incremental，absolute and composite. The incremental encoder directly outputs three groups of square wave pulses A，B and Z phases by using the photoelectric conversion principle. Phase difference between the two sets of pulses A and B is 90 for direction discrimination：It is forward direction when the Phase A pulse is ahead of Phase B，and it is the reverse direction when the Phase B pulse is ahead of Phase A，as shown in Figure 4.2.4. Phase Z is a pulse per rotation for reference point positioning.

编码器直接连接到丝杠上就可以计算出小车当前的位置。丝杠旋转一圈，编码器的 A 相和 B 相就输出 1000 个脉冲，将它们连接到 PLC 的高速计数器输入端，就可以得到小车当前的位置了。

The encoder is connected directly to the lead screw to calculate the current position of the trolley. When the lead screw is rotated for one circle，Phase A and Phase B of the encoder output 1000 pulses，which are connected to the high-speed counter input end of the PLC to get the current position of the trolley.

（二）S7-1200 的高速计数器

（Ⅱ）High speed counter of S7-1200

PLC 的普通计数器的计数过程与扫描工作方式有关，CPU 以通过每一个扫描周期读取一次被测信号的方法来捕捉被测信号的上升沿，被测信号的频率较高时，会丢失计数脉冲，因此普通计数器的最高工作频率一般仅有几十赫兹。而高速计数器（HSC）可以对发生速率快于程序循环 OB 执行速率的事件进行计数。

The technical process of the ordinary counter of PLC is related to the scanning working mode. The CPU captures the rising edge of the measured signal by reading the measured signal once every scanning cycle. When the frequency of the measured signal is high，the counting pulse will be lost. Therefore，the highest working frequency of the ordinary technical area is generally only tens of Hertz. A high-speed counter (HSC) counts events that occur at the speed faster than the program loop OB execution rate.

1. 高速计数器的功能

1. Functions of High Speed Counters

（1）HSC 有 4 种高速计数工作模式：具有内部方向控制的单相计数器、具有外部方向控制的单相计数器、具有两路时钟脉冲输入的双相计数器和 A/B 相正交计数器。每种 HSC 模式都可以使用或不使用复位输入。复位输入为 1 状态时，HSC 的实际计数值被清除。直到复位输入变为 0 状态，才能启动计数功能。

（1）The HSC has four high speed count modes：A single-phase counter with internal direction control，a single-phase counter with external direction control，a two-phase counter with two clock pulses inputs，and a A/B phase quadrature counter. The reset input can be used or not used in each HSC mode. The actual counting value of the HSC is cleared when the reset input is in the 1 state. The counting function cannot be activated until the reset input state becomes 0.

（2）某些 HSC 模式可以选用 3 种频率测量的周期（0.01s、0.1s 和 1.0s）来测量频率值。频率测量周期决定了多长时间计算和报告一次新的频率值。得到的是根据信号脉冲的计数值和测量周期计算出的频率的平均值，频率的单位为 Hz（每秒的脉冲数）。

（2）Three frequency measurement periods （0.01 s，0.1 s and 1.0 s） can be selected to measure the frequency values in some HSC modes. The frequency measurement period determines how long it takes to calculate and report a new frequency value. The average frequency calculated from the count of signal pulses and the measurement period is given in the unit of Hz （number of pulses per second）.

（3）使用扩展高速计数器指令 CTRL_HSC_EXT，可以按指定的时间周期，用硬件中断的方式测量出被测信号的周期数和精确到 μs 的时间间隔，从而计算出被测信号的周期。

（3）By using the Extended High Speed Counter instruction CTRL_HSC_EXT，the period of the signal under test can be calculated by measuring the number of periods of the signal under test and the time interval accurate to μs by means of hardware interruption in a specified time period.

2. 高速计数器的组态步骤

2. Configuration steps for high speed counters

在用户程序使用 HSC 之前，应为 HSC 组态，设置 HSC 的计数模式。某些 HSC 的参数在设备组态中初始化，以后可以用程序来修改。

Before the user program uses the HSC，set the HSC counting mode for the HSC configuration. Some HSC parameters are initialized in the device configuration and can be modified later by program.

（1）打开 PLC 的设备视图，选中其中的 CPU 模块。选中巡视窗口的"属性"选项卡左边的高速计数器 HSC1 的"常规"，用复选框选中"启用该高速计数器"。

（1）Open the device view of the PLC and select the CPU module. Select "General" for the high-speed counter HSC1 on the left side of the "Properties" tab of the Patrol window and select "Enable This High-speed Counter" with the check box.

（2）选中左边窗口的"功能"，如图 4.2.5 所示，在右边窗口设置下列参数：

（2）Select "Functions" in the left window，as shown in Figure 4.2.5，and set the following parameters in the right window：

使用"计数类型"下拉式列表，可选"计数"、"时间段"、"频率"或"运动控制"。如果设置为"时间段"和"频率"，使用"频率测量周期"下拉式列表，可以选择"0.01s""0.1s""1.0s"。

Use the Count Type drop-down list to select "Count"，"Time Period"，"Frequency"，or "Motion Control". If setting as "Time Period" and "Frequency"，use the "Frequency Measurement Period" drop-down list to select "0.01 s"，"0.1 s"，and "1.0 s".

使用"工作模式"下拉式列表，可选"单相"、"两相位"、"A/B 计数器"或"A/B 计数器四倍频"。使用"计数方向取决于"下拉式列表，可选"用户程序（内部方向控制）"或输入"外部方向控制"。用"初始计数方向"下拉式列表选择"增计数"或"减计数"。

Use the "Operating Mode" drop-down list to "select Single Phase", "Two Phase", "A/B Counter", or "A/B CounterQuadruplicated Frequency". Use the "Count Direction Depending" drop-down list to select "User Programs (Internal Direction Control)" or "enter External Direction Control." Select "Increment" or "Decrease" from the "Initial Count Direction" drop-down list.

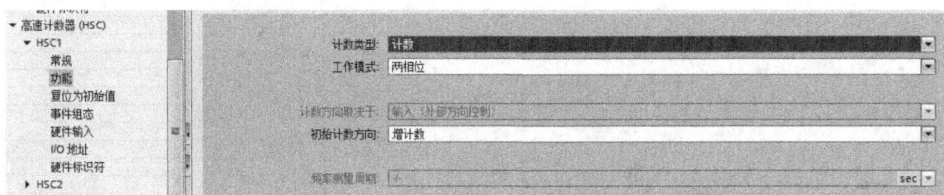

图 4.2.5　高速计数器的功能设置

Figure 4. 2. 5　Functional settings for high speed counters

（3）选中左边窗口的"复位为初始值"，如图 4.2.6 所示，可以设置"初始计数器值"和"初始参考值"。如果勾选了"使用外部复位输入"复选框，用下拉式列表选择"复位信号电平"是高电平有效还是低电平有效。

（3）Select "Reset to initial" in the left window, as shown in Figure 4. 2. 6, and set "Initial Counter Value" and "Initial Reference Value". If the "Use External Reset Input box" is checked, select "Active High or Active Low Reset Signal Level" in the drop-down list.

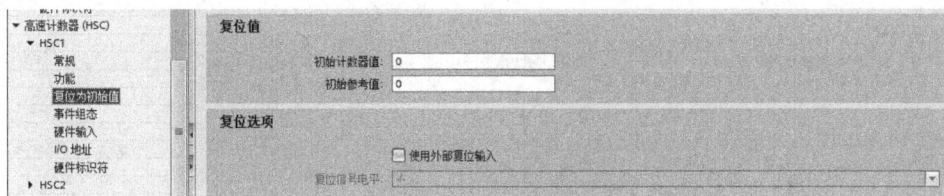

图 4.2.6　设置高速计数器的初始值与复位信号

Figure 4. 2. 6　Set the initial value of the high-speed counter and the reset signal

（4）选择左边窗口的"事件组态"，可以用右边窗口的复选框激活下列事件出现时是否产生中断，如图 4.2.7 所示。可产生的中断有：计数器值等于参考值、出现外部复位事件和出现计数方向变化事件。可以输入中断事件名称或采用默认的名称。生成硬件中断组织块 OB40 后，将它指定给计数器值等于参考值的中断事件。

（4）Select "Event Configuration" in the left window to activate whether an interrupt occurs when the following events occur by using the check boxes in the right window, as shown in Figure 4. 2. 7. The interrupts that can be generated are: The counter value is equal to the reference value, an external reset event occurs, and a count direction change event occurs. You can enter an interrupt event name or use the default name. After the hardware interrupt organization block OB40 is generated, it is assigned to an interrupt event with a counter value equal to the reference value.

（5）选中左边窗口的"硬件输入"，在右边窗口可以组态该 HSC 使用的时钟发生器输入、方向输入和复位输入的输入点。可以看到可用的最高频率，如图 4.2.8 所示。

（5）Select "Hardware Input" in the left window where you can configure the input points for the clock generator input, direction input, and reset input used by the HSC in the right window. You can see the highest frequency available, as shown in Figure 4. 2. 8.

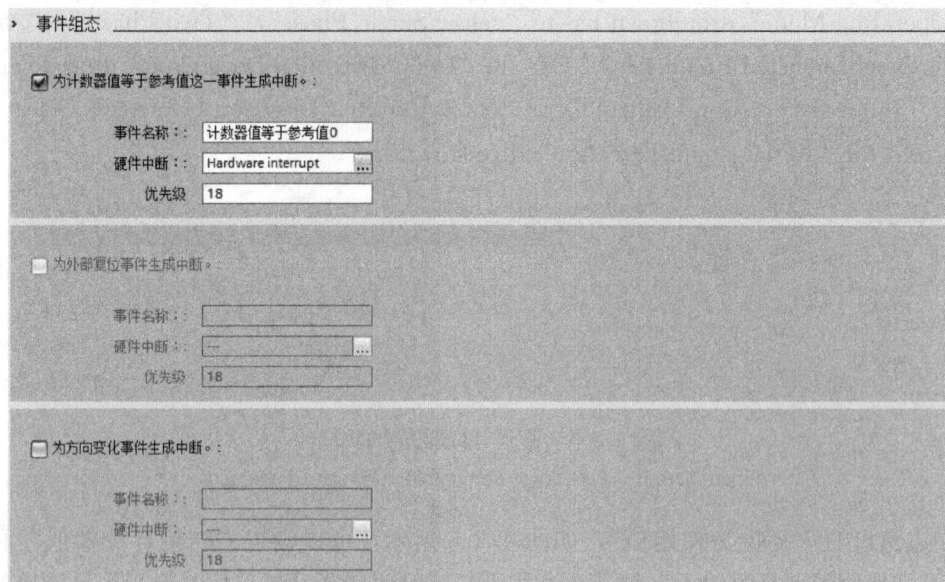

图 4.2.7　高速计数器的事件组态

Figure 4.2.7　Event configuration for high speed counters

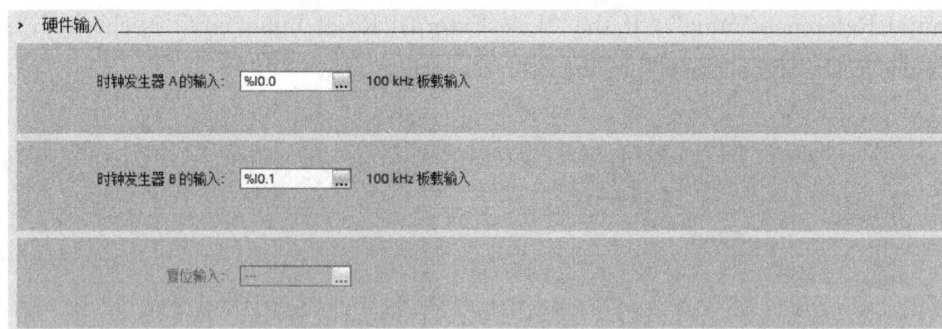

图 4.2.8　高速计数器的硬件输入

Figure 4.2.8　Hardware input for high speed counters

（6）选中左边窗口的"I/O"地址，可以在右边窗口修改 HSC 的起始地址，如图 4.2.9 所示。默认的起始地址为 1000。由于 HSC 的计数值数据类型为 DInt，占用 4 个字节，因此默认的结束地址为 1003，计数器 HSC1 的地址用 ID1000 表示。

（6）Select the "I/O" address in the left window to modify the start address of the HSC in the right window，as shown in Figure 4.2.9．The default start address is 1000．Because the HSC's count data type is DInt，which takes four bytes，the default end address is 1003，and the address of counter HSC1 is represented by ID 1000．

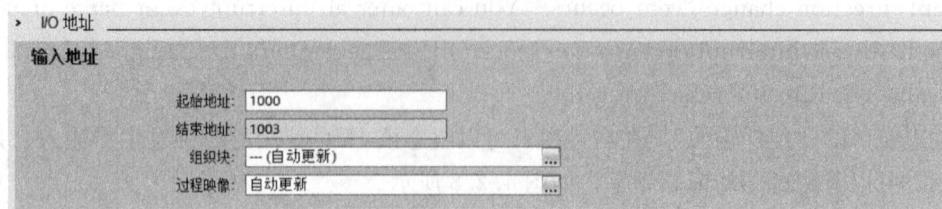

图 4.2.9　高速计数器的 I/O 地址

Figure 4.2.9　I/O address of high speed counter

（三）以预定义速度移动轴指令

（Ⅲ）Instruction to move axis at predefined speed

使用 MC_MoveVelocity 指令以指定的速度持续移动轴。使用 MC_MoveVelocity 指令，必须先启用轴。MC_MoveVelocity 指令的参数见表 4.2.1。

Use the MC_MoveVelocity instruction to continuously move the axis at the specified speed. In order to use the MC_MoveVelocity instruction，you must first enble the axis. The parameters of the MC_MoveVelocity instruction are shown in Table 4.2.1.

MC_MoveVelocity 指令的参数   表 4.2.1

Parameters of the MC_MoveVelocity instruction   Table 4.2.1

参数 Parameter	声明 Statement	数据类型 Data type	说明 Description
Axis	INPUT	TO_Axis	轴工艺对象 Shaft process object
Execute	INPUT	Bool	出现上升沿时开始任务（默认值：False） Start task on rising edge (default：False)
Velocity	INPUT	Real	指定轴运动的速度（默认值：10.0） Specify the speed at which the axis moves (default：10.0) 限值：启动/停止速度≤|Velocity|≤最大速度 Limit：Start/stop speed≤Velocity≤Maximum speed （允许 Velocity＝0.0） (Velocity＝0.0 is allowable)
Direction	INPUT	Int	指定方向： Specify direction： 0：旋转方向与参数"Velocity"中的值符号一致（默认值） 0：Rotation direction coincides with value sign in parameter "Velocity" (default) 1：正旋转方向（参数"Velocity"的值符号被忽略） 1：Positive rotation direction (value sign of parameter "Velocity" ignored) 2：负旋转方向（参数"Velocity"的值符号被忽略） 2：Negative rotation direction (value sign of parameter "Velocity" ignored)
Current	INPUT	Bool	保持当前速度 Keep Current Speed FALSE：禁用"保持当前速度" FALSE：Disable Keep Current Speed 使用参数"Velocity"和"Direction"的值（默认值） Use values for parameters "Velocity" and "Direction" (Default value) TRUE：激活"保持当前速度" TRUE：Activate Keep Current Speed 不考虑参数"Velocity"和"Direction"的值 Take no account of values for parameters "Velocity" and "Direction" 当轴继续以当前速度运动时，参数"InVelocity"返回值 TRUE The return value of parameter "InVelocity" is TRUE as the axis continues to move at the current speed
InVelocity	OUTPUT	Bool	如果"Current"＝FALSE If "Current"＝FALSE 已达到参数"Velocity"中指定的速度 The speed specified in parameter "Velocity" has been reached 如果"Current"＝TRUE If "Current"＝TRUE 轴在启动时以当前速度运动（默认值） Axis moves at current speed during startup (default)
Busy	OUTPUT	Bool	TRUE＝正在执行任务 TRUE＝Executing tasks

参数 Parameter	声明 Statement	数据类型 Data type	说明 Description
CommandAborted	OUTPUT	Bool	TRUE＝任务在执行期间被另一任务中止 TRUE＝Task stopped by another task during execution
Error	OUTPUT	Bool	FALSE：无错误 FALSE：No Errors TRUE：任务执行期间出错 TRUE：An error occurred during task execution 出错原因可在 "ErrorID" 和 "ErrorInfo" 参数中找到 The cause of the error can be found in the "ErrorID" and "ErrorInfo" parameters
ErrorID	OUTPUT	Word	参数 "Error" 的错误 ID Error ID for the "Error" parameters
ErrorInfo	OUTPUT	Word	参数 "ErrorID" 的错误信息 ID Error information ID for the "ErrorID" parameters

（四）暂停轴指令

（Ⅳ）Instruction to pause axis

使用 MC＿Halt 指令可停止所有运动并将轴切换到停止状态。停止位置未定义。为了使用 MC＿Halt 指令，必须先启用轴。MC＿Halt 指令的参数见表 4.2.2。

Use the MC＿Halt instruction to stop all movement and switch the axis to a stop state. Stop position is not defined. In order to use the MC＿Halt instruction，the axis must first be enabled. The parameters of the MC＿Halt instruction are shown in Table 4.2.2.

MC＿Halt 指令的参数　　　　　　　　表 4.2.2

Parameters of the MC＿Halt instruction　　　Table 4.2.2

参数 Parameter	声明 Statement	数据类型 Data type	说明 Description
Axis	INPUT	TO＿Axis	轴工艺对象 Shaft process object
Execute	INPUT	Bool	出现上升沿时开始任务（默认值：False） Start task on rising edge (default：False)
Done	OUTPUT	Bool	TRUE＝速度达到零 TRUE＝Speed reaches zero
Busy	OUTPUT	Bool	TRUE＝正在执行任务 TRUE＝Executing tasks
CommandAborted	OUTPUT	Bool	TRUE＝任务在执行期间被另一任务中止 TRUE＝Task stopped by another task during execution.
Error	OUTPUT	Bool	FALSE：无错误 FALSE：No Errors TRUE：任务执行期间出错 TRUE：An error occurred during task execution 出错原因可在 "ErrorID" 和 "ErrorInfo" 参数中找到 The cause of the error can be found in the "ErrorID" and "ErrorInfo" parameters
ErrorID	OUTPUT	Word	参数 "Error" 的错误 ID Error ID for the "Error" parameters
ErrorInfo	OUTPUT	Word	参数 "ErrorID" 的错误信息 ID Error information ID for the "ErrorID" parameters

三、任务实施

Ⅲ. Task Implementation

(一) 构建实时数据库

(Ⅰ) To create one real-time database

分析控制要求，可以知道需要新建三个开关型变量控制步进电机的正转、反转和停止，还需要新建一个数值型变量显示小车当前位置。实时数据库变量见表 4.2.3。

By analyzing the control requirements, we can know that it is necessary to create three new switched variables to control the normal rotation, reverse rotation and stop of the stepping motor, and to create a new numerical variable to display the current position of the trolley. Variables of real-time database as shown in Table 4.2.3.

实时数据库变量表   表 4.2.3

Variables of real-time database   Table 4.2.3

变量名 Variable name	类型 Type	报警 Alarm
正转 Positive rotation	开关型 Switch type	
反转 Negative rotation	开关型 Switch type	
停止 Stop	开关型 Switch type	
小车位置 Trolley location	数值型 Numeric type	

(二) 组态用户窗口

(Ⅱ) To configure the user window

按照项目一中介绍的操作步骤，新建监控窗口和报警窗口，并为各窗口创建合适的图形对象，最后定义各对象的动画连接。当前位置的实时显示是通过百分比填充构件实现的。需要设置其属性如图 4.2.10 所示。

Create the new monitoring window and alarm window by the steps introduced in Item I, create the proper graphic objects to each window, and finally define the animation connection with each object. Real-time display of the current position is achieved by percentage-based filling component. You need to set its properties as shown in Figure 4.2.10.

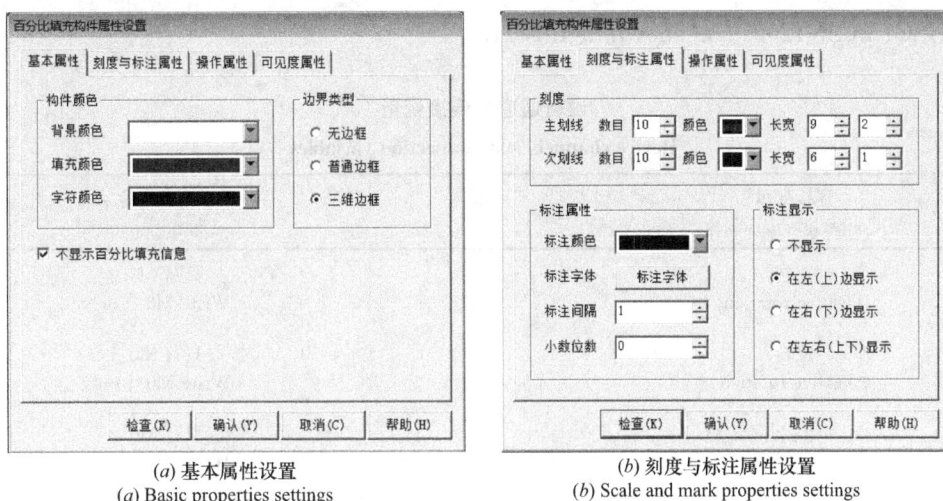

(a) 基本属性设置
(a) Basic properties settings   (b) 刻度与标注属性设置
(b) Scale and mark properties settings

图 4.2.10 百分比填充构件属性设置 （一）

Figure 4.2.10 Percentage-based filling component properties settings （Ⅰ）

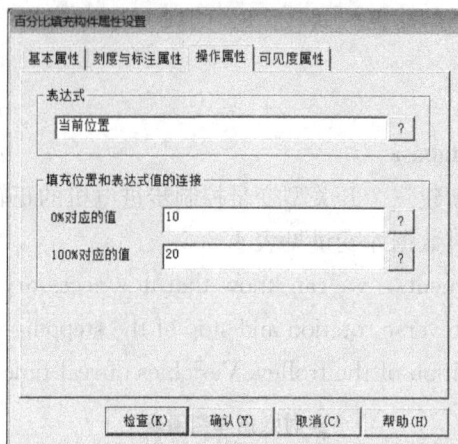

(c) 操作属性设置
(c) Operation properties settings

图 4.2.10　百分比填充构件属性设置（二）
Figure 4.2.10　Percentage-based filling component properties settings（Ⅱ）

（三）组态设备窗口
（Ⅲ）To configure the device window

按照项目一中介绍的操作步骤，组态设备窗口，按表 4.2.4 创建 PLC 和实时数据库变量的通道连接。由于"当前位置"是数值型变量，需要关联 PLC 高速计数器的值，该值是一个 32 位的整数，因此需要在新建设备通道时选择 32 位有符号二进制数的数据类型。因为高速计数器 HSC1 的地址是 ID1000，所以需要设置通道类型设置为 I 输入继电器，通道地址设置为 1000，意味着"当前位置"这个变量与 S7-1200 中高速计数器 HSC1 的计数值连接在了一起。如图 4.2.11 所示。

Configure the device window by the steps introduced in Item Ⅰ, and create the channel connection between PLC and real-time database variable according to Table 4.2.4. Because Current Position is a numeric variable，you need to associate the value of the PLC high-speed counter，which is a 32-bit integer，so you need to select the 32-bit signed binary data type when you create a new device channel. Because the address of the high-speed counter HSC1 is ID1000，it is necessary to set the channel type to I input relay and the channel address to 1000，meaning that the variable "Current Position" is connected to the count value of the high-speed counter HSC1 in S7-1200. As shown in 4.2.11.

设备通道和连接变量　　　　　　　　　　　　　　　　表 4.2.4
Device channels and connection variables　　　　　　　Table 4.2.4

连接变量 Connection variables	通道名称 Channel name
正转 Positive rotation	只写 M0.0 Write M0.0 only
反转 Negative rotation	只写 M0.1 Write M0.1 only
停止 Stop	只写 M0.2 Write M0.2 only
当前位置 Current position	只读 IDB1000 Read IDB1000 only

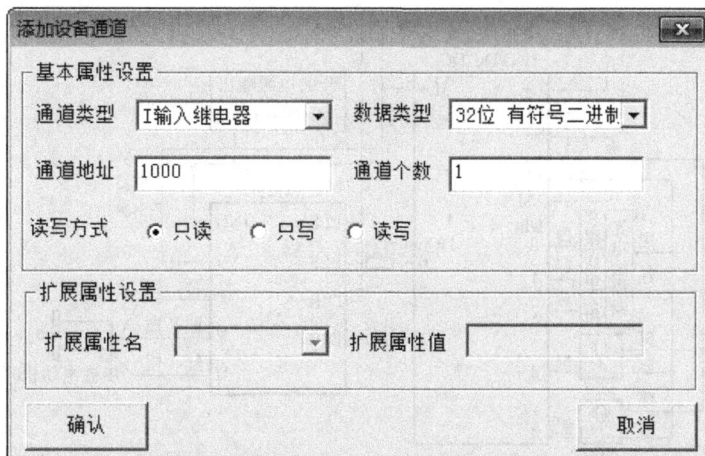

图 4. 2. 11　添加当前位置的设备通道

Figure 4. 2. 11　Add device channel at current position

触摸屏读入的是高速计数器的数值，但是需要显示的是当前位置，单位是 cm，所以要对该"当前位置"的设备通道进行处理。当前位置的计算公式为（4×高速计数器计数值/10000）+10，设置设备处理通道如图 4. 2. 12 所示。

The touch screen reads the value of the high-speed counter，but what needs to be displayed is the current position in cm，so the device channel at that "current position" is processed. The current position is calculated as（4×high-speed counter count value/10000）+10，so it is needed to set the device processing channel as shown in Figure 4. 2. 12.

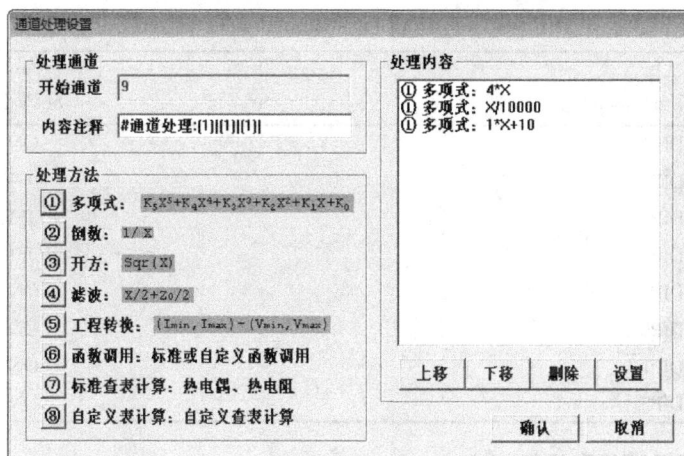

图 4. 2. 12　设备通道处理设置

Figure 4. 2. 12　Device channel processing settings

（四）连接 I/O 设备

（Ⅳ）To connect I/O device.

按照控制要求，I/O 接线图如图 4. 2. 13 所示。

Based on the control requirements，the I/O wiring diagram is shown in Figure 4. 2. 13.

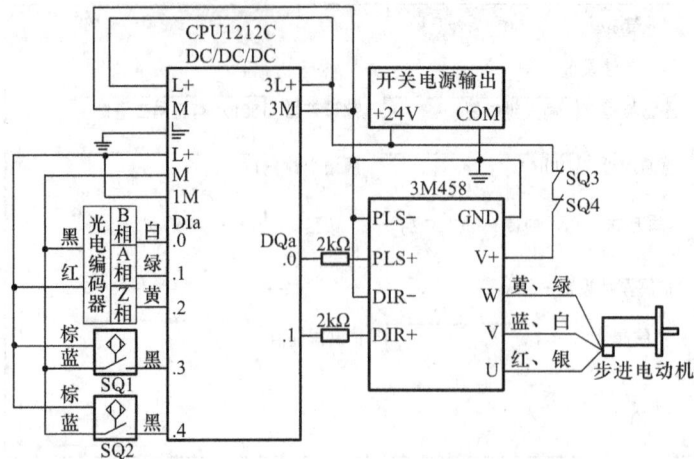

图 4. 2. 13　I/O 接线图

Figure 4. 2. 13　I/O wiring diagram

（五）调整步进电机驱动器

（Ⅴ）Adjust stepping motor driver

调整步进驱动器的细分设置为 1000 步/转，其他设置为静态电流半流和相电流 5.8A。结合表 4.1.1、4.1.2、4.1.3，设置步进电机驱动器的 8 位 DIP 见表 4.2.5。

The subdivision setting of stepping driver is adjusted to 1000 steps per rotation, and other settings are: quiescent current half-current and phase current of 5.8 A. According to Table 4.1.1, 4.1.2, 4.1.3, the 8-bit DIP setting for the stepping motor driver is shown in Table 4.2.5.

<table>
<tr><td colspan="2">步进电机驱动器的 DIP 开关设置</td><td>表 4.2.5</td></tr>
<tr><td colspan="2">DIP switch settings for stepping motor drivers</td><td>Table 4.2.5</td></tr>
<tr><td colspan="2">开关序号
Switch serial number</td><td>设置值
Set value</td></tr>
<tr><td colspan="2">DIP1</td><td>ON</td></tr>
<tr><td colspan="2">DIP2</td><td>OFF</td></tr>
<tr><td colspan="2">DIP3</td><td>OFF</td></tr>
<tr><td colspan="2">DIP4</td><td>OFF</td></tr>
<tr><td colspan="2">DIP5</td><td>ON</td></tr>
<tr><td colspan="2">DIP6</td><td>ON</td></tr>
<tr><td colspan="2">DIP7</td><td>ON</td></tr>
<tr><td colspan="2">DIP8</td><td>ON</td></tr>
</table>

（六）在 TIA 博途中新建项目和设备组态

（Ⅵ）To create a new item and device configuration in TIA Portal

按照项目一中的步骤在 TIA 博途中新建项目和设备组态。由于编程中需要用到系统存储器，因此需要在 CPU 的属性中勾选启用系统存储器字节。另外，需要启用高速计数器 HSC1 并设置其参数，如图 4.2.14 所示，其他设置均为默认值。

Create the new item and configure the device in TIA Portal by the steps in Item Ⅰ. Because system memory is required for programming, the Enable System Memory byte needs to be checked in the properties of the CPU. In addition, the high-speed counter HSC1 needs to be enabled and its parameters need to be set, as shown in Figure 4.2.14, with the other settings being the default values.

(a) 功能设置
(a) Functional settings

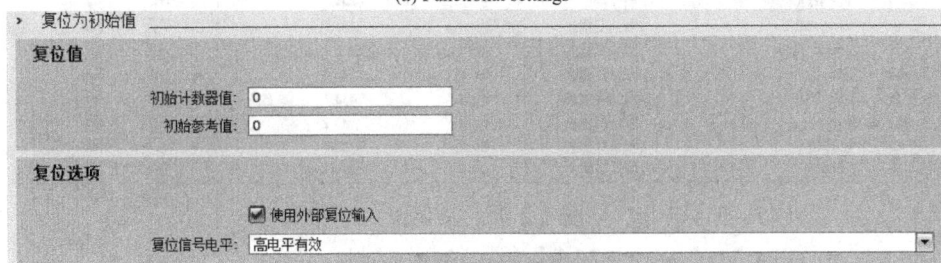

(b) 复位为初始值设置
(b) Reset to initial value settings

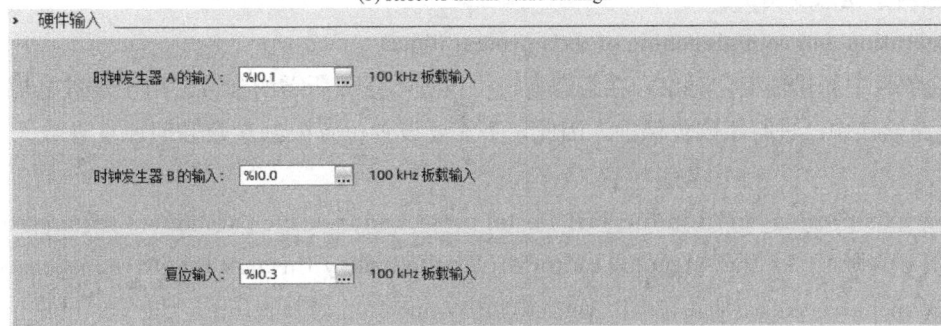

(c) 硬件输入设置
(c) Hardware input settings

图 4.2.14 高速计数器的属性设置

Figure 4.2.14 Attribute settings for high speed counters

由于 I0.0 和 I0.1 作为 HSC1 的输入端，其端口设置必须能够捕捉到高速脉冲，因此需要调整这两个输入端口的组态设置，减小输入滤波器的滤波时间。如图 4.2.15 所示。

Because I0.0 and I0.1 are the input ends to HSC1，their port settings must capture high speed pulses，the configuration settings of the two input ports need to be adjusted to reduce the filtering time of the input filter. As shown in Figure 4.2.15.

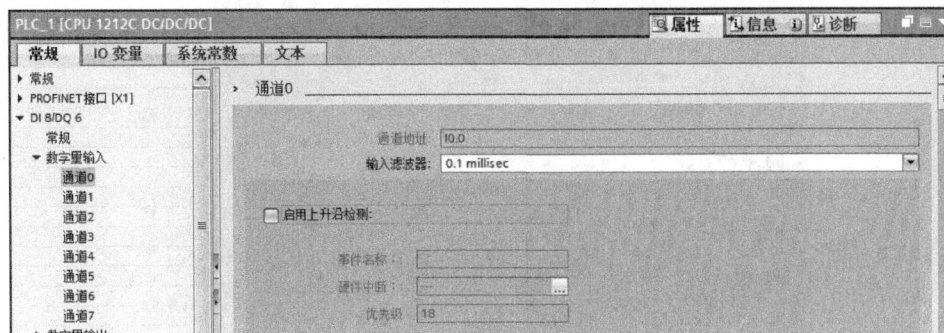

图 4.2.15 调整输入滤波时间

Figure 4.2.15 Adjust input filtering time

（七）建立变量表

（Ⅶ）To create the variable table

按照控制要求，定义变量表如图 4. 2. 16 所示。

Define the variable table as shown in Figure 4. 2. 16 based on the control requirements.

		名称	变量表	数据类型	地址	保持	在 H...	可从 ...	注释
1		System_Byte	默认变量表	Byte	%MB1	☐	☑	☑	
2		FirstScan	默认变量表	Bool	%M1.0	☐	☑	☑	
3		DiagStatusUpdate	默认变量表	Bool	%M1.1	☐	☑	☑	
4		AlwaysTRUE	默认变量表	Bool	%M1.2	☐	☑	☑	
5		AlwaysFALSE	默认变量表	Bool	%M1.3	☐	☑	☑	
6		正转	默认变量表	Bool	%M0.0	☐	☑	☑	
7		反转	默认变量表	Bool	%M0.1	☐	☑	☑	
8		停止	默认变量表	Bool	%M0.2	☐	☑	☑	
9		A点限位	默认变量表	Bool	%I0.3	☐	☑	☑	
10		B点限位	默认变量表	Bool	%I0.4	☐	☑	☑	

图 4. 2. 16　变量表

Figure 4. 2. 16　Variable table

（八）轴工艺对象组态及调试

（Ⅷ）Configuration and commissioning of axis process object

在 TIA 博途项目中新建轴工艺对象，在组态窗口中设置本任务所需的参数。由于步进电机通过传动机构带动小车，在电路连接中已进行了限位保护，因此，不需要设置位置限制参数和回原点参数。本任务需要设置的内容有：

Create a new axis process object in the TIA Portal project and set the parameters required for this task in the configuration window. As that stepping motor drives the trolley through the drive mechanism，the position limit protection has been carried out in the circuit connection，therefore，the position limit parameter need not be set，and the return parameter need not be set. The task contents to be set are as follows：

1. 常规设置

1. General setting

在常规设置中，可以设置轴的名称。驱动器选择 PTO，测量单位选择 mm。如图 4. 2. 17 所示。

In general settings，you can set the name of the axis. The driver selects the PTO and the unit of measurement selects the mm. As shown in Figure 4. 2. 17.

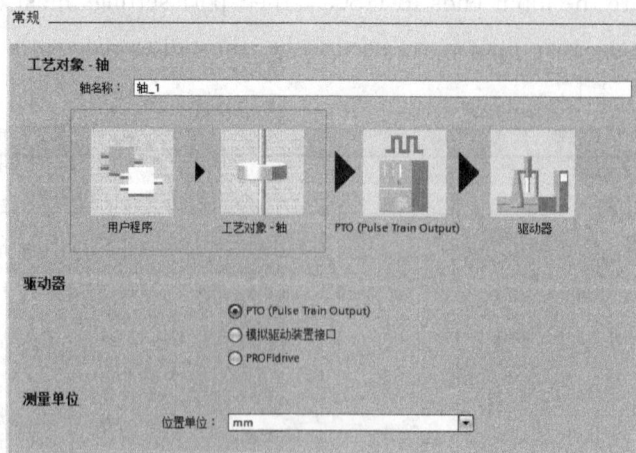

图 4. 2. 17　轴工艺对象的常规设置

Figure 4. 2. 17　General setting of shaft process object

2. 驱动器设置

启用脉冲发生器，选择 PTO 的类型为脉冲 A 和方向 B，脉冲信号由 Q0.0 提供，方向信号由 Q0.1 提供。如图 4.2.18 所示。

2. Drive settings

Enable the pulse generator and select the type of PTO as Pulse A and Direction B, the pulse signal is provided by Q0.0 and the direction signal is provided by Q0.1. As shown in Figure 4.2.18.

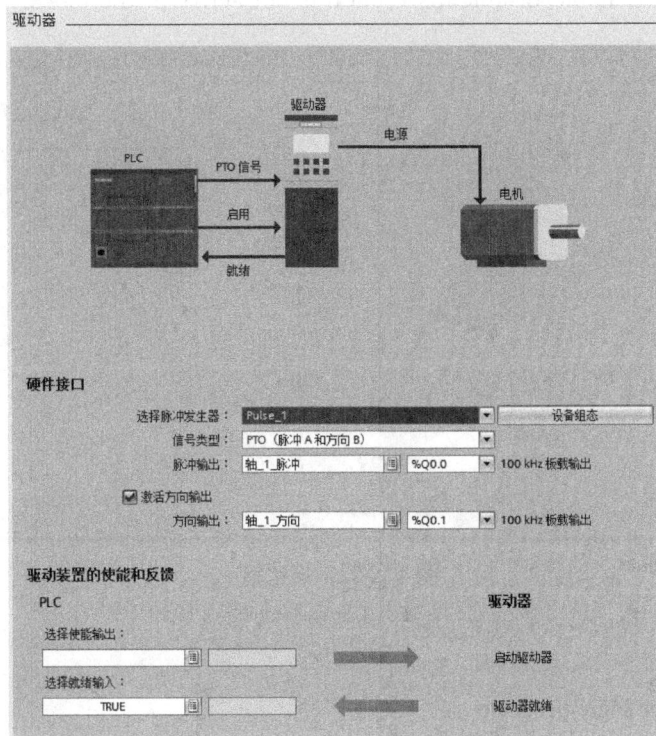

图 4.2.18 轴工艺对象的驱动器设置

Figure 4.2.18 Drive settings for shaft process object

3. 机械设置

根据步进电机驱动器的参数设置电机每转的脉冲数和位移，如图 4.2.19 所示。

3. Mechanical setting

Set the number of pulses per rotation and the displacement of the motor according to the parameters of the stepping motor driver, as shown in Figure 4.2.19.

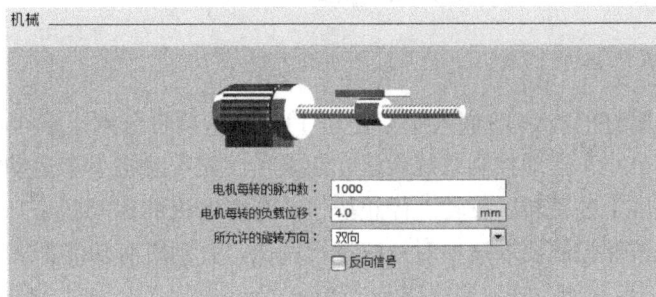

图 4.2.19 轴工艺对象的机械设置

Figure 4.2.19 Mechanical setting of shaft process object

4. 调试

选择调试工具，将 PLC 转入在线状态，可以打开轴控制面板，激活调试轴命令，对轴进行调试，如图 4.2.20 所示。注意，只有在程序块中启用/禁用轴指令没有激活的情况下才能调试轴。

4. Commissioning

Select the commissioning tool and put the PLC into online state. Open the shaft control panel, activate the commissioning shaft command and debug the shaft, as shown in Figure 4.2.20. Note that the shaft can only be debugged if the shaft enable/disable instruction is not active in the block.

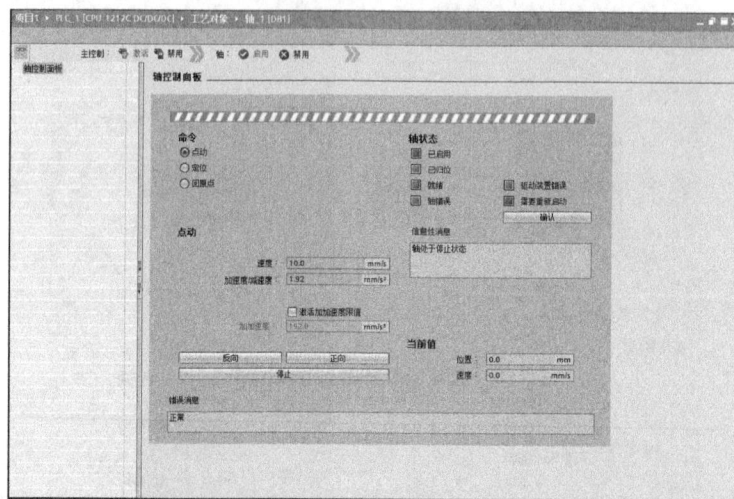

图 4.2.20　调试轴

Figure 4.2.20　Commissioning of shaft

（九）编写梯形图程序

(Ⅸ) To write the ladder logic program

先启用轴，用系统存储器中的常开位作为 MC_Power 指令的使能信号。按下正转按钮时，执行 MC_MoveVelocity 指令，速度为 4mm/s；按下反转按钮时，执行 MC_MoveVelocity 指令，速度为 -4mm/s；按下停止按钮时，执行 MC_Halt 指令，电机停止。

Enable the shaft first, and use the normally open bit in system memory as the enable signal for the MC_Power instruction. When the forward button is pressed, execute the MC_MoveVelocity instruction at a speed of 4mm/s; When the backward button is pressed, execute the MC_MoveVelocity instruction at a speed of -4mm/s; When the stop button is pressed, execute the MC_Halt instruction and the motor stops.

梯形图程序如图 4.2.21 所示。

The ladder diagram program is shown in Figure 4.2.21.

（十）调试

(Ⅹ) Debugging

完成以上步骤后，将编辑好的 MCGS 嵌入版工程和 TIA 博途项目分别下载至触摸屏和 PLC，再用编程线将二者连接起来，这样一个小车运动控制系统就建立起来了。先手动将小车运动到 A 点，当 SQ1 接通时，高速计数器复位。按下触摸屏上的正转、反转、停止按钮，观察电机的运动状态，再观察小车的位置和触摸屏上显示的位置是否相符。按照控制要求逐项对系统进行调试，直到所有功能都实现为止。

After the above steps, the embedded MCGS engineering and TIA Portal items have been edited are respectively downloaded to the touch screen and PLC, the both are connected with the programming line such that the trolley movement control system is finally created. Move the trolley manually to point A. The high-speed counter resets when SQ1 is powered on. Press the forward，reverse and stop buttons on the touch screen to observe the motion of the motor，and then observe whether the position of the trolley matches the position displayed on the touch screen. The system is debugged item by item according to the control requirements until that all functions are realized.

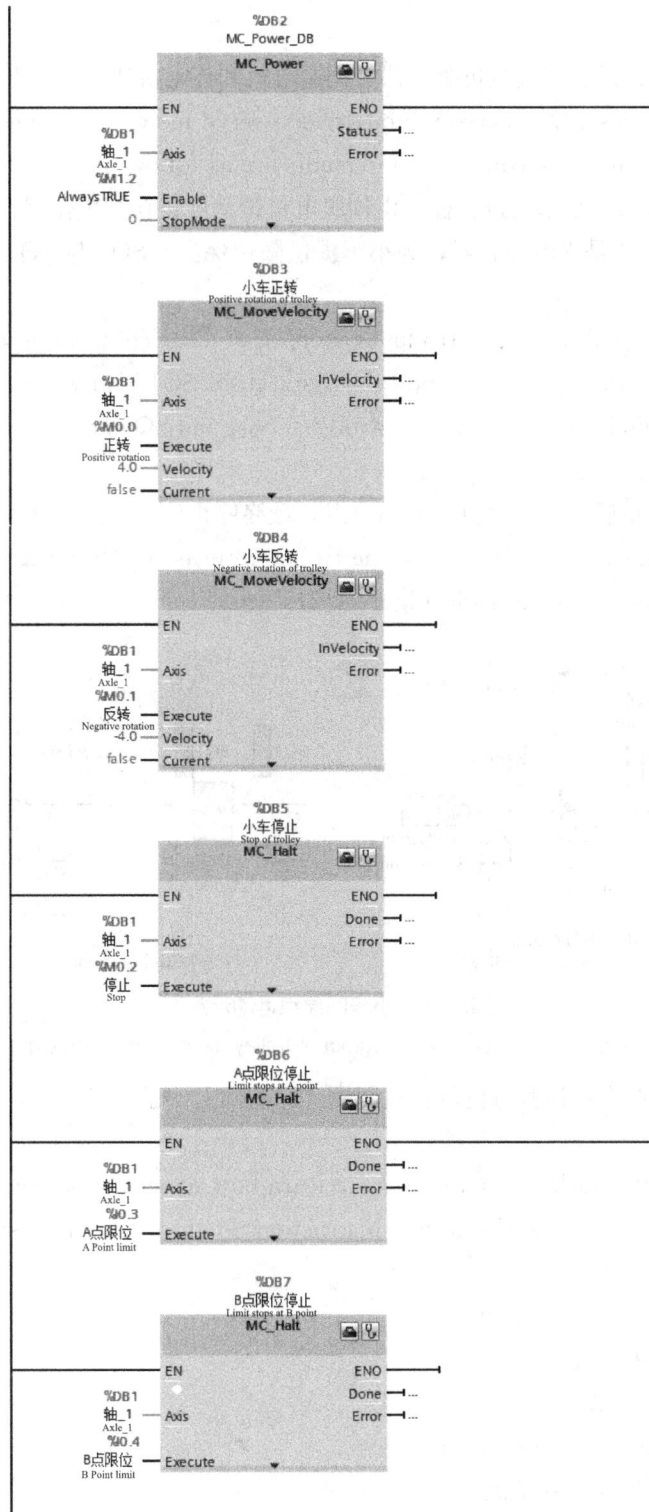

%DB2
MC_Power_DB
MC_Power

EN — ENO
%DB1
轴_1
Axle_1 — Axis — Status
— Error
%M1.2
AlwaysTRUE — Enable
0 — StopMode

%DB3
小车正转
Positive rotation of trolley
MC_MoveVelocity

EN — ENO
— InVelocity
%DB1
轴_1
Axle_1 — Axis — Error
%M0.0
正转
Positive rotation — Execute
4.0 — Velocity
false — Current

%DB4
小车反转
Negative rotation of trolley
MC_MoveVelocity

EN — ENO
— InVelocity
%DB1
轴_1
Axle_1 — Axis — Error
%M0.1
反转
Negative rotation — Execute
-4.0 — Velocity
false — Current

%DB5
小车停止
Stop of trolley
MC_Halt

EN — ENO
— Done
%DB1
轴_1
Axle_1 — Axis — Error
%M0.2
停止
Stop — Execute

%DB6
A点限位停止
Limit stops at A point
MC_Halt

EN — ENO
— Done
%DB1
轴_1
Axle_1 — Axis — Error
%0.3
A点限位
A Point limit — Execute

%DB7
B点限位停止
Limit stops at B point
MC_Halt

EN — ENO
— Done
%DB1
轴_1
Axle_1 — Axis — Error
%0.4
B点限位
B Point limit — Execute

图 4.2.21　梯形图程序

Figure 4.2.21　Ladder logic program

任务三　交流伺服电机控制小车回原点

Task Ⅲ　AC Servo Motor Controls the Trolley to Return to Origin Point

一、任务分析

Ⅰ. Task Analysis

用西门子 S7-1200 系列 PLC、伺服电机驱动器、伺服电机和传动机构控制小车回原点，控制要求如下：

Use Siemens S7-1200 series PLC，servo motor driver，servo motor and drive mechanism to control the trolley to return to origin point．The control requirements are as follows：

（1）如图 4.3.1（a）所示，小车通过丝杠，由伺服电机带动停止在某个位置，A 点为原点位置，安装有原点传感器 SQ1。SQ2 和 SQ3 是光电传感器，为小车提供限位保护。SQ4 与 SQ5 是行程开关，也为小车提供限位保护。

（1）As shown in Figure 4.3.1（a），a trolley is stopped at a certain position by a servo motor driven by a lead screw，point A is the origin position，and the origin sensor SQ1 is installed．SQ2 and SQ3 are photoelectric sensors that provide limit protection for the trolley．SQ4 and SQ5 are travel switches that also provide limit protection for the trolley.

（2）按下 SB1 回原点按钮后，小车由负方向回原点，最终停止位置如图 4.3.1（b）所示。

（2）When the SB1 origin button is pressed，the trolley returns from the negative direction to the origin point，and the final stop position is shown in Figure 4.3.1（b）.

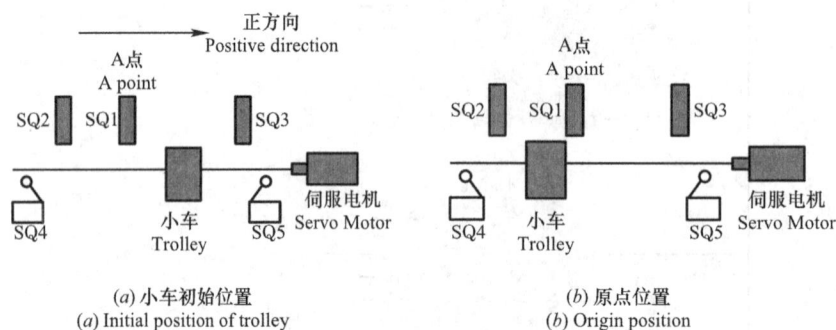

图 4.3.1　小车回原点运动示意图

Figure 4.3.1　Schematic diagram of trolley movement to its origin

为了完成本任务，需要在掌握伺服电机位置控制模式的基础上，学习 S7-1200 PLC 回原点参数的设定方法和回原点指令的使用方法。

In order to accomplish this task，it is necessary to learn how to set the parameters of the return to origin point of S7-1200 PLC and how to use the return instruction on the basis of mastering the position control mode of servo motor.

二、相关知识技能

Ⅱ. Relevant Knowledge & Skills

（一）伺服电机位置控制模式

（Ⅰ）Position control mode of servo motor

1. 伺服系统的组成及反馈控制的概念

1. Composition of servo system and concept of feedback control

"伺服"一词是跟随的意思，即被控的电机忠实地执行频繁变化的位置或速度指令，精确地控制机械系统运动的位移或角度，这种自动控制系统被称为伺服系统或随动系统。伺服系统是一个闭环的控制系统，它主要由控制元件、执行元件和反馈网络构成，如图 4.3.2 所示。以直流伺服电机为驱动元件的伺服系统叫作直流伺服系统；以交流伺服电机为驱动元件的伺服系统叫作交流伺服系统。通常，控制器需要通过伺服电机驱动器对伺服电机进行控制。

The word "servo" means follow-up，that is，the controlled motor faithfully executes frequently changing position or speed instructions to precisely control the displacement or angle of movement of the mechanical system，which is called a servo system or follow-up control system. The servo system is a closed-loop control system，which is mainly composed of control elements，actuator elements and feedback network，as shown in Figure 4.3.2. The servo system with DC servo motor as driving component is called DC servo system；The servo system with AC servo motor as driving component is called AC servo system. Typically，the controller needs to control the servo motor through the servo motor driver.

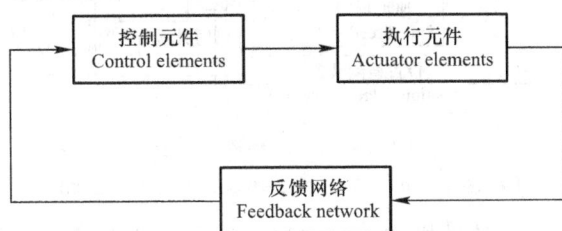

图 4.3.2　伺服控制系统的原理图

Figure 4.3.2　Schematic diagram of servo control system

步进电机控制系统是开环控制系统，如图 4.3.3 所示。

The stepping motor control system is an open-loop control system，as shown in Figure 4.3.3.

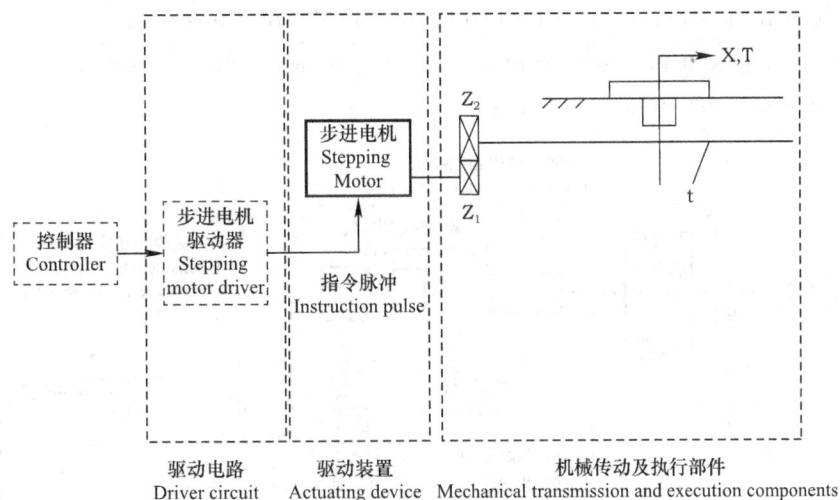

图 4.3.3　开环控制系统

Figure 4.3.3　Open-loop control system

伺服电机与伺服电机驱动器一般构成一个半闭环控制系统，如图 4.3.4 所示。通过伺服电机上的位置检测装置（通常为编码器），将伺服电机的实际位移转换成电脉冲的形式反馈到驱动器。驱动器将它与输入的位置指令进行比较，将两者的差值放大和变换，控制伺服电机向消除偏差的方向运动，直到指令位置和反馈位置的差值等于零为止。

Servo motors and servo motor drivers generally form a semi-closed-loop control system，as shown in Figure 4.3.4. The actual displacement of the servo motor is converted into electrical pulses and fed back to the

153

driver by a position detection device (usually an encoder) on the servo motor. The driver compares it with the input position instruction, amplifies and converts the difference between the two, and controls the servo motor to move in the direction of eliminating deviation until the difference between the instruction position and the feedback position is equal to zero.

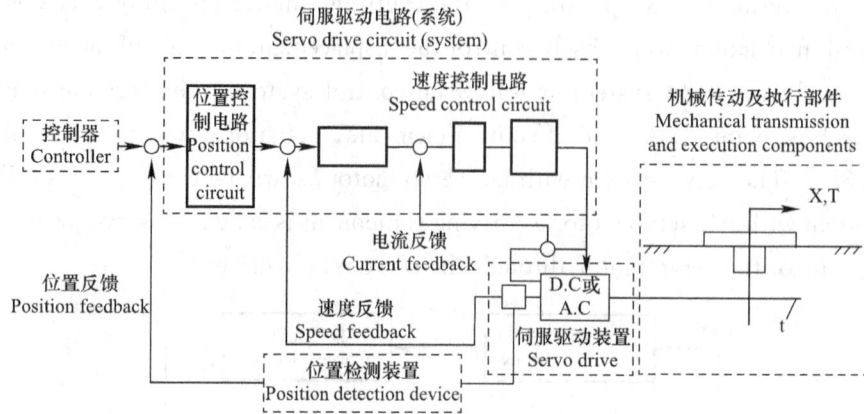

图 4.3.4　半闭环控制系统

Figure 4.3.4　Semi-closed-loop control system

如果反馈的位置信号来自于机械传动机构上的位移传感器，则构成了全闭环控制系统，如图 4.3.5 所示。与半闭环控制系统的位置反馈信号来源于伺服电机上的编码器不同，全闭环控制系统的位置反馈信号来源于位移传感器直接测量到的位置信号，因此这样的控制系统也更精确。

If the feedback position signal comes from a displacement sensor on the mechanical drive mechanism, a full closed-loop control system is formed, as shown in Figure 4.3.5. Unlike the position feedback signal of the semi-closed-loop control system which comes from the encoder of the servo motor, the position feedback signal of the full closed-loop control system comes from the position signal measured directly by the displacement sensor, so the control system is more accurate.

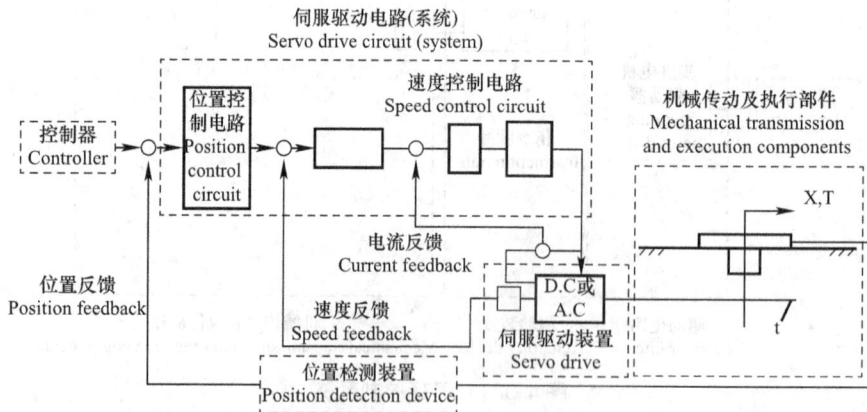

图 4.3.5　全闭环控制系统

Figure 4.3.5　Full Closed-loop control system

2. 伺服电机位置控制模式

2. Position control mode of servo motor

以台达 ASD-B2 系列驱动器为例，来说明伺服电机驱动器的控制模式。驱动器提供位置、速度、扭矩三种基本控制模式，可以用单一控制模式，即固定在一种模式控制，也可选择用混合模式来进行控制，每一种模式分两种情况，所以总共有 11 种控制模式，表 4.3.1 列出了所有的控制模式与说明。

Taking Delta ASD-B2 series drivers as an example，the control mode of servo motor drivers is explained. The driver provides three basic control modes：position，speed，and torque，which can be controlled in a single control mode，fixed in one mode，or optionally in a hybrid mode. Each mode can be controlled in two cases，so there is a total of 11 control modes. Table 4.3.1 lists all the control modes and descriptions.

伺服驱动器的控制模式 表 4.3.1

Control mode of servo driver Table 4.3.1

模式名称 Mode name	模式代号 Mode symbol	模式码 Mode code	说明 Description
位置模式 Position mode （端子输入） （Terminal input）	Pt	00	驱动器接受位置命令，控制电机至目标位置。位置命令由端子输入，信号形态为脉冲 The driver receives the position instruction and controls the motor to the target position. The position instruction is inputted from the terminal，and the signal type is pulse
位置模式 Position mode （内部寄存器输入） （Internal register input）	Pr	01	驱动器接受位置命令，控制电机至目标位置。位置命令由内部寄存器提供（共八组寄存器），可利用 DI 信号选择寄存器编号 The driver receives the position instruction and controls the motor to the target position. The position instruction is provided by an internal register（eight sets of registers）that allows the register number to be selected using the DI signal
速度模式 Speed mode	S	02	驱动器接受速度命令，控制电机至目标转速。速度命令可由内部寄存器提供（共三组寄存器），或由外部端子输入模拟电压（－10V～＋10V）。命令的选择是根据 DI 信号来选择 The driver receives the speed instruction and controls the motor to the target rotating speed. The speed instruction can be supplied from an internal register（three sets of registers）or an analog voltage can be input from an external terminal（－10V－＋10V）. The selection of instructions is based on the DI signal
速度模式 Speed mode （无模拟输入）	Sz	04	驱动器接受速度命令，控制电机至目标转速。速度命令仅可由内部寄存器提供（共三组寄存器），无法由外部端子提供。命令的选择是根据 DI 信号来选择 The driver receives the speed instruction and controls the motor to the target rotating speed. The speed instruction can only be supplied from the internal registers（three sets of registers）and cannot be supplied from the external terminals. The selection of instructions is based on the DI signal
扭矩模式 Torque mode	T	03	驱动器接受扭矩命令，控制电机至目标扭矩。扭矩命令可由内部寄存器提供（共三组寄存器），或由外部端子输入模拟电压（－10V～＋10V）。命令的选择是根据 DI 信号来选择 The driver receives the torque instruction and controls the motor to the target torque. The torque instruction can be supplied from an internal register（three sets of registers）or an analog voltage can be input from an external terminal（－10V－＋10V）. The selection of instructions is based on the DI signal
扭矩模式 Torque mode （无模拟输入） （No analog input）	Tz	05	驱动器接受扭矩命令，控制电机至目标扭矩。扭矩命令仅可由内部寄存器提供（共三组寄存器），无法由外部端子提供。命令的选择是根据 DI 信号来选择 The driver receives the torque instruction and controls the motor to the target torque. The torque instruction can only be supplied from the internal registers（three sets of registers）and cannot be supplied from the external terminals. The selection of instructions is based on the DI signal
混合模式 Mixed mode	Pt-S	06	Pt 与 S 可通过 DI 信号切换 Pt and S can be switched by DI signals
	Pt-T	07	Pt 与 T 可通过 DI 信号切换 Pt and T can be switched by DI signals.
	Pr-S	08	Pr 与 S 可通过 DI 信号切换 Pr and S can be switched by DI signals.
	Pr-T	09	Pr 与 T 可通过 DI 信号切换 Pr and T can be switched by DI signals.
	S-T	10	S 与 T 可通过 DI 信号切换 S and T can be switched by DI signals.

与位置控制有关的部分参数说明见表4.3.2。

Some of the parameters related to position control are described in Table 4.3.2.

与位置控制有关的部分参数 表 4.3.2

Some parameters related to position control Table 4.3.2

序号 S/N	参数编号 Parameter No.	参数名称 Parameter name	出厂值 Initial Value	功能和含义 Function and meaning
1	P0-02	LED 初始状态 LED initial state	0	显示电机反馈脉冲数（电子齿轮比之后） Display the number of motor feedback pulses (after the electronic gear ratio)
2	P1-00	外部脉冲列指令输入形式设定 External pulse train instruction input form setting	2	脉冲列＋符号 Pulse trains＋signs
3	P1-01	控制模式及控制命令输入源设定 Control mode and control command input source setting	00	位置控制模式（相关代码 Pt） Position control mode (related code Pt)
4	P1-44	电子齿轮比分子（N） Electronic gear ratio numerator (N)	16	按控制需求设置 Set by control requirements
5	P1-45	电子齿轮比分母（M） Electronic gear ratio denominator (M)	10	按控制需求设置，规则如下： According to the control requirements, the setting rules are as follows: 指令脉冲输入 $f1$ → $\dfrac{N}{M}$ → 位置指令 $f2$ $f2=f1\times\dfrac{N}{M}$ 指令脉冲输入比值范围：$1/50 < N/M < 25600$ Instruction pulse input ratio range: $1/50 < N/M < 25600$
6	P2-00	位置控制比例增益 Position control proportional gain	35	位置控制增益值加大时，可提升位置应答性及缩小位置控制误差量。但若设定太大时易产生振动及噪声 When the position control gain value is increased, the position responsiveness can be improved and the position control error can be reduced. However, if it is set too large, it is easy to produce vibration and noise
7	P2-02	位置控制前反馈增益 Position control feedforward gain	50	位置控制命令平滑变动时，增益值加大可改善位置跟随误差量。若位置控制命令不平滑变动时，降低增益值可降低机构的运转振动现象 When the position control instruction changes smoothly, increasing the gain value can improve the position following error. If the position control instruction does not change smoothly, reducing the gain value can reduce the vibration phenomenon of the mechanism

通过设置适当的电子齿轮比，可以得到更简单的位移和脉冲之间的换算关系。当电子齿轮比为1时，每来1个指令脉冲，编码器就反馈1个脉冲。当电子齿轮比为0.5时，则每来2个指令脉冲，编码器才反馈1个脉冲。如果伺服电机上的编码器需要1000个脉冲才转一圈，而转一圈的直线位移为3mm。在齿轮比为1时，那么1个脉冲对应的直线距离是$3\mu m$，这个比例不利于位移和脉冲之间的换算。如果将电子齿轮比调整为1:3，那么每来3000个指令脉冲，编码器反馈1000个脉冲，伺服电机旋转一圈，直线位移为3mm。这样换算下来每1个脉冲对应的直线距离是$1\mu m$，这个简单的比例关系更方便于位移和脉冲之间的换算。若按照初始值，P1-44设置为16，P1-45设置10时，脉冲数为100000旋转一周（编码器反馈脉冲为160000脉冲/转）。

By setting proper electronic gear ratio, a simple conversion relationship between displacement and pulse can be obtained. When the electronic gear ratio is 1, the encoder feeds back 1 pulse for each instruction pulse. When the electronic gear ratio is 0.5, the encoder feeds back 1 pulse for each two instruction pulses. If the encoder on the servo motor requires 1000 pulses to rotate for one cycle, the linear displacement of one cycle is 3 mm. When the gear ratio is 1, the linear distance corresponding to one pulse is 3 μm, which is not conducive to the conversion between the displacement and the pulse. If the electronic gear ratio is adjusted to 1:3, for every 3000 instruction pulses, the encoder feeds back 1000 pulses, the servo motor rotates for one cycle and the linear displacement is 3mm. In this way, the linear distance corresponding to each pulse is 1 μm. This simple proportional relation is very convenient for the conversion between the displacement and the pulse. If P1-44 is set to 16 and P1-45 is set to 10 according to initial value, the number of pulses is 100000 for each rotation (160000 pulses per rotation for encoder feedback).

(二) 回原点参数的设定方法

(Ⅱ) Setting method of return parameters

"原点"也可以叫作"参考点", "回原点"或是"寻找参考点"的作用是: 把轴实际的机械位置和S7-1200程序中轴的位置坐标统一, 以进行绝对位置定位。一般情况下, 西门子PLC的运动控制在使能绝对位置定位之前, 必须执行"回原点"或"寻找参考点"。在组态轴工艺对象的扩展参数中, 可以设置回原点参数, 回原点参数分为"主动"和"被动"两部分参数。

"Origin point" can also be called a "reference point", and the role of "return to origin point", or "search for reference point" is. The actual mechanical position of the axis is unified with the position coordinates of the axis in the S7-1200 program for absolute position positioning. Normally, Siemens PLC motion control program must perform "return to origin point" or "search for reference point" before absolute position positioning is enabled. In the extended parameters of the configuration axis process object, the return parameters can be set, which are divided into "active" and "passive" parameters.

1. 主动回原点参数设置

1. Active return parameter settings

"主动"回原点就是传统意义上的回原点或是寻找参考点。当轴触发了主动回参考点操作, 轴就会按照组态的速度去寻找原点开关信号, 并完成回原点命令。主动回原点的组态界面如图4.3.6所示。

"Active" return to the original point is the traditional sense of the return to the origin point or to search for a reference point. When the axis triggers the operation of active return to reference point, the axis searches for the origin switch signal at the configured speed and completes the return instruction. The configuration interface for active return is shown in Figure 4.3.6.

(1) 输入原点开关: 设置原点开关的DI输入点。

(1) Input origin switch: Set the DI input point of the origin switch.

(2) 选择电平: 选择原点开关的有效电平, 也就是当轴碰到原点开关时, 该原点开关对应的DI点是高电平还是低电平。

(2) Select level: Select the active level of the origin switch, that is, whether the DI point corresponding to the origin switch is high or low when the axis touches the origin switch.

(3) 允许硬件限位开关处自动反转: 如果轴在回原点的一个方向上没有碰到原点, 则需要使能该选项, 这样轴可以自动调头, 向反方向寻找原点。

(3) Allow automatic inversion at hardware limit switches: If the axis does not touch the origin in the direction of return, you need to enable this option so that the axis can automatically turn around and search for the origin point in the opposite direction.

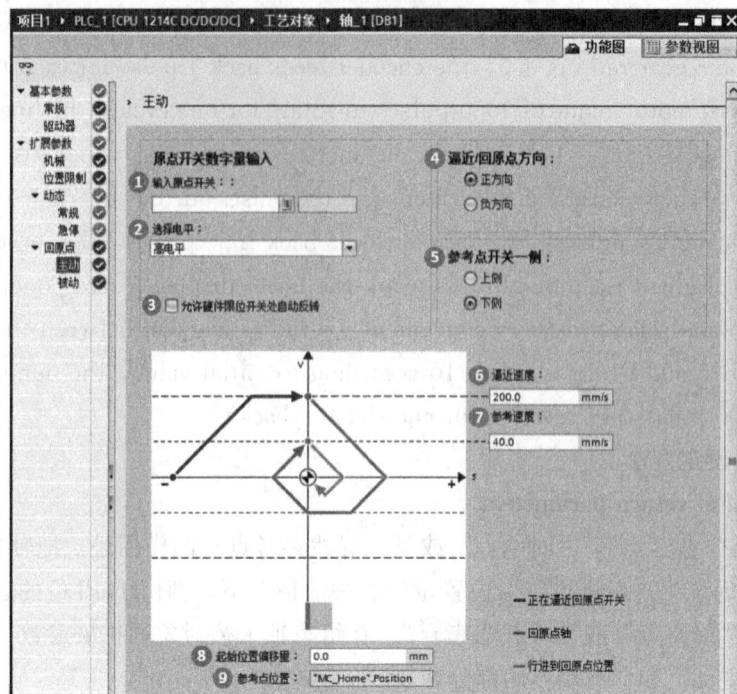

图 4.3.6　主动回原点组态界面

Figure 4.3.6　Active return configuration interface

（4）逼近/回原点方向：寻找原点的起始方向。也就是说触发了寻找原点功能后，轴是向"正方向"或是"负方向"开始寻找原点，如图 4.3.7 所示。如果知道轴和参考点的相对位置，可以合理设置"逼近/回原点方向"来缩短回原点的路径。例如，以图 4.3.7 中的负方向为例，触发回原点命令后，轴需要先运行到左边的限位开关，掉头后继续向正方向寻找原点开关。

（4）Approximation/return direction：Find the starting direction of the origin. That is，after triggering the Search for Origin function，the axis begins to find the origin in the "positive" or "negative" direction，as shown in Figure 4.3.7. If you know the relative position of the axis and the reference point，you can reasonably set the "Approximation/Return Direction" to shorten the path back to the origin. For example，with the negative direction shown in Figure 4.3.7，after triggering the return instruction，the axis needs to run to the left limit switch first，then turn around and continue to search for the origin switch in the positive direction.

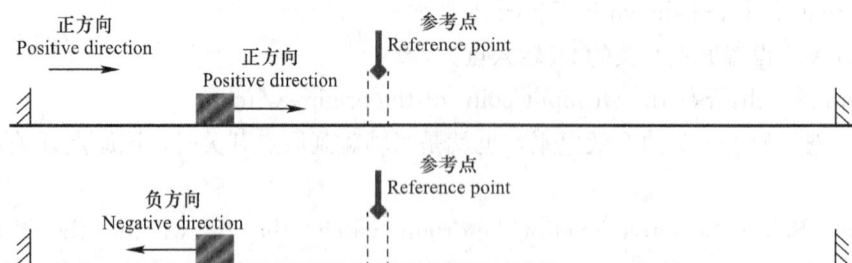

图 4.3.7　回原点方向示意图

Figure 4.3.7　Schematic diagram of return direction

（5）参考点开关一侧："上侧"指的是：轴完成回原点指令后，以轴的左边沿停在参考点开关右侧边沿。"下侧"指的是：轴完成回原点指令后，以轴的右边沿停在参考点开关左侧边沿。无论用户设置寻找原点的起始方向为正方向还是负方向，轴最终停止的位置取决于"上侧"或"下侧"，如图 4.3.8 所示。

(5) Reference point switch side: "Upper side" means: After the axis completes the return instruction, stop at the right edge of the reference point switch with the left edge of the axis. "Lower side" means: After the axis completes the return instruction, stop at the left edge of the reference point switch with the right edge of the axis. Whether the user sets the start direction of the search for origin as positive or negative, the final stop position of the axis depends on the "upper" or "lower" side, as shown in Figure 4. 3. 8.

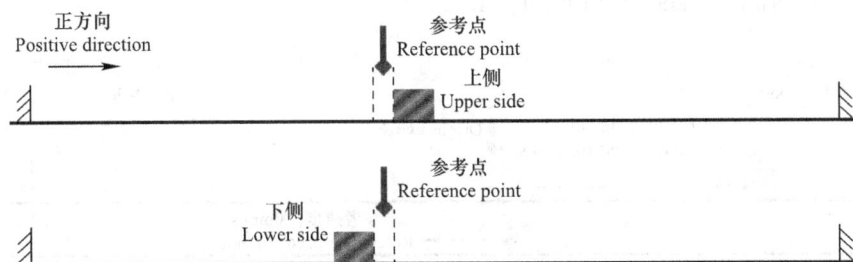

图 4.3.8 参考点上侧和下侧示意图

Figure 4. 3. 8 Schematic diagram of upper and lower sides of reference points.

（6）逼近速度：寻找原点开关的起始速度，当程序中触发了 MC _ Home 指令后，轴立即以"逼近速度"运行来寻找原点开关。

(6) Approximation speed: Find the start speed of the origin switch. When the MC _ Home instruction is triggered in the program, the axis immediately runs at "approximation speed" to find the origin switch.

（7）参考速度：最终接近原点开关的速度，当轴第一次碰到原点开关有效边沿后运行的速度，也就是触发了 MC _ Home 指令后，轴立即以"逼近速度"运行来寻找原点开关，当轴碰到原点开关的有效边沿后轴从"逼近速度"切换到"参考速度"来最终完成原点定位。"参考速度"要小于"逼近速度"，"参考速度"和"逼近速度"都不宜设置的过快。在可接受的范围内，设置较慢的速度值。

(7) Reference speed: Speed approaching the origin switch, it means the running speed when the axis touches the effective edge of the origin switch for the first time; that is, when the MC _ Home instruction is triggered, the axis immediately runs at the "approximation speed" to find the origin switch, and when the axis touches the effective edge of the origin switch, the axis switches from the "approximation speed" to the "reference speed" to finally complete the origin positioning. "Reference Speed" should be less than "Approximation Speed", "Reference Speed" and "Approximation Speed" should not be set too fast. Set the slower speed value within an acceptable range.

（8）起始位置偏移量：该值不为零时，轴会在距离原点开关一段距离（该距离值就是偏移量）停下来，把该位置标记为原点位置值。该值为零时，轴会停在原点开关边沿处。

(8) Start position offset: When this value is not zero, the axis stops at certain distance away from the origin switch (the distance value is the offset) and the position is marked as the origin position value. When this value is zero, the axis stops at the edge of the origin switch.

（9）参考点位置：该值就是（8）中的原点位置值。

(9) Reference point position: This value is the origin position value in (8).

根据轴与原点开关的相对位置，分成四种情况：轴在原点开关负方向侧，轴在原点开关的正方向侧，轴刚执行过回原点指令，轴在原点开关的正下方。下面以一个例子来说明主动回原点的执行过程。

According to the relative position of the axis and the origin switch, it can be divided into four cases: The axis is on the negative side of the origin switch and the axis is on the positive side of the origin switch. The axis has just performed the return instruction and the axis is directly below the origin switch. An example is given below to illustrate the execution process of an active return instruction.

例子：逼近速度＝10.0mm/s，参考速度＝2.0mm/s，正方向回原点，参考点开关上侧。当轴在原点开关负方向时，回原点过程如图 4.3.9 所示。

Example：Approximation speed＝10.0mm/s, reference speed＝2.0mm/s, positive return to origin, upper side of reference point switch. When the axis is in the negative direction of the origin switch, the process of return to the origin is shown in Figure 4.3.9.

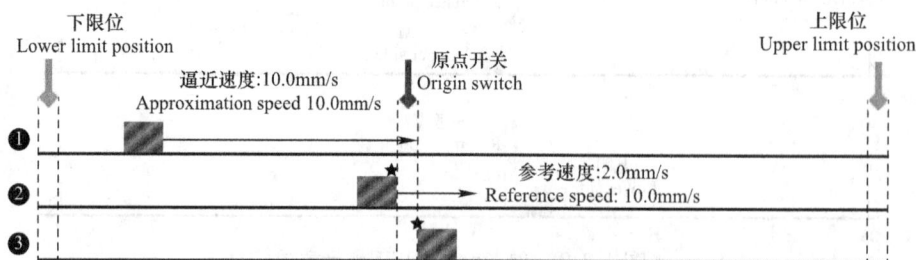

图 4.3.9　轴在原点开关负方向

Figure 4.3.9　Axis is at negative direction of origin switch

（10）当程序以 Mode＝3 触发 MC_Home 指令时，轴立即以"逼近速度 10.0mm/s"向右（正方向）运行寻找原点开关；

（10）When the program triggers the MC_Home instruction with Mode＝3, the axis immediately runs to the right side (positive direction) at the "approximation speed of 10.0 mm/s" to search for the origin switch;

（11）当轴碰到参考点的有效边沿，切换运行速度为"参考速度 2.0mm/s"继续运行；

（11）When the axis touches the effective edge of the reference point, switch the operation speed to "reference speed 2.0 mm/s" to continue operation;

（12）当轴的左边沿与原点开关有效边沿重合时，轴完成回原点动作。

（12）When the left edge of the axis coincides with the active edge of the origin switch, the axis completes the return action.

当轴在原点开关正方向时，回原点过程如图 4.3.10 所示。

When the axis is in the positive direction of the origin switch, the process of return to the origin is shown in Figure 4.3.10.

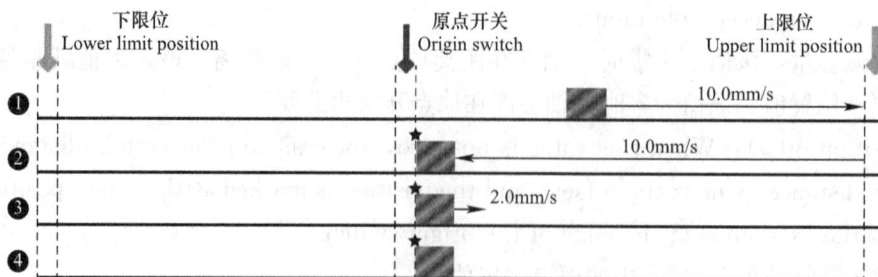

图 4.3.10　轴在原点开关正方向

Figure 4.3.10　Axis is at positive direction of origin switch

（13）当轴在原点开关的正方向（右侧）时，触发主动回原点指令，轴会以"逼近速度"运行直到碰到右限位开关，如果在这种情况下，用户没有使能"允许硬件限位开关处自动反转"选项，则轴因错误取消回原点动作并按急停速度使轴制动；如果用户使能了该选项，则轴将以组态的减速度减速（不是以紧急减速度）运行，然后反向运行，反向继续寻找原点开关；

(13) When the axis is in the positive direction of the origin switch (right side), the active return instruction is triggered and the axis runs at an "approximation speed" until the right limit switch is encountered. In this case, if the user does not enable the "Allow automatic reverse at the hardware limit switch" option, the axis cancels the return action due to an error and brakes the axis at the emergency stop speed; If this option is enabled by the user, the axis will decelerate at the configured deceleration (not at the emergency deceleration), then run in the reverse direction, continuing to search for the origin switch in the reverse direction;

(14) 当轴掉头后, 继续以"逼近速度"向负方向寻找原点开关的有效边沿;

(14) Continue to search for the effective edge of the origin switch in the negative direction at the "approximation speed" after the axis turns around;

(15) 原点开关的有效边沿是右侧边沿, 当轴碰到原点开关的有效边沿后, 将速度切换成"参考速度"最终完成定位。

(15) The effective edge of the origin switch is the right edge, and when the axis touches the effective edge of the origin switch, the speed is switched to the "reference speed" to finally complete the positioning.

当轴已经回原点, 再次触发回原点指令时, 其回原点的过程如图 4.3.11 所示。

When the axis has returned to the origin and the return instruction is triggered again, the process of return is shown in Figure 4.3.11.

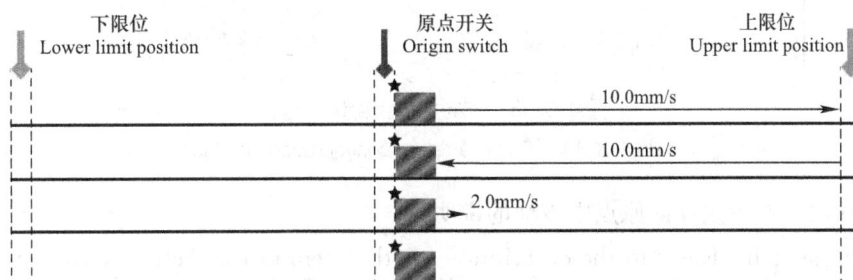

图 4.3.11　轴已经回原点再次触发回原点指令

Figure 4.3.11　The axis has returned to the origin and the return instruction is triggered again

当轴在原点开关正下方, 其回原点的过程如图 4.3.12 所示。

When the axis is directly below the origin switch, the process of return is shown in Figure 4.3.12.

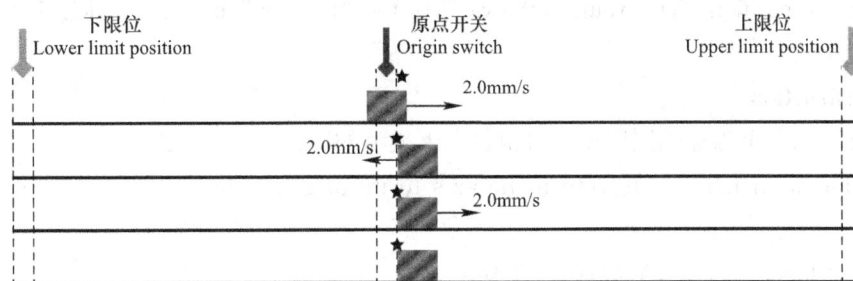

图 4.3.12　轴在原点开关正下方

Figure 4.3.12　The axis is directly below the origin switch

2. 被动回原点参数设置

2. Passive return parameter settings

被动回原点指的是轴在运行过程中碰到原点开关, 轴的当前位置将设置为回原点位置值。被动回原点的组态界面如图 4.3.13 所示。

Passive return means that the axis touches the origin switch during operation and the current position of the axis is set as return position value. The configuration interface for passive return is shown in Figure 4. 3. 13.

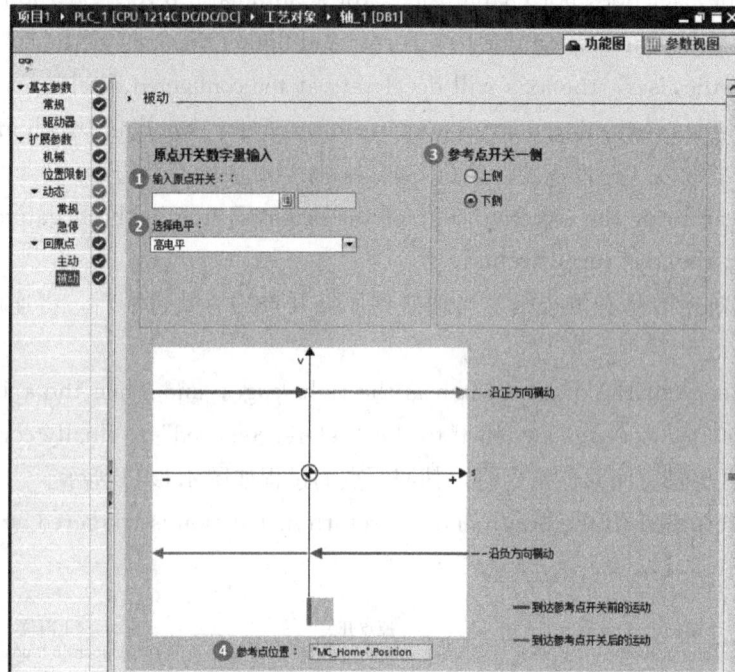

图 4. 3. 13　被动回原点组态界面

Figure 4. 3. 13　Passive return configuration interface

（1）输入原点开关：参考主动回原点中该项的说明。

（1）Input origin switch：Refer to the description for this item in the Active Return.

（2）选择电平：参考主动回原点中该项的说明。

（2）Select level：Refer to the description for this item in the Active Return.

（3）参考点开关一侧：参考主动回原点中第 5 项的说明。

（3）Reference point switch side：Refer to the description for item 5 in the Active Return.

（4）参考点位置：该值是 MC ＿ Home 指令中"Position"管脚的数值。

（4）Reference point position：This value is the value of the "Position" pin in the MC ＿ Home instruction.

（三）回原点指令

（Ⅲ）Return instruction

使用 MC ＿ Home 指令可以让轴回原点。可执行以下类型的归位：

Use the MC ＿ Home instruction to return the axis to its origin. The following types of homing can be performed：

（1）主动归位（Mode＝3），自动执行归位步骤。

（1）Active homing（Mode＝3），automatically performing the homing step.

（2）被动归位（Mode＝2），被动归位期间，运动控制指令 MC ＿ Home 不会执行任何归位运动。用户必须通过其他运动控制指令，执行这一步骤中所需的往返运动。检测到归位开关时，轴即归位。

（2）Passive homing（Mode＝2），during which the motion control instruction MC ＿ Home does not perform any homing motion. The user must perform the required round-trip motion in this step through other motion control instructions. When a homing switch is detected，the axis is homed.

（3）直接绝对归位（Mode＝0），将当前的轴位置设置为参数"Position"的值。

（3）Direct absolute return（Mode＝0），setting the current axis position to the value of parameter "Position".

（4）直接相对归位（Mode＝1），将当前轴位置的偏移值设置为参数"Position"的值。

（4）Direct relatively return（Mode＝1），setting deviation value of the current axis position to the value of parameter "Position".

为了使用 MC_Home 指令，轴工艺对象必须正确组态，且轴已启用。MC_Home 指令的参数见表 4.3.3。

In order to use the MC_Home instruction，the axis process object must be configured correctly and the axis enabled. The parameters of the MC_Home instruction are shown in Table 4.3.3.

MC_Home 指令的参数 表 4.3.3
Parameters of the MC_Home instruction Table 4.3.3

参数 Parameter	声明 Statement	数据类型 Data type	说明 Description
Axis	INPUT	TO_Axis	轴工艺对象 Shaft process object
Execute	INPUT	Bool	出现上升沿时开始任务（默认值：False） Start task on rising edge（default：False）
Position	INPUT	Real	Mode＝0、2 和 3 时，完成归位操作之后，轴的绝对位置 When Mode＝0，2，and 3，the absolute position of the axis after the homing operation is completed Mode＝1 时，对当前轴位置的修正值 Correction value of the current axis position when Mode＝1
Mode	INPUT	Int	归位模式，参照前述归位类型 Homing mode，referring to the aforementioned homing type
Done	OUTPUT	Bool	TRUE＝命令已完成 TRUE＝Instruction completed
Busy	OUTPUT	Bool	TRUE＝正在执行任务 TRUE＝Executing tasks
CommandAborted	OUTPUT	Bool	TRUE＝任务在执行期间被另一任务中止 TRUE＝Task stopped by another task during execution
Error	OUTPUT	Bool	FALSE：无错误 FALSE：No Errors TRUE：任务执行期间出错 TRUE：An error occurred during task execution 出错原因可在"ErrorID"和"ErrorInfo"参数中找到 The cause of the error can be found in the "ErrorID" and "ErrorInfo" parameters.
ErrorID	OUTPUT	Word	参数"Error"的错误 ID Error ID for the "Error" parameters
ErrorInfo	OUTPUT	Word	参数"ErrorID"的错误信息 ID Error information ID for the "ErrorID" parameters

三、任务实施
Ⅲ. Task Implementation

（一）连接 I/O 设备

（Ⅰ）To connect I/O device

按照控制要求，I/O 接线图如图 4.3.14 所示。

Based on the control requirements，the I/O wiring diagram is shown in Figure 4.3.14.

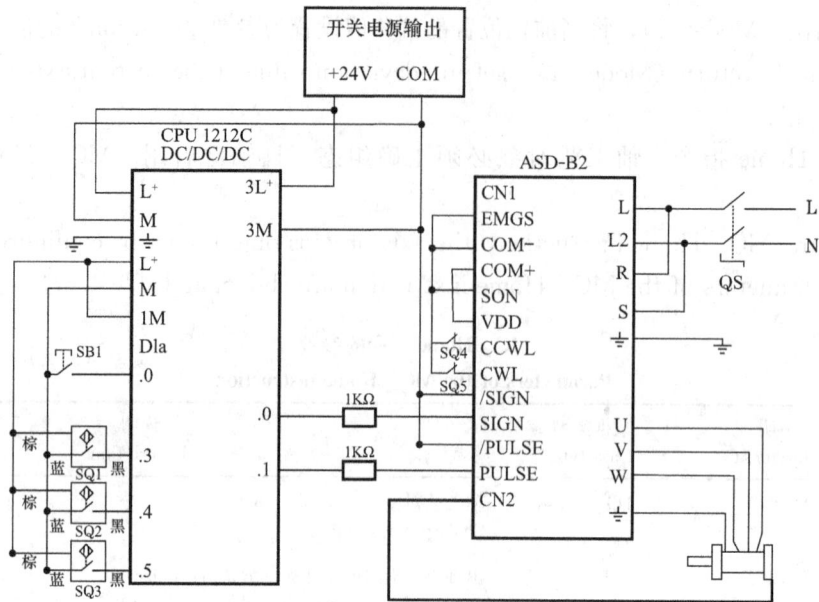

图 4.3.14 I/O 接线图

Figure 4.3.14 I/O wiring diagram

(二) 调整伺服电机驱动器

(Ⅱ) Adjust servo motor driver

调整伺服电机驱动器的参数见表 4.3.4。

The parameters for adjusting the servo motor driver is shown in Table 4.3.4.

伺服电机的参数设置 表 4.3.4

Parameter setting of servo motor **Table 4.3.4**

参数 Parameter	设置值 Set value
P0-02	0
P1-00	102
P1-01	000
P1-44	40
P1-45	1
P2-00	35
P2-02	50

(三) 在 TIA 博途中新建项目和设备组态

(Ⅲ) To create a new item and device configuration in TIA Portal

按照项目一中的步骤在 TIA 博途中新建项目和设备组态。由于编程中需要用到系统存储器，因此需要在 CPU 的属性中勾选启用系统存储器字节。

Create the new item and configure the device in TIA Portal by the steps in Item I. Because system memory is required for programming，the Enable System Memory byte needs to be checked in the properties of the CPU.

（四）建立变量表

（Ⅳ）To create the variable table

按照控制要求，定义变量表如图 4.3.15 所示。

Define the variable table as shown in Figure 4.3.15 based on the control requirements.

		名称	变量表	数据类型	地址	保持	在 H...	可从...	注释
		PLC 变量							
1		System_Byte	默认变量表	Byte	%MB1	☐	☑	☑	
2		FirstScan	默认变量表	Bool	%M1.0	☐	☑	☑	
3		DiagStatusUpdate	默认变量表	Bool	%M1.1	☐	☑	☑	
4		AlwaysTRUE	默认变量表	Bool	%M1.2	☐	☑	☑	
5		AlwaysFALSE	默认变量表	Bool	%M1.3	☐	☑	☑	
6		启动回原点按钮	默认变量表	Bool	%I0.0	☐	☑	☑	

图 4.3.15 变量表

Figure 4.3.15 Variable table

（五）轴工艺对象组态及调试

（Ⅴ）Configuration and commissioning of axis process object

在 TIA 博途项目中新建轴工艺对象，在组态窗口中设置本任务所需的参数。由于伺服电机的端子连接了左右限位的开关，如果小车碰到开关，伺服电机立即停止，因此不需要设置位置限制参数，但是因为本任务的内容为回原点，因此需要设置回原点参数。本任务需要设置的内容有：

Create a new axis process object in the TIA Portal project and set the parameters required for this task in the configuration window. Since the servo motor terminals are connected with the left and right limit switches，the servo motor stops immediately if the trolley touches the switch，so there is no need to set the position limit parameter，but because the content of this task is about the return to the origin point，the return parameter needs to be set. The task contents to be set are as follows：

1. 常规设置，如图 4.3.16 所示

1. General setting as shown in Figure 4.3.16

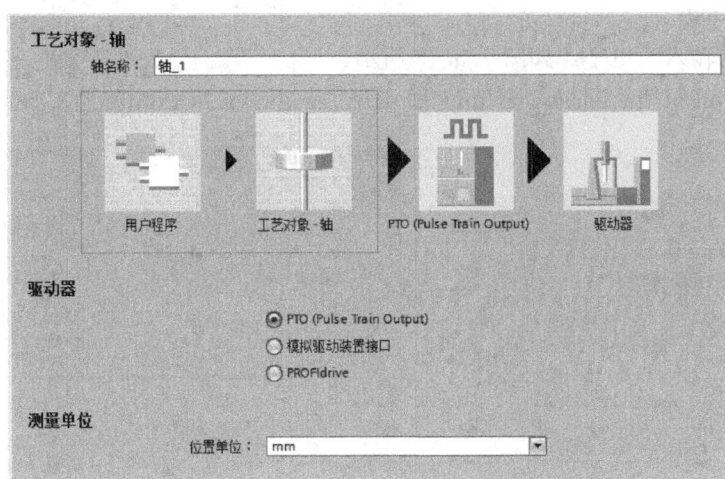

图 4.3.16 轴工艺对象的常规设置

Figure 4.3.16 General setting of shaft process object

2. 驱动器设置，如图 4.3.17 所示

2. Driver setting as shown in Figure 4.3.17

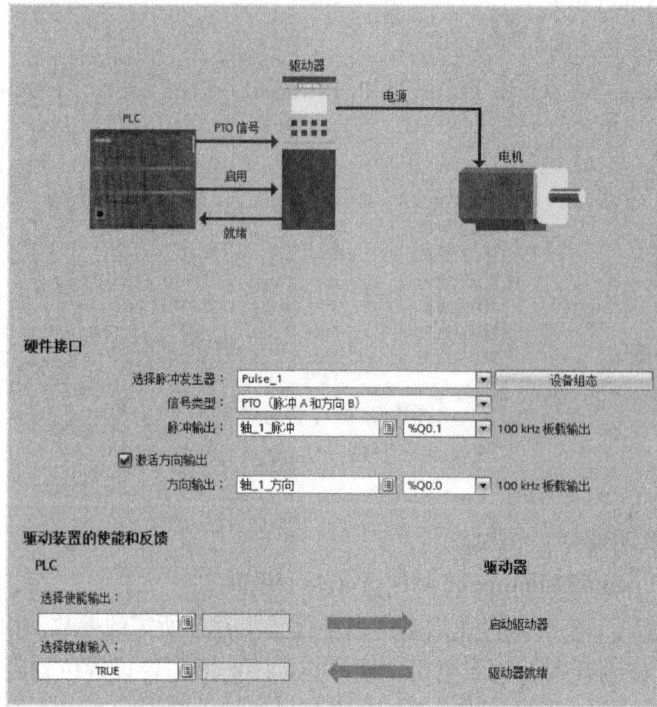

图 4.3.17　轴工艺对象的驱动器设置

Figure 4.3.17　Drive settings for shaft process object

3. 机械设置

3. Mechanical setting

根据步进电机驱动器的参数设置电机每转的脉冲数和位移，如图 4.3.18 所示。

Set the number of pulses per rotation and the displacement of the motor according to the parameters of the stepping motor driver，as shown in Figure 4.3.18.

4. 位置限制设置

4. Position limit settings

设置硬件限位开关的输入及电平，如图 4.3.19 所示。

Set the input and level of the hardware limit switch，as shown in Figure 4.3.19.

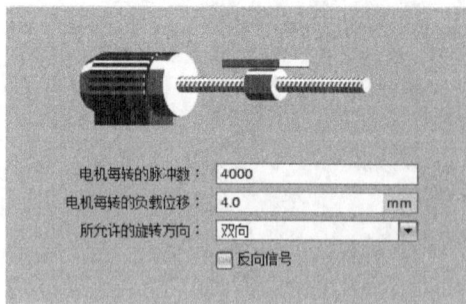

图 4.3.18　轴工艺对象的机械设置

Figure 4.3.18　Mechanical setting of shaft process object

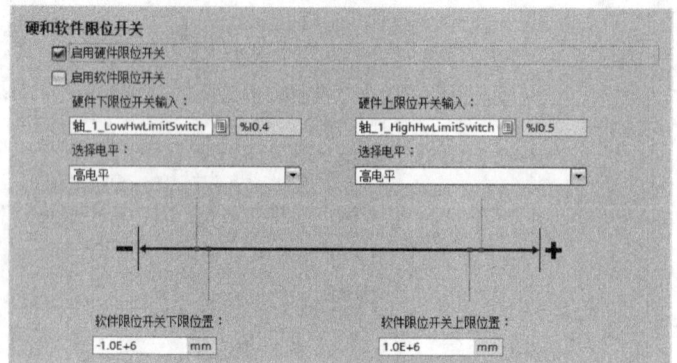

图 4.3.19　轴工艺对象的位置限制设置

Figure 4.3.19　Position limit settings for axis process objects

5. 主动回原点设置

5. Active return setting

分析控制要求，设置逼近/回原点的方向为负方向，参考点开关一侧选择下侧。由于伺服电机的外部接线有限位设置，因此不设置允许硬件限位开关处反转，如图 4.3.20 所示。

Analyze the control requirements，set the direction of approximation/return as negative direction，and select the lower side as one side of reference point switch. Because the external wiring of the servo motor is set with position limit switch，no setting is allowed for reverse rotation at the hardware limit switch，as shown in Figure 4.3.20.

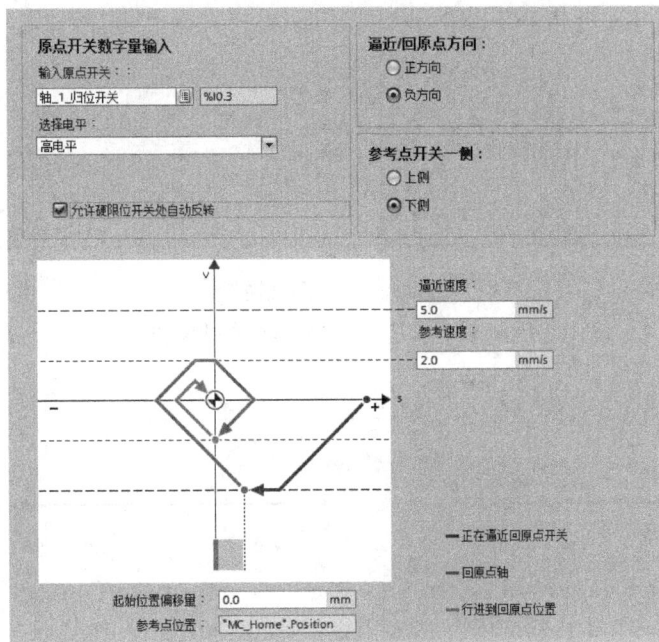

图 4.3.20　主动回原点设置

Figure 4.3.20　Active return setting.

6. 调试

6. Commissioning

选择调试工具，将 PLC 转入在线状态，可以打开轴控制面板，激活调试轴命令，对轴进行回原点功能的调试，如图 4.3.21 所示。注意，只有在程序块中启用/禁用轴指令没有激活的情况下才能调试轴。

Select the commissioning tool and put the PLC into online state. Open the shaft control panel，activate the commissioning shaft command and debug the shaft function of returning to the origin，as shown in Figure 4.3.21. Note that the shaft can only be debugged if the shaft enable/disable instruction is not active in the block.

（六）编写梯形图程序

（Ⅵ）To write the ladder logic program

先启用轴，用系统存储器中的常开位作为 MC_Power 指令的使能信号。设置回原点指令的归位模式为 3，当按下启动回原点按钮时，执行回原点指令。

Enable the shaft first，and use the normally open bit in system memory as the enable signal for the MC_Power instruction. Set the homing mode of the return instruction as 3，and execute the return instruction when the Return button is pressed.

梯形图程序如图 4.3.22 所示。

The ladder diagram program is shown in Figure 4.3.22.

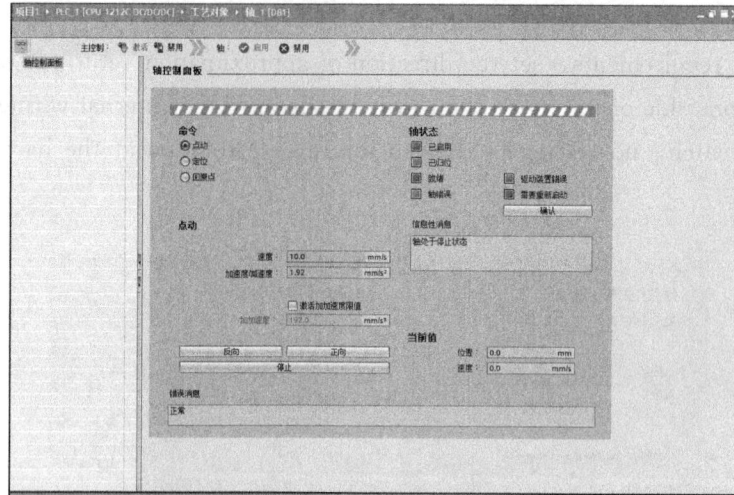

图 4.3.21 调试轴

Figure 4.3.21 Commissioning of shaft

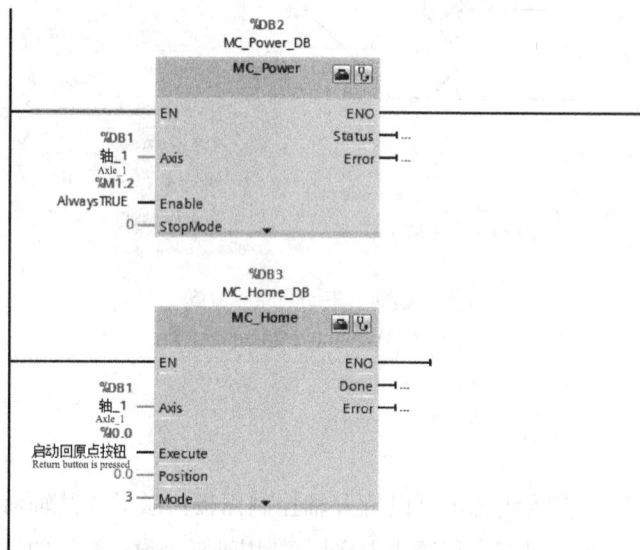

图 4.3.22 梯形图程序

Figure 4.3.22 Ladder logic program

（七）调试

（Ⅶ）Debugging

完成以上步骤后，先手动将小车调整到前图 4.3.1 (a) 所示位置，按下启动回原点按钮，观察小车的运动情况，不断调试，直到小车能停在原点开关位置为止。

After completing the above steps, first manually adjust the trolley to the position shown in Figure 4.3.1 (a), press the Return button, observe the movement of the trolley, and continuously make commissioning until the trolley can stop at the position of origin switch.

任务四 交流伺服电机的小车相对定位运动控制

Task Ⅳ Relative Positioning Motion Control of Trolley for AC Servo Motor

一、任务分析
Ⅰ. Task Analysis

用西门子 S7-1200 系列 PLC、伺服电机驱动器、伺服电机和传动机构对小车进行控制，控制要求如下：

Use Siemens S7-1200 series PLC, servo motor driver, servo motor and drive mechanism to control the trolley. The control requirements are as follows:

（1）小车通过丝杠由伺服电机带动停止在某个位置，A 点为原点位置，安装有原点传感器 SQ1。SQ2 和 SQ3 是光电传感器，为小车提供限位保护。SQ4 与 SQ5 是行程开关，也为小车提供限位保护。

(1) A trolley is driven by a servo motor to stop at a certain position through a lead screw, a point A is the origin position, and an origin sensor SQ1 is installed. SQ2 and SQ3 are photoelectric sensors that provide limit protection for the trolley. SQ4 and SQ5 are travel switches that also provide limit protection for the trolley.

（2）按下 SB1 按钮后，小车可从任意位置回到原点 A，停在如图 4.4.1 所示的位置。接着，小车开始按照一定规则做直线运动：先以 10mm/s 的速度向正方向运行 25mm，接着再以 8mm/s 的速度向正方向运行 45mm，停留 5s 后，再以 15mm/s 的速度向反方向运行 60mm，最后以 5mm/s 的速度回原点 A。伺服电机每旋转一圈，小车行进 4mm。

(2) When the SB1 button is pressed, the trolley can return to origin A from any position and stop at the position shown in Figure 4.4.1. Then the trolley began to move in a straight line according to certain rules: At the speed of 10mm/s, it runs 25mm in the positive direction, then at the speed of 8mm/s, it runs 45mm in the positive direction, after staying for 5s, it runs 60mm in the reverse direction at the speed of 15mm/s, and finally it returns to the origin A at the speed of 5mm/s. For each rotation of the servo motor, the trolley travels 4mm.

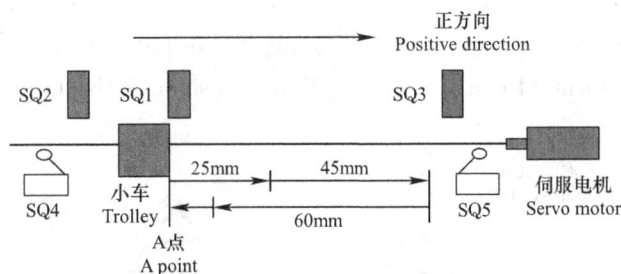

图 4.4.1 小车运动示意图

Figure 4.4.1 Schematic diagram of trolley movement

（3）按下 SB3 按钮，小车减速停止。再次按下 SB1 后，小车重新回到原点开始上述运行过程。

(3) Press the SB3 button and the trolley will slow down and stop. When SB1 is pressed again, the trolley returns to its origin position and begins the operation described above.

为了完成本任务，需要在理解相对定位和绝对定位的基础上学习 S7-1200 PLC 以相对方式定位轴指令的使用方法。

In order to accomplish this task, it is necessary to learn how to use the S7-1200 PLC relative positioning axis instruction based on the understanding of relative positioning and absolute positioning.

二、相关知识技能

Ⅱ. Relevant Knowledge & Skills

（一）相对定位和绝对定位

（Ⅰ）Relative positioning and absolute positioning

相对定位是以当前位置为起点，以指定的方向运动指定的位移的定位，与当前位置有关，与原点（参考点）的位置无关；绝对定位是以原点（参考点）为基准，运动到指定位置的定位，与当前位置无关，与起点位置有关。

Relative positioning: take the current position as the start point and move for specified displacement in the specified direction, which is related to the current position, but irrelated to the position of the origin point (reference point); Absolute positioning: move to a specified position based on the origin point (reference point), which is irrelated to the current position, but related to the starting position.

类比地理学中，海拔是指地面某个地点或者地理事物高出或者低于海平面的垂直距离，也就是说，地球上所有的高度都可以用海平面作为参考点来衡量。例如，一座山山顶的高度是海拔150m，山脚的高度是海拔50m，这是采用绝对定位来表达的，它们都是以参考点（也就是海平面）为基准来衡量的。通过比较两个点海拔数值的大小，可以知道它们谁更高，也就可以确定移动的方向。使用绝对定位时，必须指定一个参考点。假如有一个登山者在山脚，需要爬到山顶，那么以他现在的位置来看，他需要向上攀登100m才能到达，这就是相对定位。相对定位是以（登山者）当前的位置为参考点，以指定的方向（向上攀登）和指定的位移（100m）来进行定位的方法。

In terms of geography, altitude is the vertical distance above or below sea level at a point on the surface of the earth, that is, all heights on the earth can be measured by sea level as a reference point. For example, the height of a mountain top is 150m above sea level and the height of its foot is 50m above sea level, which is expressed in absolute positioning and is measured by reference point (i. e. sea level). By comparing the altitude values of the two points, you can know which is higher and determine the direction of movement. When using absolute positioning, you must specify a reference point. If there is a climber at the foot of the mountain, who need to climb to the top of the mountain, then in his current position, he needs to climb up 100m to reach the top, this is the relative positioning. Relative positioning is a method of positioning in a specified direction (climbing up) and a specified displacement (100m), using the current position of the (climber) as a reference point.

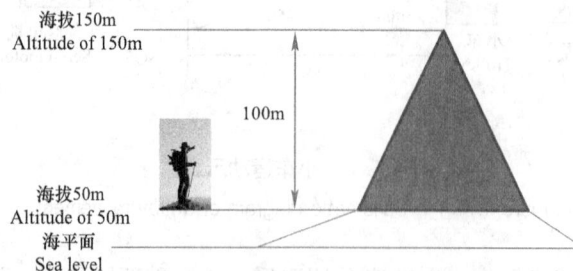

图4.4.2 相对定位与绝对定位的类比

Figure 4.4.2 Analogy between relative positioning and absolute positioning

（二）以相对方式定位轴指令

（Ⅱ）Axis positioning instructions in relative manner

通过运动控制指令MC_MoveRelative，启动相对于起始位置的定位运动。

The positioning movement relative to the start position is initiated by the motion control instruction MC _ MoveRelative.

为了使用 MC _ MoveRelative 指令，轴工艺对象必须正确组态，且轴已启用。MC _ MoveRelative 指令的参数见表 4.4.1。

In order to use the MC _ MoveRelative instruction，the axis process object must be configured correctly and the axis enabled. The parameters of the MC _ MoveRelative instruction is shown in Table 4.4.1.

MC _ MoveRelative 指令的参数 表 4.4.1

The parameters of the MC _ MoveRelative instruction Table 4.4.1

参数 Parameter	声明 Statement	数据类型 Data type	说明 Description
Axis	INPUT	TO _ Axis	轴工艺对象 Shaft process object
Execute	INPUT	Bool	出现上升沿时开始任务（默认值：False） Start task on rising edge (default：False)
Distance	INPUT	Real	定位操作的移动距离（默认值：0.0） Movement distance for positioning operation (default：0.0) 限值 Limit $-1.0e^{12} \leqslant$ 距离 $\leqslant 1.0e^{12}$ $-1.0e^{12} \leqslant Distance \leqslant 1.0e^{12}$
Velocity	INPUT	Real	轴的速度（默认值：10.0） Shaft speed (Default value：10.0) 由于所组态的加速度和减速度以及要途经的距离等原因，不会始终保持这一速度 This speed is not always maintained due to the acceleration and deceleration of the configuration and the distance to be traveled 限值 Limit 启动/停止速度 $\leqslant$ Velocity $\leqslant$ 最大速度 Start/stop speed $\leqslant$ Velocity $\leqslant$ Maximum speed
Done	OUTPUT	Bool	TRUE＝目标位置已到达 TRUE＝Target position reached
Busy	OUTPUT	Bool	TRUE＝正在执行任务 TRUE＝Executing tasks
CommandAborted	OUTPUT	Bool	TRUE＝任务在执行期间被另一任务中止 TRUE＝Task stopped by another task during execution
Error	OUTPUT	Bool	FALSE：无错误 FALSE：No Errors TRUE：任务执行期间出错 TRUE：An error occurred during task execution 出错原因可在"ErrorID"和"ErrorInfo"参数中找到 The cause of the error can be found in the "ErrorID" and "ErrorInfo" parameters
ErrorID	OUTPUT	Word	参数"Error"的错误 ID Error ID for the "Error" parameters
ErrorInfo	OUTPUT	Word	参数"ErrorID"的错误信息 ID Error information ID for the "ErrorID" parameters

三、任务实施

Ⅲ．Task Implementation

（一）连接 I/O 设备

（Ⅰ）To connect I/O device.

按照控制要求，I/O 接线图如图 4.4.3 所示。

Based on the control requirements，the I/O wiring diagram is shown in Figure 4. 4. 3.

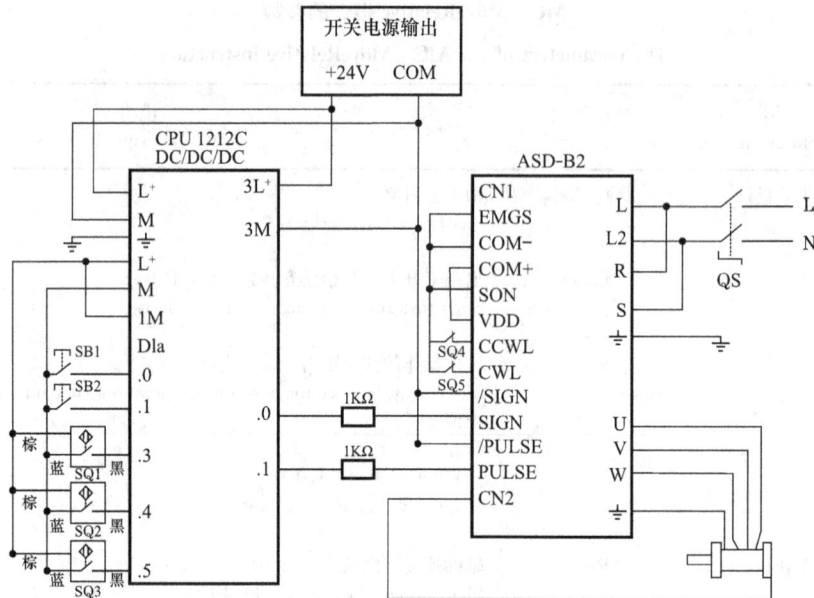

图 4.4.3　I/O 接线图

Figure 4. 4. 3　I/O wiring diagram

（二）调整伺服电机驱动器

（Ⅱ）Adjust servo motor driver

调整伺服电机驱动器的参数如表 4.4.2 所示。

The parameters for adjusting the servo motor driver is shown in Table 4. 4. 2.

伺服电机的参数设置　　　　　表 4. 4. 2

Parameter setting of servo motor　　　　　Table 4. 4. 2

参数 Parameter	设置值 Set value
P0-02	0
P1-00	102
P1-01	000
P1-44	40
P1-45	1
P2-00	35
P2-02	50

（三）在 TIA 博途中新建项目和设备组态

（Ⅲ）To create a new item and device configuration in TIA Portal

按照项目一中的步骤在 TIA 博途中新建项目和设备组态。由于编程中需要用到系统存储器，因此需要在 CPU 的属性中勾选启用系统存储器字节。

Create the new item and configure the device in TIA Portal by the steps in Item I. Because system memory is required for programming，the Enable System Memory byte needs to be checked in the properties of the CPU.

（四）建立变量表

（Ⅳ）To create the variable table

按照控制要求，定义变量表如图 4.4.4 所示。

Define the variable table as shown in Figure 4.4.4 based on the control requirements.

		名称	变量表	数据类型	地址	保持	在 H...	可从 ...	注释
1		System_Byte	默认变量表	Byte	%MB1		☑	☑	
2		FirstScan	默认变量表	Bool	%M1.0		☑	☑	
3		DiagStatusUpdate	默认变量表	Bool	%M1.1		☑	☑	
4		AlwaysTRUE	默认变量表	Bool	%M1.2		☑	☑	
5		AlwaysFALSE	默认变量表	Bool	%M1.3		☑	☑	
6		SB1启动按钮	默认变量表	Bool	%I0.0		☑	☑	
7		Sb2暂停按钮	默认变量表	Bool	%I0.1		☑	☑	

图 4.4.4　变量表

Figure 4.4.4　Variable table

（五）轴工艺对象组态及调试

（Ⅴ）Configuration and commissioning of axis process object

在 TIA 博途项目中新建轴工艺对象，在组态窗口中设置本任务所需的参数。本任务需要设置的内容有：

Create a new axis process object in the TIA Portal project and set the parameters required for this task in the configuration window. The task contents to be set are as follows：

1. 常规设置，如图 4.4.5 所示

1. General setting as shown in Figure 4.4.5

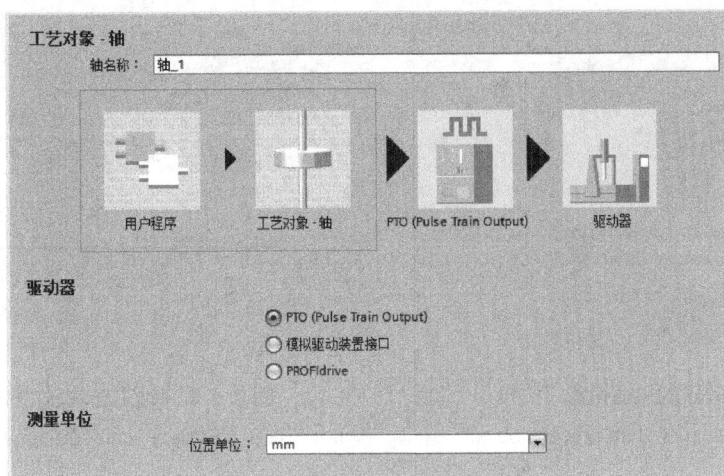

图 4.4.5　轴工艺对象的常规设置

Figure 4.4.5　General setting of shaft process object

2. 驱动器设置，如图 4.4.6 所示

2. Driver setting as shown in Figure 4.4.6

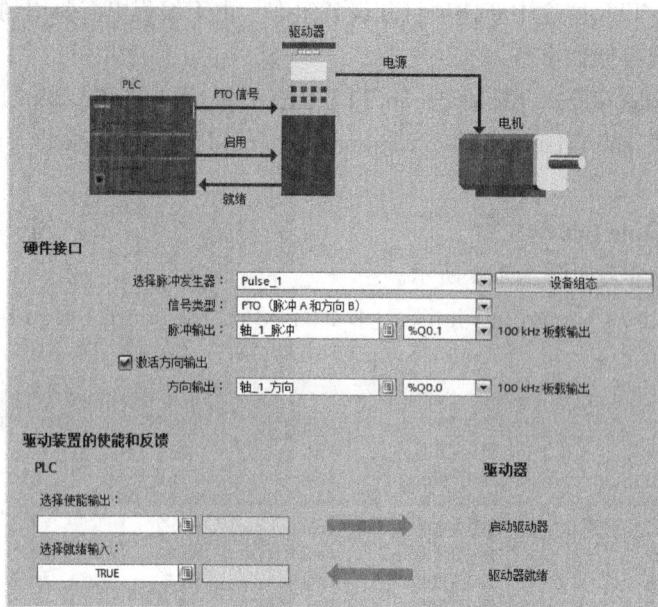

图 4.4.6　轴工艺对象的驱动器设置

Figure 4.4.6　Drive settings for shaft process object

3. 机械设置

3. Mechanical setting

根据步进电机驱动器的参数设置电机每转的脉冲数和位移，如图 4.4.7 所示。

Set the number of pulses per rotation and the displacement of the motor according to the parameters of the stepping motor driver, as shown in Figure 4.4.7.

4. 位置限制设置

4. Position limit settings

设置硬件限位开关的输入及电平，如图 4.4.8 所示。

Set the input and level of the hardware limit switch, as shown in Figure 4.4.8.

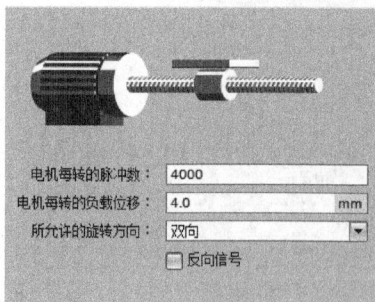

图 4.4.7　轴工艺对象的机械设置

Figure 4.4.7　Mechanical setting of shaft process object

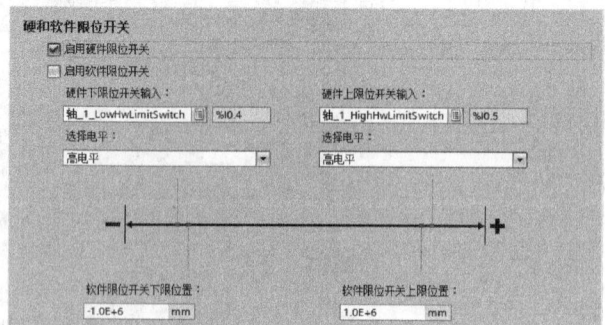

图 4.4.8　轴工艺对象的位置限制设置

Figure 4.4.8　Position limit settings for axis process objects

5. 主动回原点设置

5. Active return setting

分析控制要求，设置逼近/回原点的方向为负方向，参考点开关一侧选择下侧。由于伺服电机的外部接线有限位设置，因此不设置允许硬件限位开关处反转，如图 4.4.9 所示。

Analyze the control requirements，set the direction of approximation/return as negative direction，and select the lower side as one side of reference point switch. Because the external wiring of the servo motor is set with position limit switch，no setting is allowed for reverse rotation at the hardware limit switch，as shown in Figure 4.4.9.

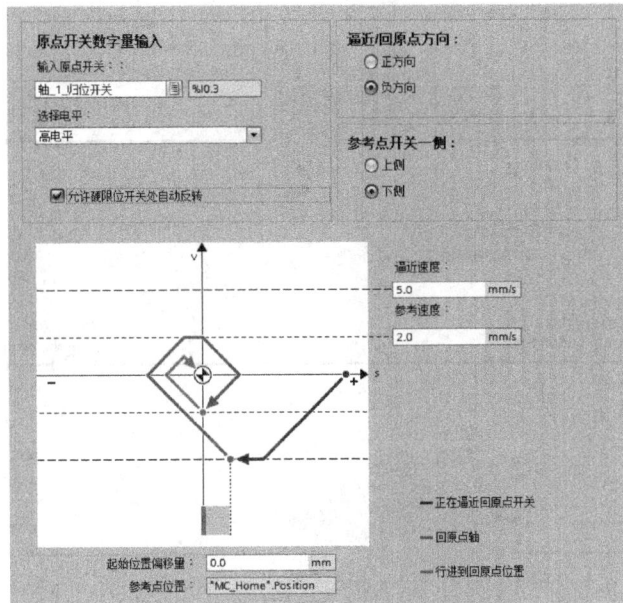

图 4.4.9　主动回原点设置

Figure 4.4.9　Active return setting

6. 调试

6. Commissioning

选择调试工具，将 PLC 转入在线状态，打开轴控制面板，激活调试轴命令，对轴进行回原点功能的调试，如图 4.4.10 所示。注意，只有在程序块中启用/禁用轴指令没有激活的情况下才能调试轴。

Select the commissioning tool and put the PLC into online state. Open the shaft control panel，activate the commissioning shaft command and debug the shaft function of returning to the origin，as shown in Figure 4.4.10. Note that the shaft can only be debugged if the shaft enable/disable instruction is not active in the block.

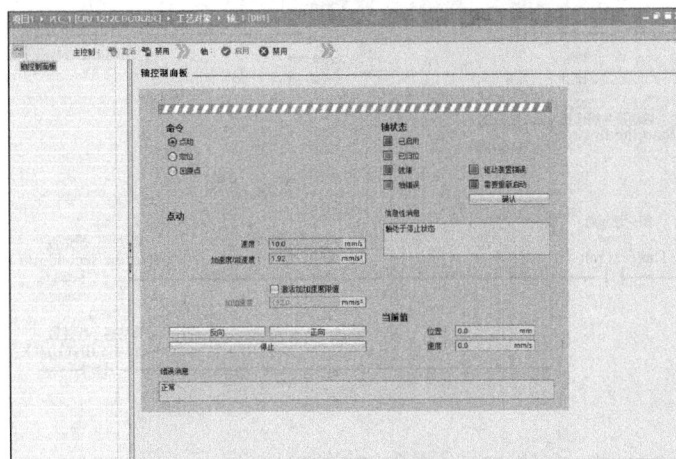

图 4.4.10　调试轴

Figure 4.4.10　Commissioning of shaft

（六）编写梯形图程序
（Ⅵ）To write the ladder logic program

梯形图程序如图 4.4.11 所示。

The ladder diagram program is shown in Figure 4.4.11.

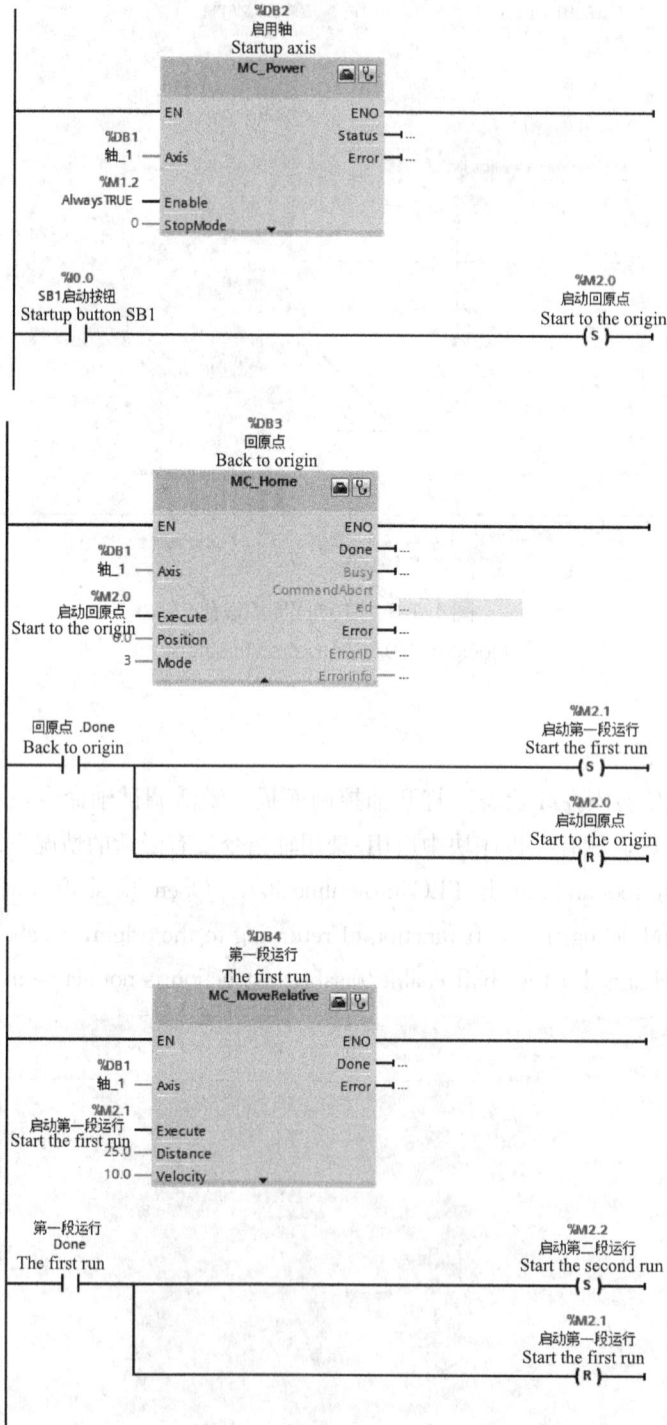

图 4.4.11　梯形图程序（一）

Figure 4.4.11　Ladder logic program（一）

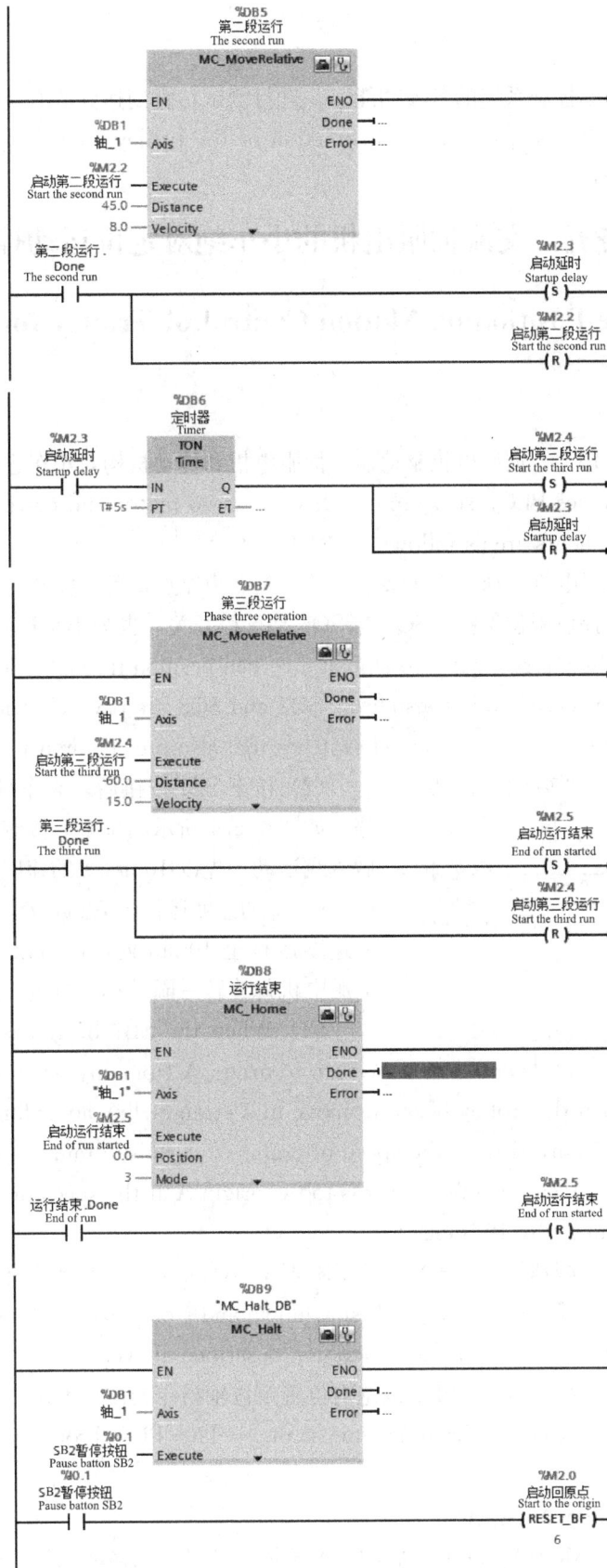

图 4.4.11 梯形图程序（二）

Figure 4.4.11 Ladder logic program（二）

（七）调试

（Ⅶ）**Debugging**

完成以上步骤后，对小车控制系统的各项功能一一进行测试，直到达到所有控制要求为止。

After completing the above steps, test each function of the trolley control system one by one until all control requirements are met.

任务五 交流伺服电机的小车绝对定位运动控制

Task Ⅴ Absolute Positioning Motion Control of Trolley for AC Servo Motor

一、任务分析

Ⅰ. Task Analysis

用西门子 S7-1200 系列 PLC、伺服电机驱动器、伺服电机和传动机构对小车进行控制，控制要求如下：

Use Siemens S7-1200 series PLC, servo motor driver, servo motor and drive mechanism to control the trolley. The control requirements are as follows:

（1）小车通过丝杠由伺服电机带动停止在某个位置，A点为原点位置，安装有原点传感器 SQ1。SQ2 和 SQ3 是光电传感器，为小车提供限位保护。SQ4 与 SQ5 是行程开关，也为小车提供限位保护。

（1）A trolley is driven by a servo motor to stop at a certain position through a lead screw, a point A is the origin position, and an origin sensor SQ1 is installed. SQ2 and SQ3 are photoelectric sensors that provide limit protection for the trolley. SQ4 and SQ5 are travel switches that also provide limit protection for the trolley.

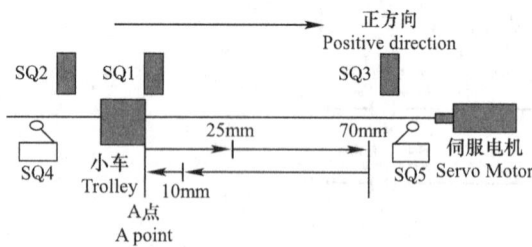

图 4.5.1 小车运动示意图

Figure 4.5.1 Schematic diagram of trolley movement

（2）按下 SB1 按钮后，小车可从任意位置回到原点 A，停在如图 4.5.1 所示的位置。接着，小车开始按照一定规则做直线运动：先以 10mm/s 的速度运行至 25mm 处，接着再以 8mm/s 的速度运行至 70mm 处，停留 5s 后，再以 15mm/s 的速度运行至 10mm 处，最后以 5mm/s 的速度回原点 A。伺服电机每旋转一圈，小车行进 4mm。

（2）When the SB1 button is pressed, the trolley can return to origin A from any position and stop at the position shown in Figure 4.5.1. Then the trolley began to move in a straight line according to certain rules: At the speed of 10mm/s, it runs 25mm, then at the speed of 8mm/s, it runs 70mm, after staying for 5s, it runs 10mm at the speed of 15mm/s, and finally it returns to the origin A at the speed of 5mm/s. For each rotation of the servo motor, the trolley travels 4mm.

（3）按下 SB2 按钮，小车减速停止。再次按下 SB1 后，小车重新回到原点开始上述运行过程。

（3）Press the SB2 button and the trolley will slow down and stop. When SB1 is pressed again, the trolley returns to its origin position and begins the operation described above.

为了完成本任务，需要学习 S7-1200 PLC 以绝对位置定位轴指令的使用方法。

To accomplish this task, you need to learn how to use the S7-1200 PLC absolute positioning axis instruction.

二、相关知识技能

Ⅱ. Relevant Knowledge & Skills

通过运动控制指令 MC_MoveAbsolute，启动轴定位运动，以将轴移动到某个绝对位置。

The axis positioning movement is initiated to move the axis to an absolute position via the motion control instruction MC_MoveAbsolute.

为了使用 MC_MoveAbsolute 指令，轴工艺对象必须正确组态，且轴已启用并回到原点。MC_Move-Relative 指令的参数见表 4.5.1。

In order to use the MC_MoveAbsolute instruction，the axis process object must be configured correctly and the axis enabled and returned to its origin. The parameters of the MC_MoveRelative instruction are shown in Table 4.5.1.

MC_MoveAbsolute 指令的参数   表 4.5.1

Parameters of the MC_MoveAbsolute instruction   Table 4.5.1

参数 Parameter	声明 Statement	数据类型 Data type	说明 Description
Axis	INPUT	TO_Axis	轴工艺对象 Shaft process object
Execute	INPUT	BOOL	出现上升沿时开始任务（默认值：False） Start task on rising edge（default：False）
Position	INPUT	Real	绝对目标位置（默认值：0.0） Absolute target location（default：0.0） 限值 Limit $-1.0e^{12} \leqslant$ 距离 $\leqslant 1.0e^{12}$ $-1.0e^{12} \leqslant Distance \leqslant 1.0e^{12}$
Velocity	INPUT	Real	轴的速度（默认值：10.0） Shaft speed（Default value：10.0） 由于所组态的加速度和减速度以及要途经的距离等原因，不会始终保持这一速度 This speed is not always maintained due to the acceleration and deceleration of the configuration and the distance to be traveled 限值 Limit 启动/停止速度 $\leqslant$ Velocity $\leqslant$ 最大速度 Start/stop speed $\leqslant$ Velocity $\leqslant$ Maximum speed
Done	OUTPUT	Bool	TRUE=达到绝对目标位置 TRUE= absolute target position reached
Busy	OUTPUT	Bool	TRUE=正在执行任务 TRUE=Executing tasks
CommandAborted	OUTPUT	Bool	TRUE=任务在执行期间被另一任务中止 TRUE=Task stopped by another task during execution
Error	OUTPUT	BOOL	FALSE：无错误 FALSE：No Errors TRUE：任务执行期间出错 TRUE：An error occurred during task execution 出错原因可在 "ErrorID" 和 "ErrorInfo" 参数中找到 The cause of the error can be found in the "ErrorID" and "ErrorInfo" parameters
ErrorID	OUTPUT	Word	参数 "Error" 的错误 ID Error ID for the "Error" parameters
ErrorInfo	OUTPUT	Word	参数 "ErrorID" 的错误信息 ID Error information ID for the "ErrorID" parameters

三、任务实施
Ⅲ．Task Implementation
（一）连接 I/O 设备
（Ⅰ）To connect I/O device.

按照控制要求，I/O 接线图如图 4.5.2 所示。

Based on the control requirements, the I/O wiring diagram is shown in Figure 4.5.2.

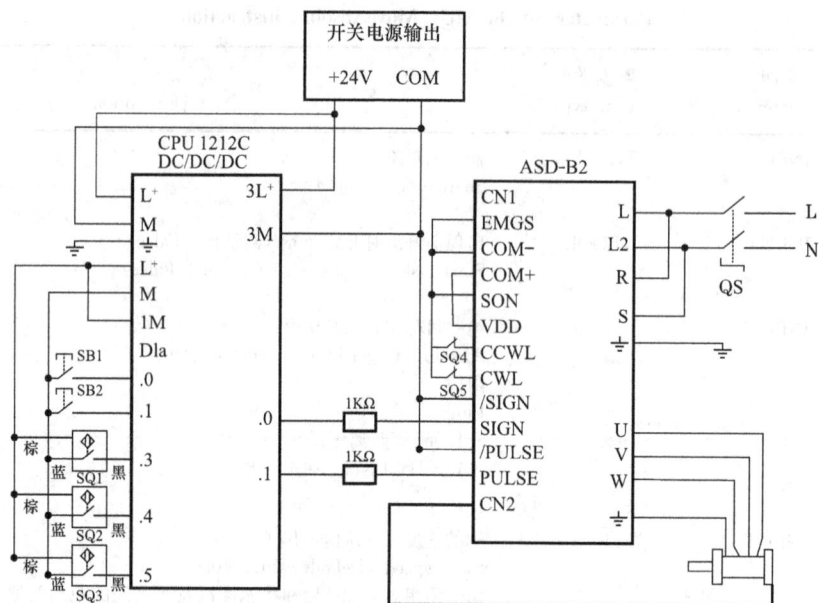

图 4.5.2　I/O 接线图

Figure 4.5.2　I/O wiring diagram

（二）调整伺服电机驱动器
（Ⅱ）Adjust servo motor driver

调整伺服电机驱动器的参数见表 4.5.2。

The parameters for adjusting the servo motor driver are shown in Table 4.5.2.

伺服电机的参数设置　　　　　　　　　　　　　　　　表 4.5.2

Parameter setting of servo motor　　　　　　　　　　Table 4.5.2

参数 Parameter	设置值 Set value
P0-02	0
P1-00	102
P1-01	000
P1-44	40
P1-45	1
P2-00	35
P2-02	50

（三）在 TIA 博途中新建项目和设备组态

（Ⅲ）To create a new item and device configuration in TIA Portal

按照项目一中的步骤在 TIA 博途中新建项目和设备组态。由于编程中需要用到系统存储器，因此需要在 CPU 的属性中勾选启用系统存储器字节。

Create the new item and configure the device in TIA Portal by the steps in Item Ⅰ. Because system memory is required for programming，the Enable System Memory byte needs to be checked in the properties of the CPU.

（四）建立变量表

（Ⅳ）To create the variable table

按照控制要求，定义变量表如图 4.5.3 所示。

Define the variable table as shown in Figure 4.5.3 based on the control requirements.

		名称	变量表	数据类型	地址	保持	在 H...	可从 ...	注释
1		System_Byte	默认变量表	Byte	%MB1	☐	☑	☑	
2		FirstScan	默认变量表	Bool	%M1.0	☐	☑	☑	
3		DiagStatusUpdate	默认变量表	Bool	%M1.1	☐	☑	☑	
4		AlwaysTRUE	默认变量表	Bool	%M1.2	☐	☑	☑	
5		AlwaysFALSE	默认变量表	Bool	%M1.3	☐	☑	☑	
6		SB1启动按钮	默认变量表	Bool	%I0.0	☐	☑	☑	
7		Sb2暂停按钮	默认变量表	Bool	%I0.1	☐	☑	☑	

图 4.5.3 变量表

Figure 4.5.3 Variable table

（五）轴工艺对象组态及调试

（Ⅴ）Configuration and commissioning of axis process object

在 TIA 博途项目中新建轴工艺对象，在组态窗口中设置本任务所需的参数。本任务需要设置的内容有：

Create a new axis process object in the TIA Portal project and set the parameters required for this task in the configuration window. The task contents to be set are as follows：

1. 常规设置，如图 4.5.4 所示

1. General setting as shown in Figure 4.5.4

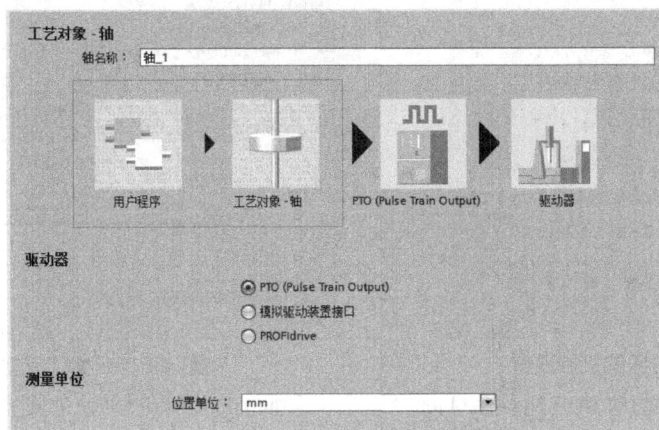

图 4.5.4 轴工艺对象的常规设置

Figure 4.5.4 General setting of shaft process object

2. 驱动器设置，如图 4.5.5 所示

2. Driver setting as shown in Figure 4.5.5

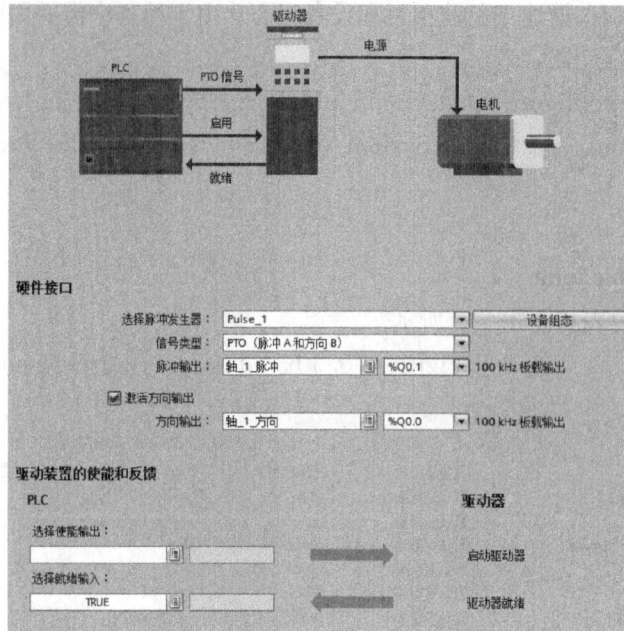

图 4.5.5　轴工艺对象的驱动器设置

Figure 4.5.5　Drive settings for shaft process object

3. 机械设置

3. Mechanical setting

根据步进电机驱动器的参数设置电机每转的脉冲数和位移，如图 4.5.6 所示。

Set the number of pulses per rotation and the displacement of the motor according to the parameters of the stepping motor driver，as shown in Figure 4.5.6.

4. 位置限制设置

4. Position limit settings

设置硬件限位开关的输入及电平，如图 4.5.7 所示。

Set the input and level of the hardware limit switch，as shown in Figure 4.5.7.

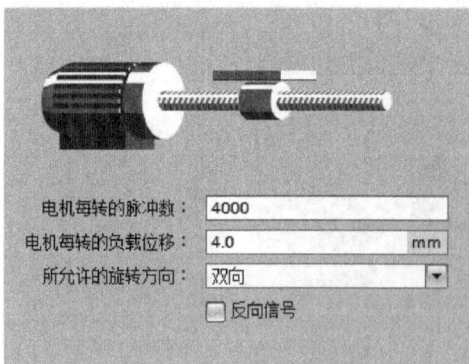

图 4.5.6　轴工艺对象的机械设置

Figure 4.5.6　Mechanical setting of shaft process object

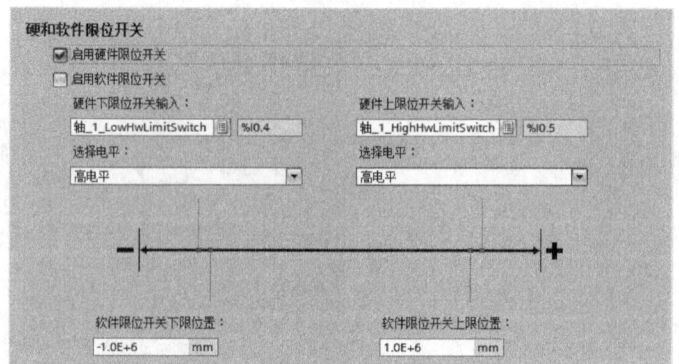

图 4.5.7　轴工艺对象的位置限制设置

Figure 4.5.7　Position limit settings for axis process objects

5. 主动回原点设置

5. Active return setting

分析控制要求，设置逼近/回原点的方向为负方向，参考点开关一侧选择下侧。由于伺服电机的外部接线有限位设置，因此不设置允许硬件限位开关处反转，如图4.5.8所示。

Analyze the control requirements，set the direction of approximation/return as negative direction，and select the lower side as one side of reference point switch. Because the external wiring of the servo motor is set with position limit switch，no setting is allowed for reverse rotation at the hardware limit switch，as shown in Figure 4.5.8.

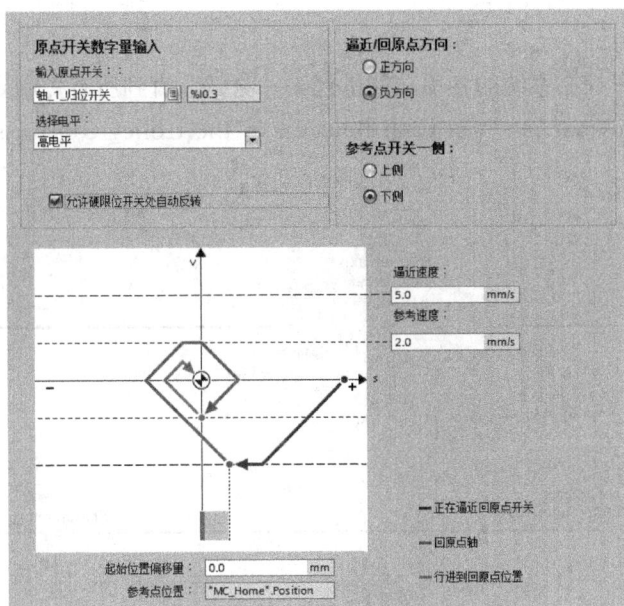

图4.5.8　主动回原点设置

Figure 4.5.8　Active return setting

6. 调试

6. Commissioning

选择调试工具，将PLC转入在线状态，可以打开轴控制面板，激活调试轴命令，对轴进行回原点功能的调试，如图4.5.9所示。注意，只有在程序块中启用/禁用轴指令没有激活的情况下才能调试轴。

Select the commissioning tool and put the PLC into online state. Open the shaft control panel，activate the commissioning shaft command and debug the shaft function of returning to the origin，as shown in Figure 4.5.9. Note that the shaft can only be debugged if the shaft enable/disable instruction is not active in the block.

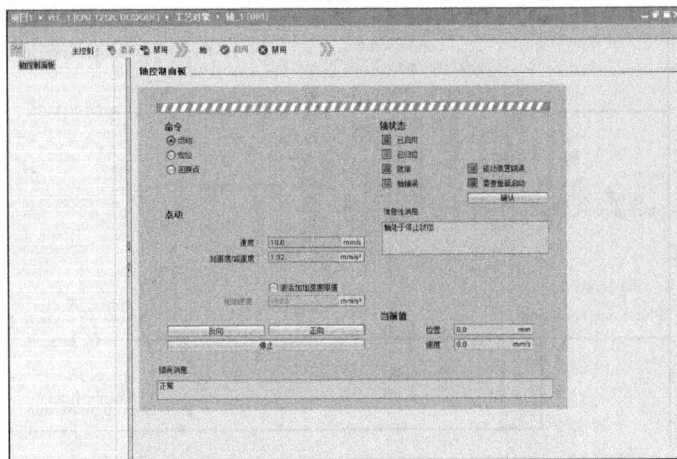

图4.5.9　调试轴

Figure 4.5.9　Commissioning of shaft

（六）编写梯形图程序

（Ⅵ）**To write the ladder logic program**

梯形图程序如图 4.5.10 所示。

The ladder diagram program is shown in Figure 4.5.10.

（七）调试

（Ⅶ）**Debugging**

完成以上步骤后，对小车控制系统的各项功能进行一一测试，直到所有控制要求达到为止。

After completing the above steps，test each function of the trolley control system one by one until all control requirements are met.

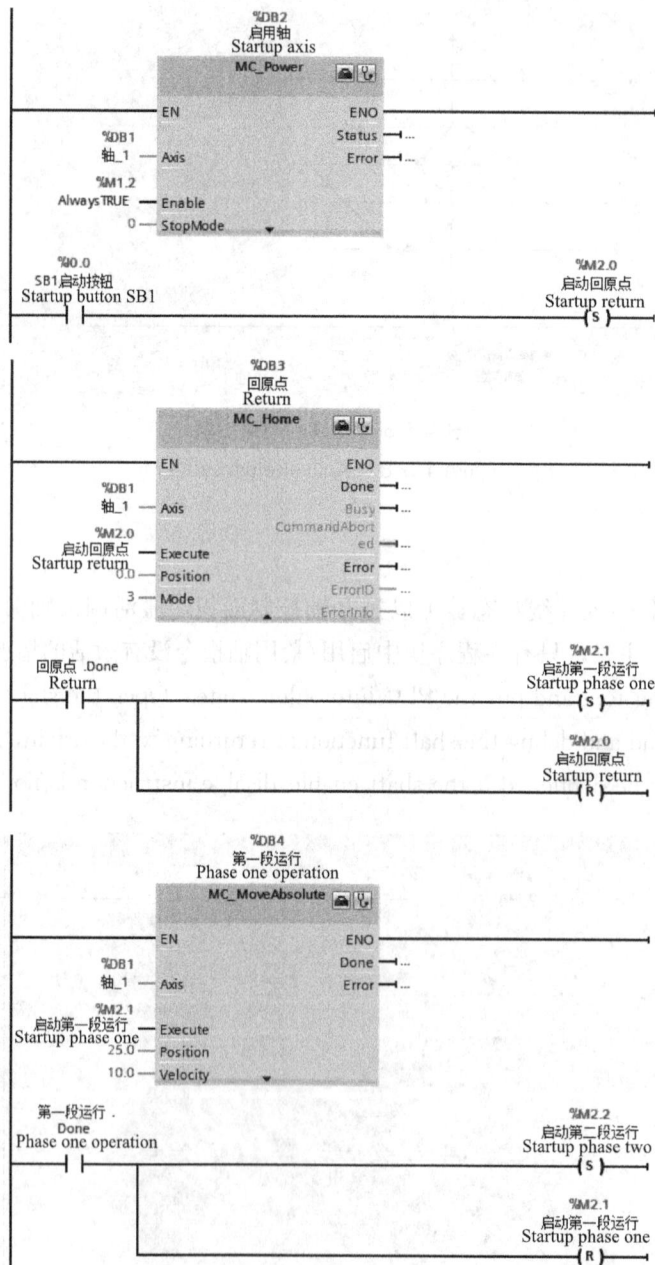

图 4.5.10　梯形图程序（一）

Figure 4.5.10　Ladder logic program（Ⅰ）

%DB5
第二段运行
The second run
MC_MoveAbsolute

EN   ENO
                     Done — ...
%DB1   Error — ...
轴_1 —— Axis
%M2.2
启动第二段运行
Start the second run —— Execute
         70.0 —— Position
          8.0 —— Velocity ▼

第二段运行   %M2.3
The second run   启动延时
Done   Startup delay
—| |—————————————————————————————————————(S)

                                      %M2.2
                                      启动第二段运行
                                      Start the second run
                 —————————————————————(R)

%DB6
定时器
Timer

%M2.3   TON   %M2.4
启动延时   Time   启动第三段运行
Startup delay   Start the third run
—| |————————— IN   Q ———————————(S)
        T#5s— PT   ET — ...
                                      %M2.3
                                      启动延时
                                      Startup delay
                 —————————————————————(R)

%DB7
第三段运行
The third run
MC_MoveAbsolute

EN   ENO
                     Done — ...
%DB1   Error — ...
轴_1 —— Axis
%M2.4
启动第三段运行
Start the third run —— Execute
         10.0 —— Position
         15.0 —— Velocity ▼

第三段运行   %M2.5
Done   启动运行结束
The third run   End of run started
—| |—————————————————————————————————————(S)

                                      %M2.4
                                      启动第三段运行
                                      Start the third run
                 —————————————————————(R)

%DB8
运行结束
End of run
MC_MoveAbsolute

EN   ENO
                     Done — ...
%DB1   Error — ...
轴_1 —— Axis
%M2.5
启动运行结束
End of run started —— Execute
          0.0 —— Position
          5.0 —— Velocity ▼

运行结束 .Done   %M2.5
End of run   启动运行结束
                                      End of run started
—| |—————————————————————————————————————(R)

%DB9
MC_Halt_DB
MC_Halt

EN   ENO
                     Done — ...
%DB1   Error — ...
轴_1 —— Axis
%I0.1
SB2暂停按钮
Pause button SB2 —— Execute ▼

%I0.1
SB2暂停按钮   %M2.0
Pause button SB2   启动回原点
                                      Start to the origin
—| |—————————————————————————————————————(RESET_BF)
                                              6

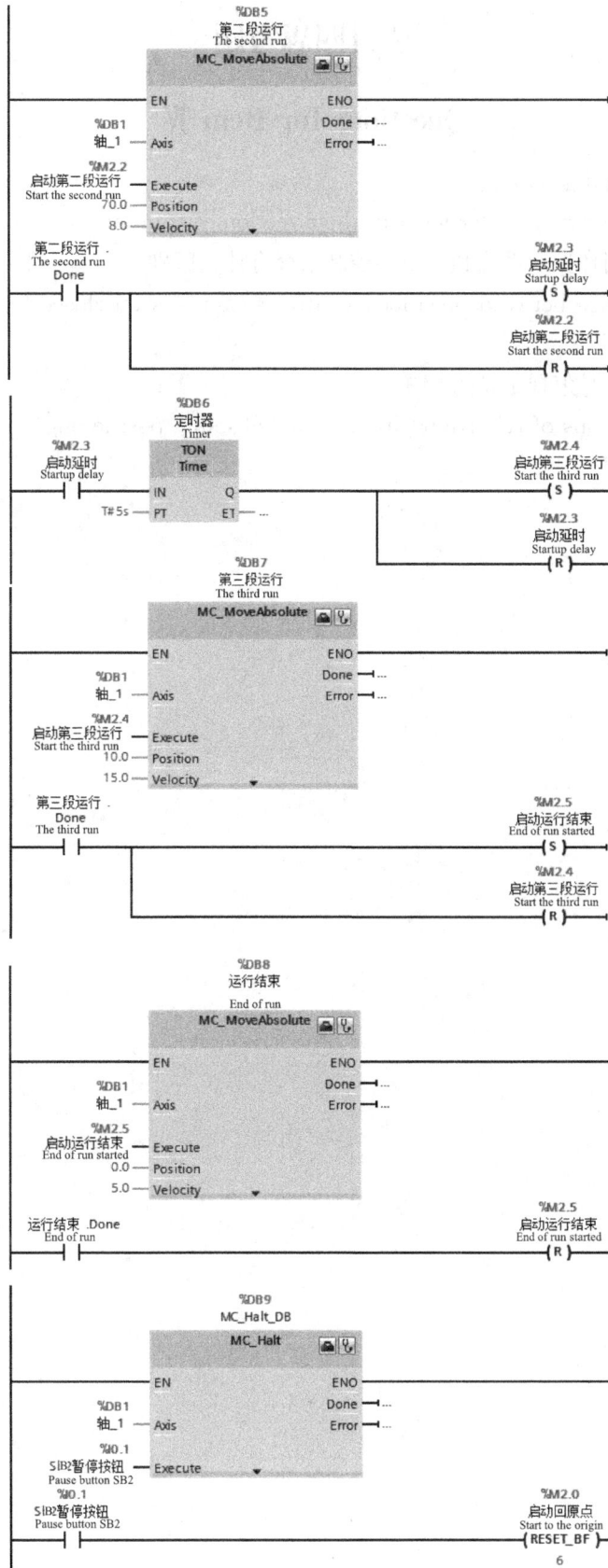

图 4.5.10　梯形图程序（二）

Figure 4.5.10　Ladder logic program（Ⅱ）

项目四思考题

Questions for Item Ⅳ

1. 光电编码器是怎样分辨旋转方向的？

1. How does a photoelectric encoder identify the direction of rotation?

2. 开环控制系统、半闭环控制系统和全闭环控制系统有什么区别？

2. What is the difference between open-loop control system，semi-closed-loop control system and full closed-loop control system?

3. 相对定位与绝对定位是怎样定义的？

3. What are the definitions of relative positioning and absolute positioning?

项目五 数控机床电气故障维修
Item V Electrical Trouble Maintenance of CNC Machine Tool

任务一 数控机床概述及基本操作

Task I General Introduction and Basic Operation of CNC Machine Tools

一、任务分析
I. Task Analysis

在进行数控机床的维修过程中，往往需要对数控机床的相关功能、参数、PMC 信号及程序进行调阅确认，以便判断分析故障产生的原因。这就要求维修维护人员具有以下数控机床基本操作能力：

In the maintenance process of CNC machine tools, it is often necessary to read and confirm the relevant functions, parameters, PMC signals and programs of CNC machine tools in order to judge and analyze the causes of the failure. This requires maintenance personnel to have the following basic operating capabilities of CNC machine tools:

（1）熟悉数控系统界面、MDI 面板及软键的功能作用。

（1）Familiar with CNC system interface, MDI panel and soft key function.

（2）能够通过机床面板调阅数控系统参数、PMC 程序及参数。

（2）Able to access CNC system parameters, PMC programs and parameters through the machine panel.

（3）能够对机床进行一般的操作并利用 MDI 模式进行简单机床功能测试。

（3）Able to operate the machine tool in general and use MDI mode to test the function of simple machine tool.

为了完成本任务，我们选择 YL-558 型数控车床实验台作为实训设备，该实验台配备有 FANUC 0i Mate-TD 数控系统、三菱 E740 通用变频器（主轴驱动）、四工位电动刀架及十字滑台。通过完成本任务，可以学会如何通过机床面板调阅系统参数、PMC 程序、PMC 参数，并能利用 MDI 模式进行简单机床功能测试。

In order to accomplish this task, we chose YL-558 CNC experimental platform as the training equipment. The experimental platform is equipped with FANUC 0i Mate-TD CNC system, Mitsubishi E740 universal frequency converter (spindle drive), four-station electric tool holder and cross slide table. Through the completion of this task, we can learn how to access system parameters, PMC programs, PMC parameters through the machine panel, and can use MDI mode for simple machine tool function testing.

二、相关知识技能

Ⅱ. Relevant Knowledge & Skills

（一）数控机床概述

（Ⅰ）Overview of CNC machine tools

1. 数控机床简介

1. Introduction to CNC machine tools

数控机床是一种装有程序控制系统的自动化机床，能够根据编好的加工程序控制机档运行加工工件。它是集机械、微电子、传感器、计算机、液压气动等多门高新技术为一体的高度自动化设备。

CNC machine tool is an automatic machine tool equipped with program control system, which can control the machine operation and process workpieces according to the programmed machining program. It isone highly-automated equipment combining the machinery, microelectronics, sensors, computers, hydro-pneumatic and other high-technologies.

与传统机床相比，数控机床具有以下特点：

Compared with traditional machine tools, CNC machine tools have the following characteristics:

（1）高度柔性：数控机床主要根据加工程序来进行零件加工，它可以完成很多传统机床无法加工的工艺。因此，数控机床适用于经常需要更换加工工艺的场合，例如单件、小批量产品的生产及新产品的开发，从而缩短了生产准备周期，节省了大量工艺装备费用。

（1）Highly flexible: CNC machine tool is mainly used to process parts based on the machining program, it can complete the process that many traditional machine tools cannot achieve. Therefore, CNC machine tools are suitable for the occasions where the machining process needs to be changed frequently, such as the production of single-piece and small-batch products and the development of new products, thus shortening the production preparation cycle and saving a lot of process equipment costs.

（2）加工精度高：数控机床的加工精度一般可达 0.05～0.1mm，因为它是按数字控制信号形式控制，系统每输出一个脉冲信号，机床移动部件移动一个脉冲当量，而且其进给过程的反向间隙及螺距误差可由系统进行补偿，因此数控机床定位精度比较高。

（2）High machining accuracy: The machining precision of CNC machine tool can reach 0.05—0.1 mm generally, because it is controlled in digital control signal form, every pulse signal is output by the system, the moving part of the machine tool moves one pulse equivalent, and the reverse clearance and pitch error of the feed process can be compensated by the system, so the positioning precision of CNC machine tool is relatively high.

（3）加工质量稳定可靠：同一台机床加工同一批零件，在相同加工条件下，使用相同的刀具和加工程序，刀具的移动轨迹完全相同，零件一致性好，质量稳定。

（3）Processing quality is stable and reliable: Under the same machining conditions, if using the same cutting tools and machining procedures, the cutting tool movement path is identical, parts consistency is good, and quality is stable when machining the same batch of parts with the same machine tool.

（4）生产效率高：与刀库配合加工，可实现一台机床进行多道工序的连续加工，减少半成品工序间周转时间，提高生产率。

（4）Higher productivity: Matching with the tool changer, one machine can realize the continuous machining of multiple processes, which reduces the turnover time between processes for semi-finished products, and improves the productivity.

（5）改善劳动条件：操作者只需做编辑加工程序、装卸工件、观测加工过程和成品检验等工作，劳动强度大大降低。

(5) Improve working conditions: Operators only need to do editing and processing procedures, loading and unloading workpieces, observation and processing and finished products inspection work, so labor intensity is greatly reduced.

2. 数控机床的组成

2. Composition of CNC machine tools

数控机床一般由计算机数控系统和机床本体两部分组成。其中计算机数控系统是由输入/输出装置、计算机数控装置、可编程控制器、主轴驱动系统和进给伺服驱动系统等组成的一个整体系统，如图 5.1.1 所示。

CNC machine tool is generally composed of computer CNC system and machine body. The CNC system consists of input/output device, CNC device, programmable controller, spindle drive system and feed servo drive system, as shown in Figure 5.1.1.

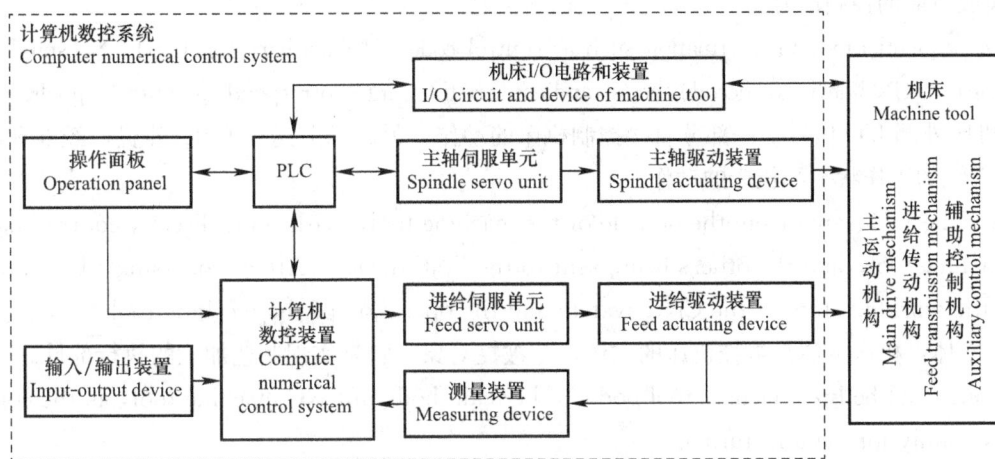

图 5.1.1　数控机床组成框图

Figure 5.1.1　Block diagram of CNC machine tool

（1）输入/输出装置：输入/输出装置用来完成数控加工程序的输入/输出、参数设定和状态显示等。

(1) Input output device: The input/output device is used to complete the input/output, parameter setting and status display of CNC machining program.

（2）数控装置：数控装置又称 CNC 装置，是数控机床的核心，它负责把输入装置送来的脉冲信号进行编译、运算和逻辑处理，然后将处理后的信息指令输出给伺服系统，让设备各部分进行规定的、有序的动作。

(2) Numerical control device: CNC device, also called numerical control device, is the core of CNC machine tool. It is responsible for compiling, calculating and logic processing the pulse signal from the input device, and then outputting the processed information instructions to the servo system, so that all parts of the equipment can perform prescribed and orderly actions.

（3）伺服驱动系统：伺服驱动系统把来自数控装置的微弱脉冲信号进行整形、放大后，转换成机床运动部件的机械位移。

(3) Servo drive system: The servo drive system shapes and amplifies the weak pulse signal from the CNC device, and then converts it into the mechanical displacement of the moving parts of the machine tool.

（4）位置反馈装置：位置反馈装置负责对数控机床各运动部件的实际位移进行检测，并把检测结果转换成电信号反馈给数控装置。数控系统将反馈信号与指令信号进行比较产生误差信号，以此控制机床纠正误差。

（4）Position feedback device：Position feedback device detects the actual displacement of each moving part of CNC machine tool，and transforms the detection result into electrical signal and then feed back signal to CNC device．CNC system compares the feedback signal with the instruction signal to generate error signal，so as to control the machine to correct the error.

（5）辅助控制装置：现代数控机床常用 PLC 与机床 I/O 电路和装置（由继电器、电磁阀、行程开关、接触器等组成）构成辅助控制装置，共同完成以下任务。

（5）Auxiliary control device：PLC and I/O circuits and devices（including relays，solenoid valves，travel switches，contactors，etc.）are commonly used in modern CNC machine tools to form auxiliary control devices，which will accomplish the following tasks together.

① 接受 CNC 装置的控制代码 M（辅助功能）、S（主轴功能）、T（刀具功能）等顺序动作信息，对其进行译码、转换成对应的控制信号。

① Receive sequential motion information such as control codes M（auxiliary function），S（spindle function），T（tool function）of CNC devices，decode them，and convert them into corresponding control signals.

② 接受机床外部 I/O 信号，一部分直接控制机床的动作，另一部分送往 CNC 装置，经处理后，输出指令控制 CNC 系统的工作状态和机床的动作。

② Receive the I/O signal from the outside of the machine tool，with some directly controlling the operation of the machine tool，and the others being sent to the CNC device．After processing，the output instruction controls the working state of the CNC system and the operation of the machine tool.

（6）机床本体：机床本体是数控机床的主体，是被控对象，它是实现制造加工的执行部件。

（6）Machine tool body：Machine tool body is the main body of CNC machine tool，is the controlled object，and it is mainly for manufacturing.

（二）数控机床的分类

（Ⅱ）Classification of CNC machine tools

数控机床的种类很多，可以按不同的方法对数控机床进行分类：按工艺用途分类和按控制方式分类等。

There are many kinds of CNC machine tools，which can be classified according to different methods：Classification by process use and by control mode，etc.

1. 按工艺用途分类

1. Classification by process use

① 普通数控机床：普通数控机床一般指在加工工艺过程中的一个工序上实现数字控制的自动化机床，如数控铣床、数控车床、数控钻床、数控磨床与数控齿轮加工机床等。普通数控机床在自动化程度上还不够完善，刀具的更换与零件的装夹仍需人工来完成。

① General CNC machine tools：General CNC machine tools generally refer to the automatic machine tools that realize digital control in a process of machining，such as CNC milling machine，CNC lathe，CNC drilling machine，CNC grinding machine and CNC gear machining machine．General CNC machine tools are not perfect in automation，tool replacement and parts clamping still need to be completed manually.

② 加工中心：加工中心是带有刀库和自动换刀装置的数控机床，它将数控铣床、数控镗床、数控钻床的功能组合在一起，零件在一次装夹后，可以将其大部分加工面进行铣削。

② Machining center：Machining center is a kind of CNC machine tool with tool changer and automatic tool changer．It combines the functions of CNC milling machine，CNC boring machine and CNC drilling machine. After the parts are clamped once，most of the machined surfaces can be milled.

2. 按控制方式分类

2. Classification by control mode

① 开环控制数控机床：这类机床不带位置检测反馈装置，通常用步进电机作为执行机构。输入数据经过数控系统的运算，发出脉冲指令，使步进电机转过一个步距角，再通过机械传动机构转换为工作台的直线移动，移动部件的移动速度和位移量由输入脉冲的频率和脉冲个数所决定，如图 5.1.2 所示。

① CNC machine tools with open-loop control：This kind of machine tool is not equipped with the position detection feedback device，and usually uses the stepping motor as the actuator. The input data are calculated by the numerical control system，and the pulse instruction is issued to make the stepping motor rotate through a step angle，and then converted into linear movement of the worktable by the mechanical transmission mechanism. The moving speed and displacement of the moving parts are determined by the frequency and the number of the input pulses，as shown in Figure 5.1.2.

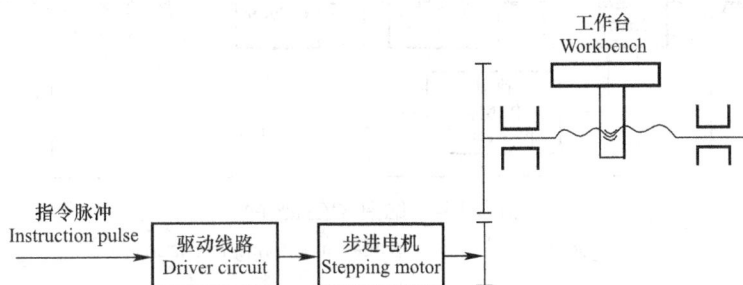

图 5.1.2　触摸屏监控画面

Figure 5.1.2　Monitoring interface of touch screen

② 半闭环控制数控机床：在电机的端头安装检测元件（一般为光电编码器），通过检测电机旋转角度来间接检测移动部件的位移，然后反馈到数控系统中。由于大部分机械传动环节未包括在系统闭环环路内，因此可获得更稳定的控制特性。其控制精度虽不如全闭环控制数控机床，但调试比较方便，因而被广泛采用，如图 5.1.3 所示。

② CNC machine tools with semi-closed loop control：A detection element （usually a photoelectric encoder） is installed at the end of the motor，and the displacement of the moving part is indirectly detected by detecting the rotation angle of the motor，and then fed back to the CNC system. Since most of the mechanical transmission links are not included in the closed loop of the system，more stable control characteristics can be obtained. Although its control accuracy is not as good as that of full closed-loop CNC machine tool，it is more convenient to debug and is widely used，as shown in Figure 5.1.3.

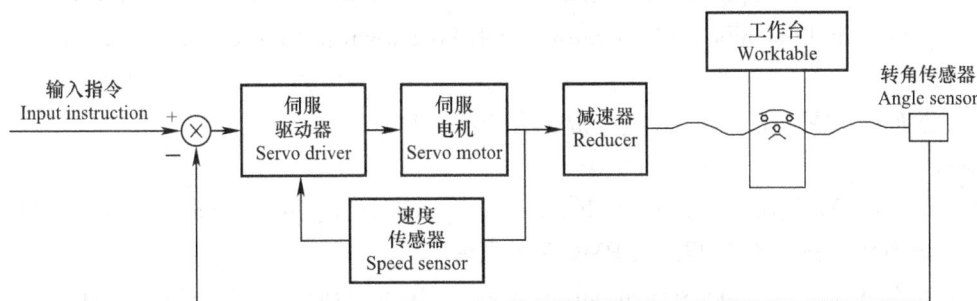

图 5.1.3　触摸屏监控画面

Figure 5.1.3　Monitoring interface of touch screen

③ 全闭环控制数控机床：这类数控机床带有位置检测反馈装置，其位置检测反馈装置采用直线位移检测元件，直接安装在机床的移动部件上，将测量结果直接反馈到数控装置中，通过反馈可消除从电机到机床移动部件整个机械传动链中的传动误差，最终实现精确定位，如图5.1.4所示。

③ CNC machine tools with full closed-loop control：A CNC machine tool of this type is provided with a position detection feedback device, the position detection feedback device uses linear displacement detection element and is directly installed on the moving part of the machine tool. The measurement result is directly fed back to the CNC device. The transmission error in the whole mechanical transmission chain from the motor to the moving part of the machine tool can be eliminated through the feedback, and the accurate positioning can be realized finally, as shown in Figure 5.1.4.

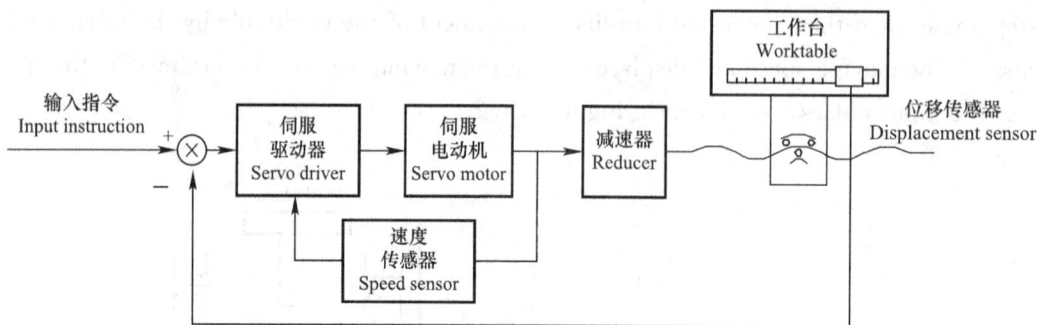

图 5.1.4 触摸屏监控画面

Figure 5.1.4 Touch screen monitoring interface

三、任务实施
Ⅲ. Task Implementation

（一）认识 FANUC 0i Mate-TD 数控系统面板各按键及功能
（Ⅰ）Understanding of the keys and functions of FANUC 0i Mate-TD CNC system panel

数控机床及数控系统品牌很多，各个品牌又有不同的规格型号，所以数控机床的数控系统操作面板各有不同，这里仅介绍 FANUC 0i Mare-TD 数控系统及相关基本操作。

CNC machine tools and CNC system brands are various, each brand has its own specifications and models, so CNC system operator panels of CNC machine tools are different. Here we only introduce FANUC 0i Mare-TD CNC system and related basic operations.

1. 数控系统面板

1. CNC system panel

FANUC 0i 系列数控系统面板主要由三部分组成，即 CRT 显示屏、MDI 面板（编辑面板）和机床操作面板，如图 5.1.5～图 5.1.8 所示。

FANUC 0i series CNC system panel is mainly composed of three parts, namely CRT display screen, MDI panel (edit panel) and machine tool operator panel, as shown in Figure 5.1.5～5.1.8.

2. MDI 面板

MDI 面板按键布局如图 5.1.9 所示，其按键功能见表 5.1.1。

2. MDI panel

The key layout of the MDI panel is shown in Figure 5.1.9, and the key functions are shown in Table 5.1.1.

（二）数控机床系统参数、PMC 程序及 PMC 参数调阅
（Ⅱ）Access to CNC machine tool system parameters, PMC program and PMC parameters

数控机床要想正常、精准地运行和加工零件，除了机械和电气部件外，还必须保证系统参数、PMC 参数及 PMC 程序设定和编写正确。因此，作为维修人员必须掌握这些数据的调阅方法。

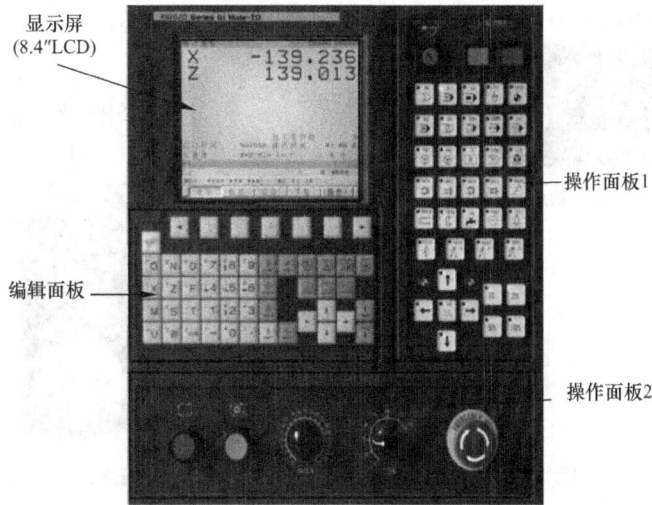

图 5.1.5　配 FANUC 0i 系统的机床面板

Figure 5.1.5　Machine tool panel with FANUC 0i system

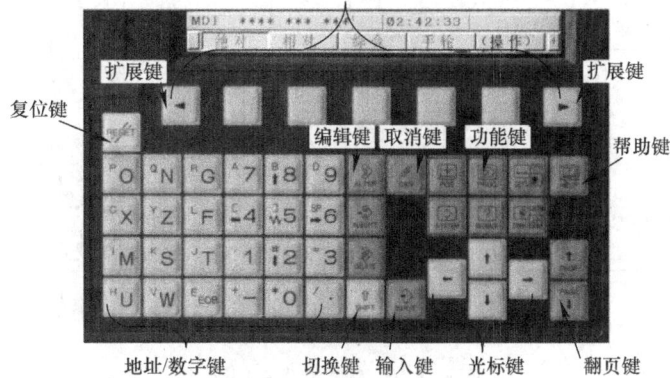

图 5.1.6　FANUC 0i 系列数控系统 MDI 面板

Figure 5.1.6　FANUC 0i series CNC system MDI panel

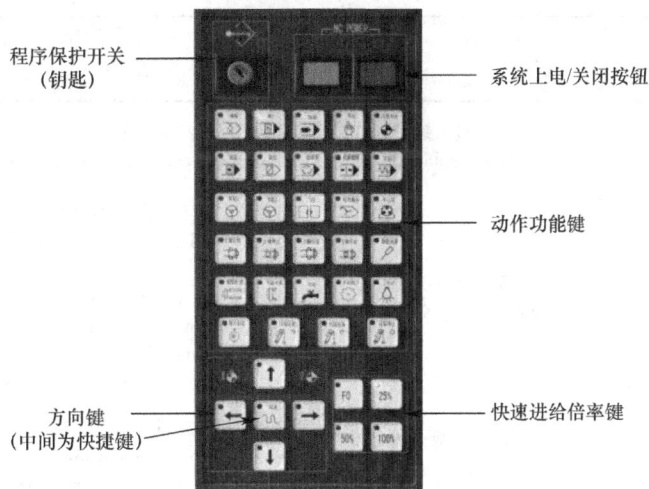

图 5.1.7　数控机床操作面板

Figure 5.1.7　CNC machine tool operator panel

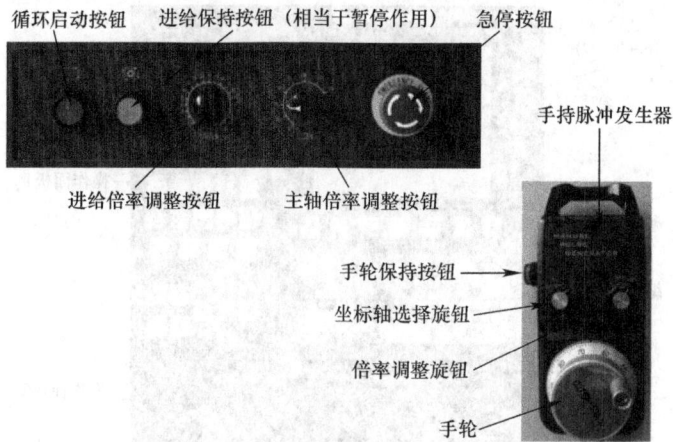

循环启动按钮　进给保持按钮（相当于暂停作用）　急停按钮

手持脉冲发生器

手轮保持按钮

坐标轴选择旋钮

倍率调整旋钮

手轮

进给倍率调整按钮　主轴倍率调整按钮

图 5.1.8　数控机床操作面板及手持脉冲发生器

Figure 5.1.8　CNC machine tool operator panel and hand-held pulse generator

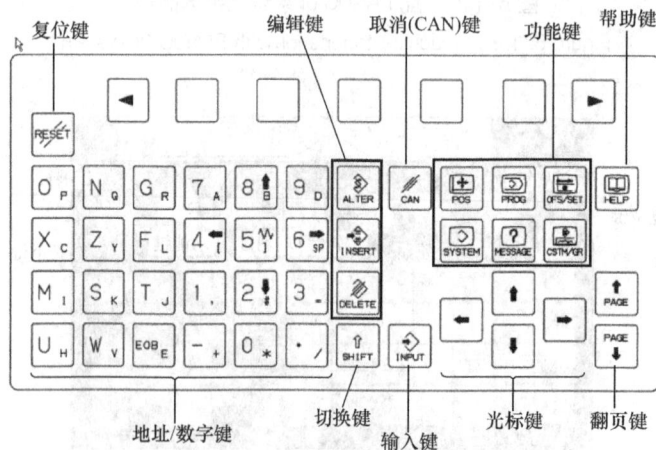

复位键　　编辑键　取消(CAN)键　功能键　帮助键

地址/数字键　　切换键　　输入键　　光标键　　翻页键

图 5.1.9　MDI 键盘布局图

Figure 5.1.9　MDI keyboard layout diagram

MDI 按键功能　　　　　　　　　　　　　　　表 5.5.1

MDI key functions　　　　　　　　　　　　　Table 5.5.1

序号 S/N	按键符号 Key symbol	名称 Name	功能说明 Function description
1	POS	位置显示键 Position display key	显示刀具的坐标位置 Display the coordinate position of the cutting tool
2	PROG	程序显示键 Program display key	在"edit"模式下显示存储器内的程序；在"MDI"模式下，输入和显示 MDI 数据；在"AOTO"模下显示当前待加工或者正在加工的程序 Display the program in the memory under "edit" mode; input and display the MDI data under "MDI" mode; display the program currently waiting to be processed or being processed under "AOTO" mode
3	OFFSET SETTING	参数设定/显示键 Parameter setting/display key	设定并显示刀具补偿值工件坐标系已经及宏程序变量 Set and display the cutting tool's offset value, the workpiece coordinate system and the macro program variable
4	SYSTEM	系统显示键 System display key	系统参数设定与显示，以及自诊断功能数据显示等 Set and display system parameters, and display the self-diagnostic function data, etc

序号 S/N	按键符号 Key symbol	名称 Name	功能说明 Function description
5	MESSAGE	报警信息显示键 Alarm information display key	显示 NC 报警信息 Display the NC alarm information
6	CUSTOM GRAPH	图形显示键 Graphic display key	显示刀具轨迹等图形 Display the cutter path and other graphics
7	RESET	复位键 Reset key	用于所有操作停止或解除报警，CNC 复位 Stop the operation or cancel the alarm or reset CNC
8	HELP	帮助键 Help key	提供与系统相关的帮助信息 Provide system-related help information
9	DELETE	删除键 Delete key	在"Edit"模式下，删除已输入的字及 CNC 中存在的程序 Under "Edit" mode, delete the words entered and the programs that exist in the CNC
10	INPUT	输入键 Input key	加工参数等数值的输入 Input numerical values such as machining parameters
11	CAN	取消键 Cancel key	清除输入缓冲器中的文字或者符号 Clear the text or symbols in the input buffer
12	INSERT	插入键 Insert key	在"Edit"模式下，在光标后输入的字符 Under "Edit" mode, input the character behind the cursor
13	ALTER	替换键 Alternate key	在"Edit"模式下，替换光标所在位置的字符 Under "Edit" mode, replace the character at the cursor position
14	SHIFT	上档键 Shift key	用于输入处在上档位置的字符 Enter the characters that are in the upper position
15	PAGE↑ PAGE↓	光标翻页键 Page turning key	向上或者向下翻页 Page up or page down
16	(程序编辑键盘)	程序编辑键 Program editing key	用于 NC 程序的输入 Input the NC program
17	(光标移动键)	光标移动键 Cursor movement key	用于改变光标在程序中的位置 Change the position of the cursor in the program

In order to run the CNC machine tool and process machine parts normally and accurately, besides mechanical and electrical components, the system parameters, PMC parameters and PMC program must be set and edited correctly. Therefore, the maintenance personnel must grasp this data access method.

1. 数控机床系统参数的调阅步骤

1. Steps of accessing CNC machine tool system parameters

在数控系统正常上电的情况下，按下 MDI 键盘功能键"SYSTEM"，系统将进入参数设定界面，如图 5.1.10 所示。

When the CNC system is powered up normally, press the MDI keyboard function key "SYSTEM" to enter the parameter setting interface, as shown in Figure 5.1.10.

在参数画面中，通过在 MDI 键盘输入想要查阅的参数号，然后按"Number Search"软键让显示画面直接跳转到该参数号，如图 5.1.11 所示。

In the parameter screen, enter the parameter number you want to search on the MDI keyboard, and then press the "Number Search" softkey to jump to the parameter number directly, as shown in Figure 5.1.11.

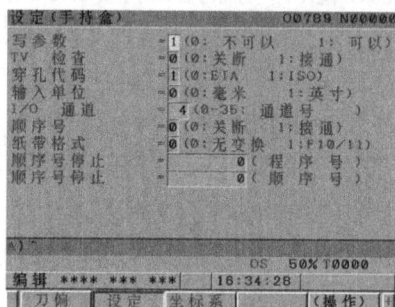

图 5.1.10　参数画面

Figure 5.1.10　Parameter screen

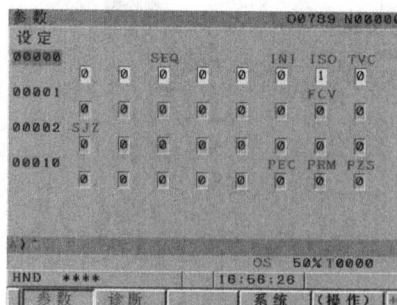

图 5.1.11　参数号搜索画面

Figure 5.1.11　Parameter number search screen

2. PMC 程序的调阅步骤

2. Steps of accessing PMC procedures

首选按"SYSTEM"功能键进入参数画面，此时系统软键如图 5.1.12 所示。再按最右边扩展键三次，来到 PMC 画面选择界面，如图 5.1.13 所示。

First press the "SYSTEM" function key to enter the parameter screen, to see the system soft key as shown in Figure 5.1.12. Then press the rightmost extension key three times to enter the PMC screen selection interface, as shown in Figure 5.1.13.

图 5.1.12　参数画面对应的软键

Figure 5.1.12　Soft key corresponding to parameter screen

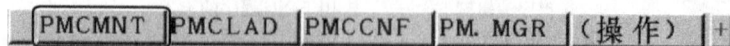

图 5.1.13　PMC 画面选择对应的软键

Figure 5.1.13　Soft key corresponding to PMC screen selection

（1）按下"PMCMNT"软件进入 PMC 维护画面，此画面显示的是 PMC 各 I/O 信号的状态信息，如图 5.1.14 所示。

196

(1) Press "PMCMNT" softkey to enter the PMC maintenance screen，which displays the status information of each PMC I/O signal，as shown in Figure 5.1.14.

（2）按下"PMCLAD"软件进入 PMC 梯形图列表画面主要显示梯形图结构，如图 5.1.15 所示。

(2) Press "PMCLAD" softkey to enter the PMC diagram chart list screen，which mainly displays the ladder diagram structure，as shown in Figure 5.1.15.

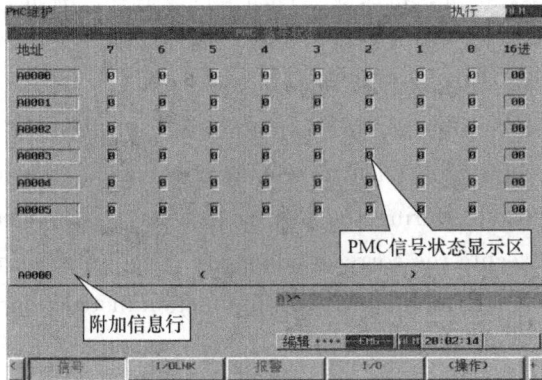

图 5.1.14　PMC 状态信息表画面

Figure 5.1.14　PMC status information table screen

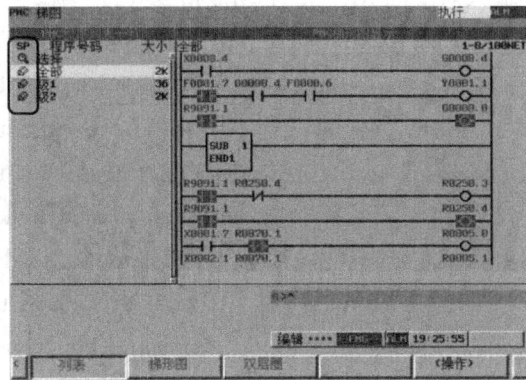

图 5.1.15　PMC 梯形图列表画面

Figure 5.1.15　PMC ladder chart list screen

在"SP"区选择文件夹后，按下"梯形图"软件，进入到对应文件夹内的梯形图监控画面，如图 5.1.16 所示。

After selecting the folder in the "SP" area，press the "ladder diagram" soft key to enter the ladder diagram monitoring interface in the corresponding folder，as shown in Figure 5.1.16.

（3）在 PMC 画面，单击"PMCMNT"并单击"＋"可以进入 PMC 参数画面，如图 5.1.17 所示。在 PMC 参数主要是 PMC 梯形图中的各种寄存器、定时器、计数器、辅助继电器的设定值等。

(3) On the PMC screen，click "PMCMNT" and click "＋" to enter the PMC parameter screen，as shown in Figure 5.1.17. The PMC parameters are mainly set values of various registers，timers，counters，auxiliary relays and so on in the PMC ladder diagram.

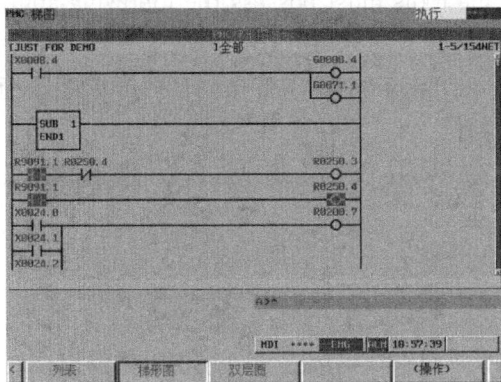

图 5.1.16　PMC 梯形图监控画面

Figure 5.1.16　PMC ladder diagram monitoring interface

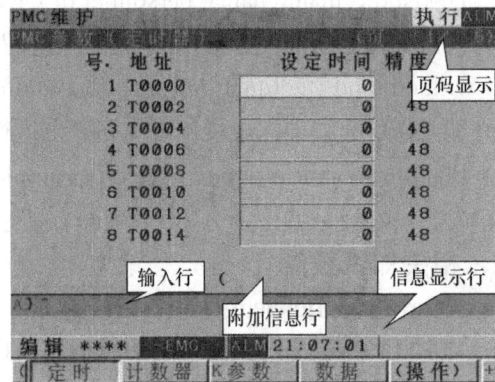

图 5.1.17　PMC 参数画面

Figure 5.1.17　PMC parameters screen

（三）利用 MDI 模式进行简单机床功能测试

（Ⅲ）Simple machine tool function testing using MDI mode

MDI 又称为手动数据输入模式，在该模式下可以输入简短的指令（最多 10 行程序段），按下起动按键可以自动执行命令。一般可配合手动模式（手轮）用于找正、对刀、检测等工作。

MDI is also known as manual data entry mode，where you can enter simple instructions（up to 10 lines of code）and press the Start button to automatically execute instructions. It can be used in manual mode（with handwheel）for alignment，knife alignment，detection and so on.

在维修过程中常常须要对机床的各种功能进行测试，如主轴、刀具等功能是否正常的测试。在测试过程中，只要输入对应的 M、S、T 功能指令并按"循环启动"按键，机床就会自动执行该指令。例如主轴功能检测时的操作只需在 MDI 模式下输入"M03 S1000；"然后按"插入"键，再按"循环启动"按键，主轴就会按正方向旋转起来，如图 5.1.18 所示。

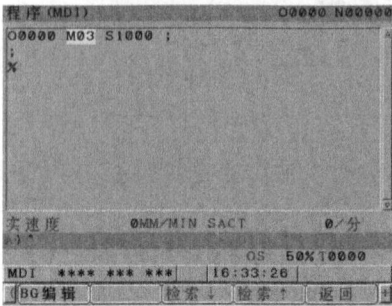

图 5.1.18　MDI 画面

Figure 5.1.18　MDI screen

In the maintenance process，it is often necessary to test the various functions of the machine tool，such as whether the spindle，cutter functions are normal or not. During the testing process，the machine tool will automatically execute the corresponding M，S，T function instructions as long as the corresponding M，S，T function instructions are input and the Cycle Start button is pressed. For example，for spindle function detection，you can simply enter "M03 S1000" in MDI mode；Then press "Insert" and press "Cycle Start"，and the spindle will rotate in a positive direction，as shown in Figure 5.1.18.

任务二　数控机床的硬件连接及电路分析

Task Ⅱ　Hardware Connection and Circuit Analysis of CNC Machine Tool

一、任务分析

Ⅰ. Task Analysis

数控机床大多数的电气故障都是因为硬件连接电路松动、老化等现象引起的。因此，数控机床维修维护人员必须具备以下能力：

Most of the electrical faults of CNC machine tools are caused by the looseness and aging of hardware connection circuits. Therefore，maintenance personnel of CNC machine tools must possess the following abilities：

（1）了解数控系统面板各接口的功能、含义及连接去向。

（1）Understand the function，meaning and connection direction of each interface of CNC system panel.

（2）能熟练、正确地完成硬件连接并开机进行功能检查。

（2）Be able to skillfully and correctly complete the hardware connection and carry out function check in start state.

二、相关知识技能

Ⅱ. Relevant Knowledge & Skills

（一）FANUC 数控系统典型硬件的结构及接口

（Ⅰ）Structure and interface of typical hardware of FANUC CNC system

1. 数控系统的组成及特点

1. Composition and characteristics of CNC system

一套完整 FANUC0iMateTD 数控系统一般包括：数控系统本身、主轴驱动单元、主轴电机、FANUC 0i 系统用 I/O 单元、进给伺服放大器、伺服电机、机床操作面板。如图 5.2.1 所示。

A complete FANUC0i MateTD CNC system generally includes：CNC system itself，spindle drive unit， spindle motor，I/O unit for FANUC 0i system，feed servo amplifier，servo motor，machine tool operator panel. As shown in Figure 5.2.1.

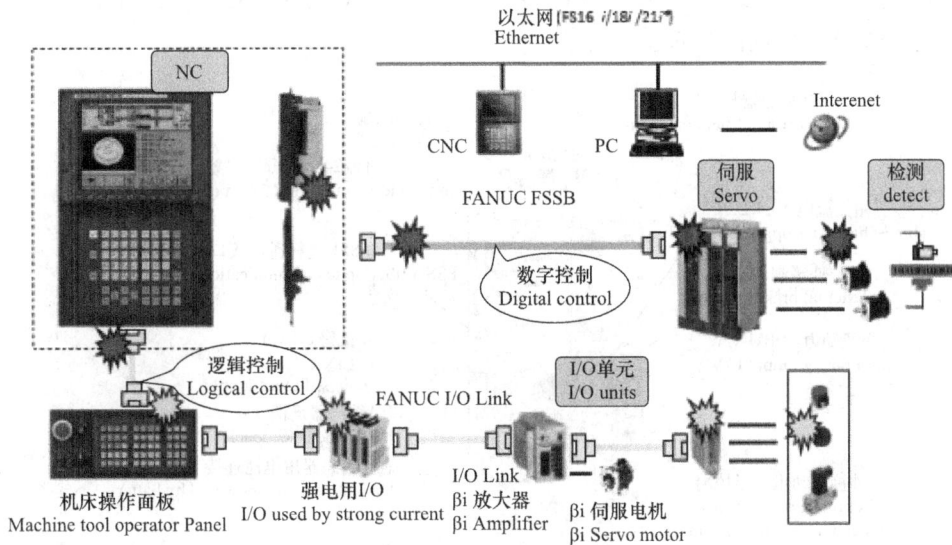

图 5.2.1 硬件图

Figure 5.2.1 Hardware diagram

2. 系统端口功能及连接

2. System port functions and connections

（1）数控系统端口：CP1、FU1、CA114、JA2、JD36A、JD36B、JA40、JA41、JD51A、CD38A、COP10A、JGA、FAN0、FAN1。如图 5.2.2 所示。

（1）CNC system port：CP1，FU1，CA114，JA2，JD36A，JD36B，JA40，JA41，JD51A，CD38A，COP10A，JGA，FAN0，FAN1. As shown in Figure 5.2.2.

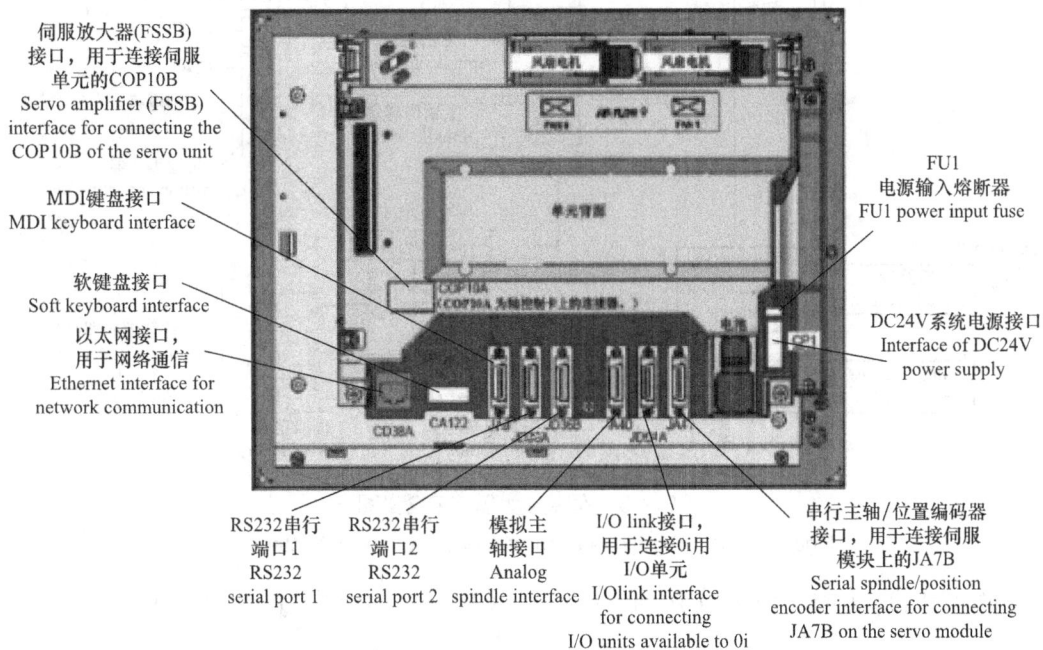

图 5.2.2 系统面板接口含义图

Figure 5.2.2 Significance diagram of system panel interface

199

（2）伺服单元端口：L1—L3、U—W、DCC/DCP、CX29、CX30、CXA20、CXA19A、CX19B、COP10A、COP10B、JX5、CX5X。如图 5.2.2 所示。

（2）Servo unit port：L1—L3，U—W，DCC/DCP，CX29，CX30，CXA20，CXA19A，CX19B，COP10A，COP10B，JX5，CX5X. As shown in Figure 5.2.2.

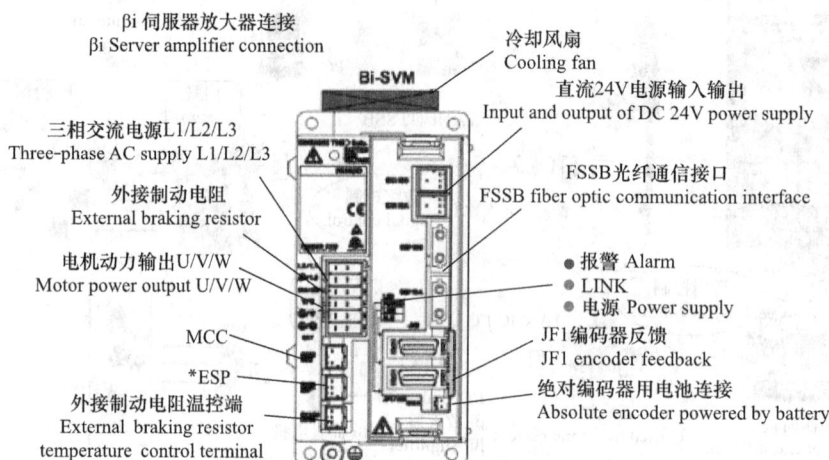

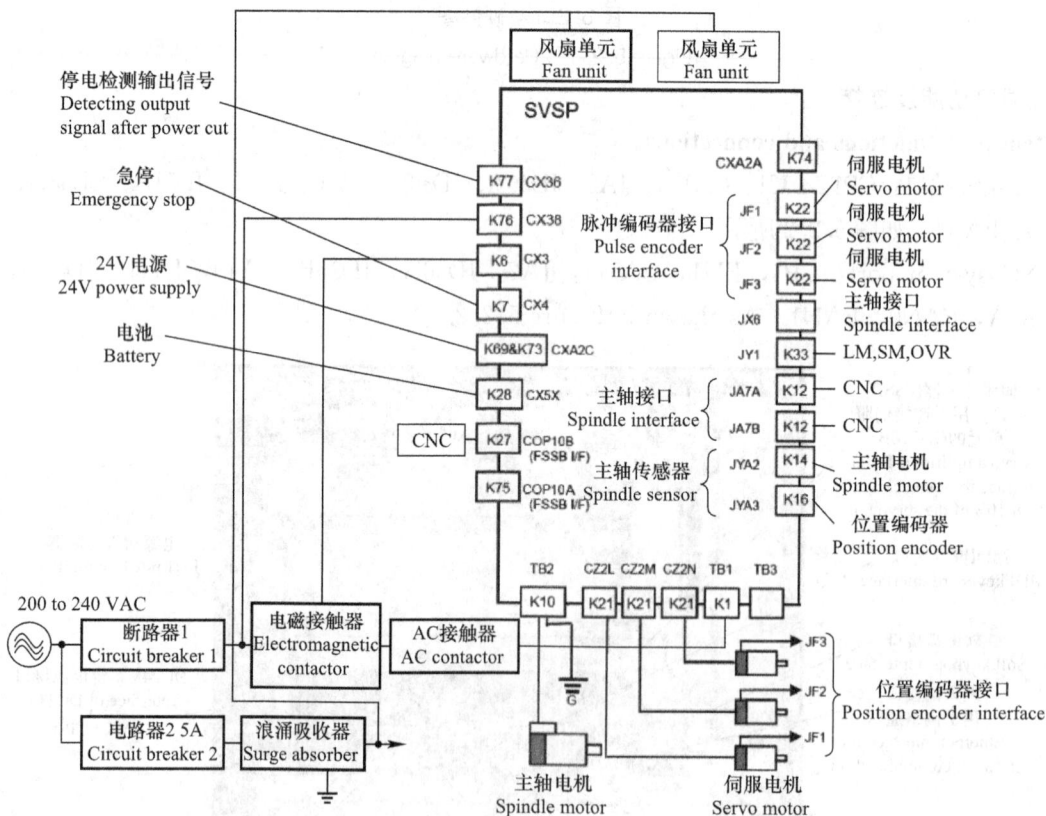

图 5.2.3 伺服放大器接口含义图

Figure 5.2.3 Significance diagram of servo amplifier interface

（3）I/O 模块端口：CP1、CP2、CP104、CP105、CP106、CP107、JA3、JD1A、JD1B。如图 5.2.4 所示。

（3）I/O module port：CP1，CP2，CP104，CP105，CP106，CP107，JA3，JD1A，JD1B. As shown in Figure 5.2.4.

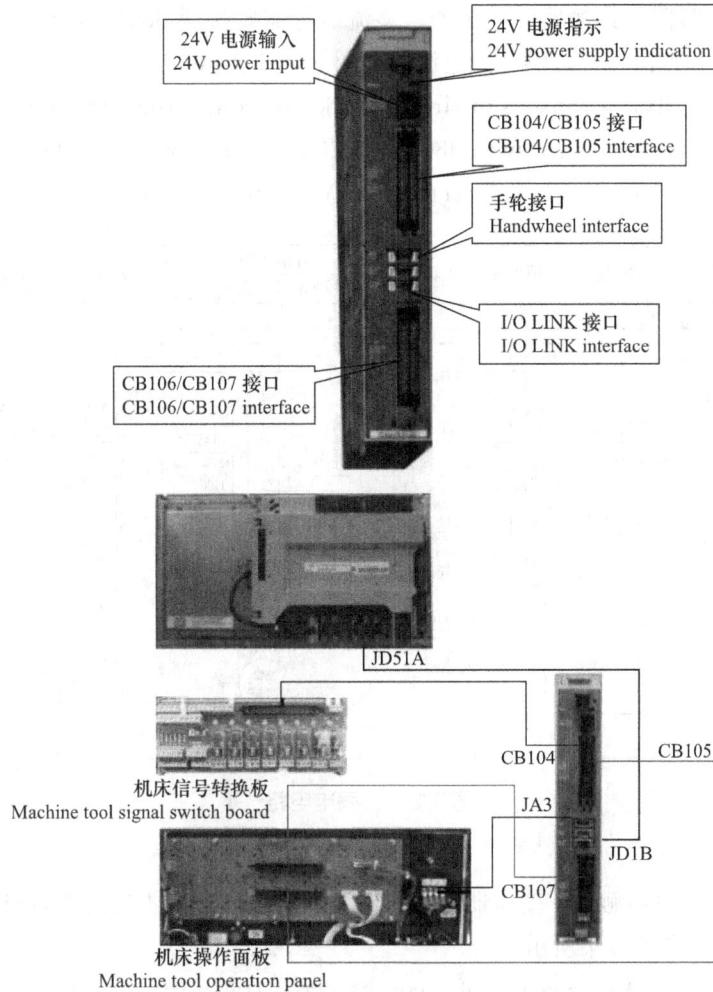

图 5. 2. 4 I/O 接口定义图

Figure 5. 2. 4 I/O interface definition diagram

（二）数控机床电气控制系统的构成（图 5. 2. 5）

（Ⅱ）Component of electrical control system of CNC machine tool（Figure 5. 2. 5）

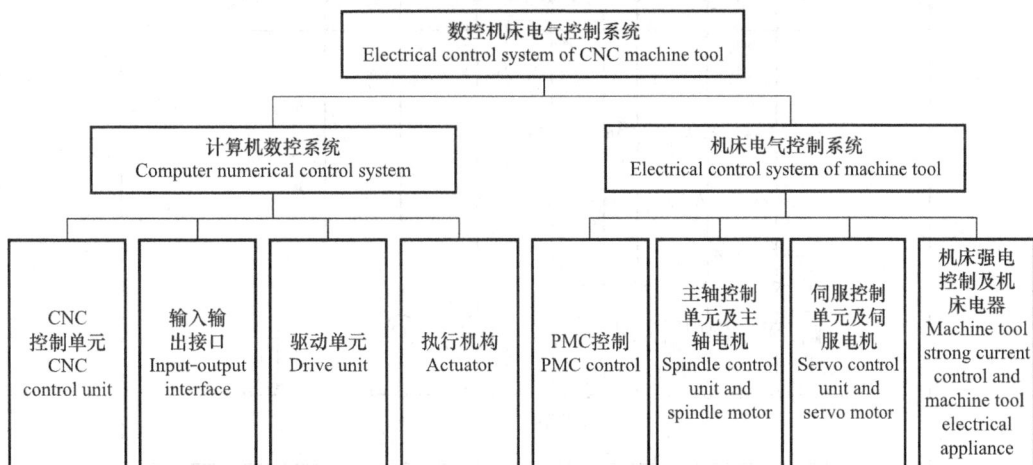

图 5. 2. 5 数控机床电气构成图

Figure 5. 2. 5 Electrical configuration diagram of CNC machine tool

1. 主电路：一般由断路器、接触器、变压器、整流器等需要接通 380V 或 220V 强电源的元器件组成，主要为了给电机、变频器和整流器等提供动力电源。如图 5.2.6 所示。

1. Main circuit: Generally, it consists of circuit breakers, contactors, transformers, rectifiers and other components that need to be connected with 380V or 220V strong power supply, mainly to provide power supply for motors, frequency converters and rectifiers. As shown in Figure 5.2.6.

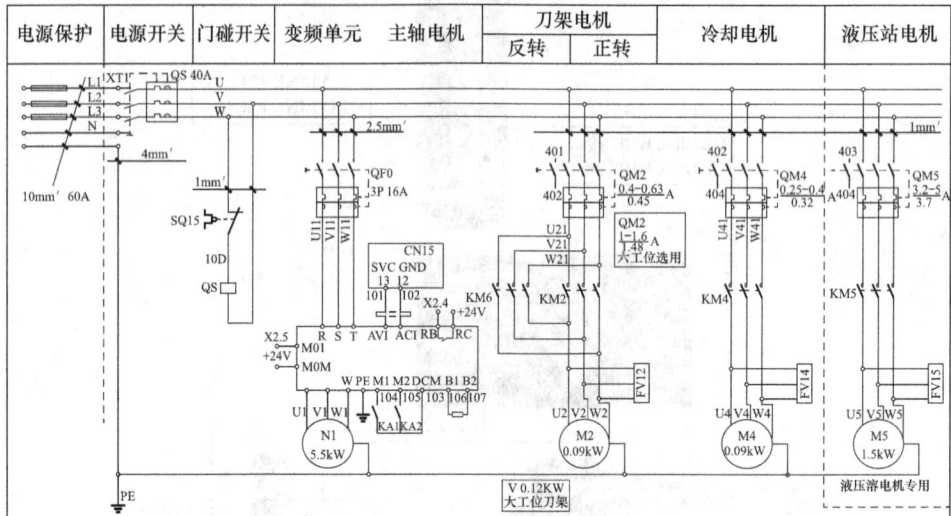

图 5.2.6　强电电路

Figure 5.2.6　Strong power circuit

2. 控制电路：一般由继电器触点、变压器，接触器线圈等组成，主要是把微弱控制信号用来控制主电路动作，相当于放大作用。如图 5.2.7 所示。

2. Control circuit: Generally it consists of relay contacts, transformers, contactor coils and so on. The weak control signal is mainly used to control the main circuit action, which plays the role of amplification. As shown in Figure 5.2.7.

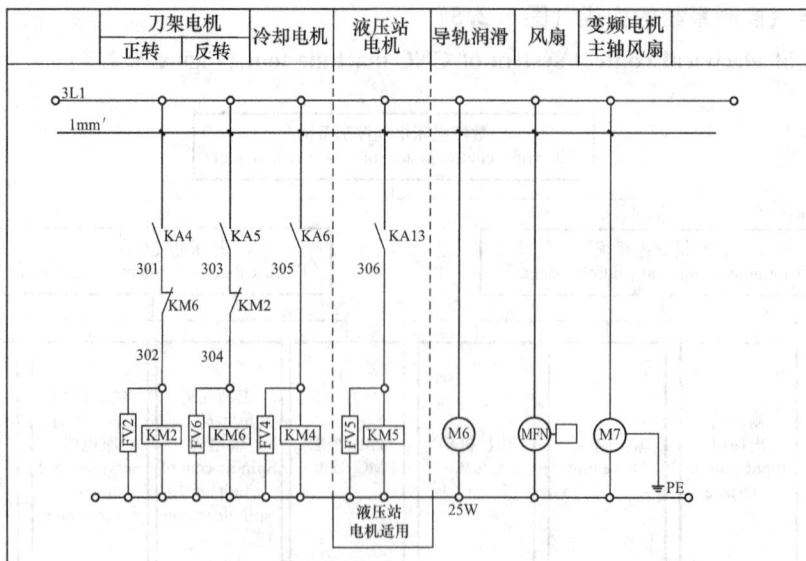

图 5.2.7　控制回路

Figure 5.2.7　Control loop

3. PMC 控制信号回路：一般由按钮、小型继电器、各类传感器、二极管等组成，主要为了对 I/O 接口进行收发信号。如图 5.2.8 所示。

3. PMC control signal loop: Generally it consists of the button, small relay, various sensors, diodes and other components, mainly to send and receive signals for I/O interface. As shown in Figure 5.2.8.

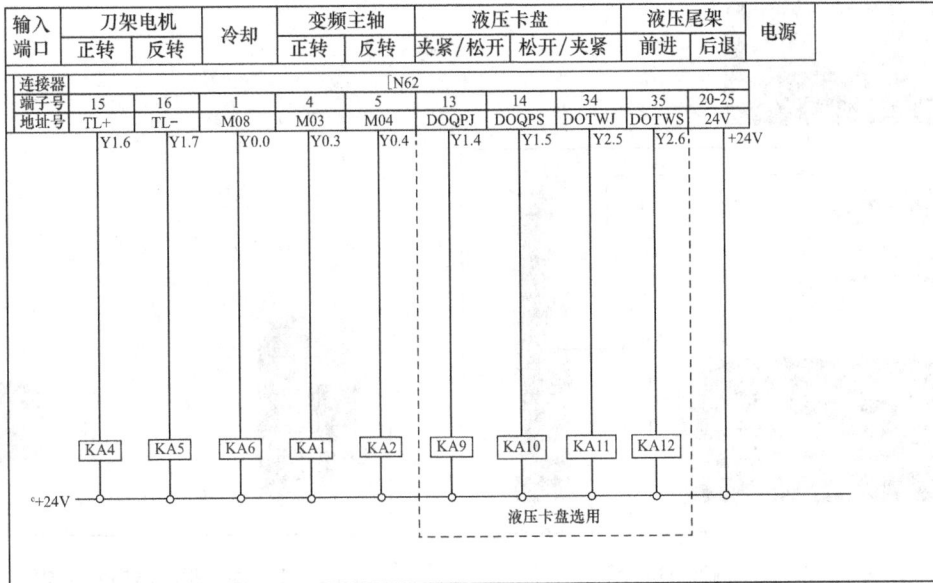

输入端口	刀架电机		冷却	变频主轴		液压卡盘			液压尾架		电源
	正转	反转		正转	反转	夹紧/松开	松开/夹紧		前进	后退	
连接器					N62						
端子号	15	16	1	4	5	13	14	34	35	20-25	
地址号	TL+	TL-	M08	M03	M04	DOQPJ	DOQPS	DOTWJ	DOTWS	24V	
	Y1.6	Y1.7	Y0.0	Y0.3	Y0.4	Y1.4	Y1.5	Y2.5	Y2.6		+24V
	KA4	KA5	KA6	KA1	KA2	KA9	KA10	KA11	KA12		
+24V							液压卡盘选用				

图 5.2.8 I/O 连接图

Figure 5.2.8 I/O connection diagram

三、任务实施
Ⅲ. Task Implementation

本次任务主要是了解各端口的作用与定义，并在实训台上进行硬件连接。连接图如下（也可以参考电缆图）：

This task purpose is mainly to understand the role and definition of each port, and carry out hardware connection on the training bench. Connection diagram As follows (See cable diagram):

（1）完成系统、X 轴放大器、Z 轴放大器的 FSSB 总线的连接。如图 5.2.9 所示。

(1) Complete the connection of the FSSB bus of the system, X-axis amplifier and Z-axis amplifier. As shown in Figure 5.2.9.

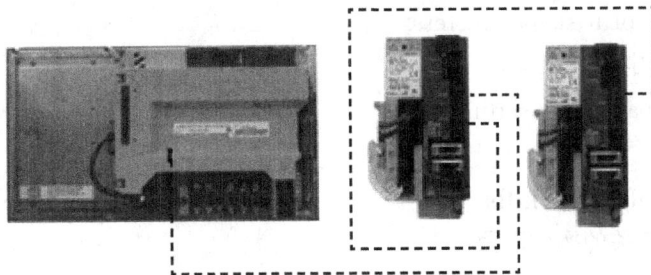

图 5.2.9 伺服系统连接图

Figure 5.2.9 Servo system connection diagram

（2）完成 I/O LINK 的连接。如图 5.2.10 所示。

(2) Complete the I/O LINK connection. As shown in Figure 5.2.10.

（3）完成伺服电机、伺服放大器的连接如图 5.2.11 所示。

（3）Complete the connection of servo motor and servo amplifier as shown in Figure 5.2.11.

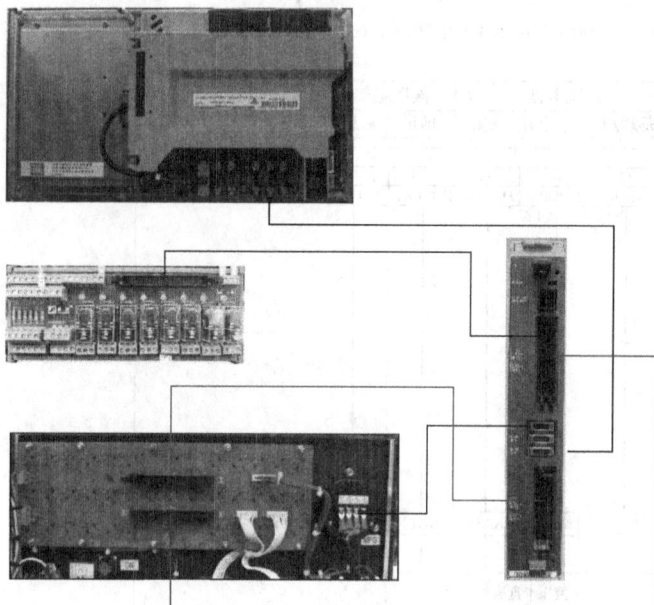

图 5.2.10　I/O 硬件连接图
Figure 5.2.10　I/O hardware connection diagram

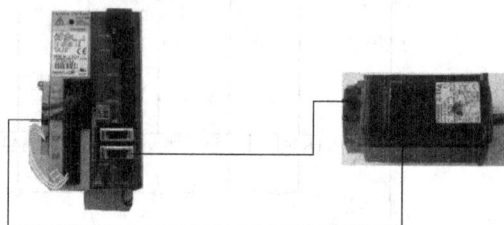

图 5.2.11　伺服驱动器与伺服电机硬件连接图
Figure 5.2.11　Hardware connection diagram
of servo driver and servo motor

任务三　数控机床基本参数设定

Task Ⅲ　Basic Parameter Setting of CNC Machine Tool

一、任务分析

Ⅰ. Task Analysis

（1）了解 Fanuc 系统参数设定画面。

（1）Understand the Fanuc system parameter setting screen.

（2）掌握基本参数的含义。

（2）Grasp the meaning of basic parameters.

（3）了解基本参数的设定。

（3）Understand basic parameter settings.

二、相关知识技能

Ⅱ. Relevant Knowledge & Skills

（一）和机床加工操作有关的操作画面

（Ⅰ）Screen operations related to machine tool processing

1. 回零点方式

1. Return-to-zero mode

回零点方式，主要是进行机床机械坐标系的设定，选择回零方式，用机床操作面板上各轴返回参考点用的开关使刀具沿参数（1006♯5）指定的方向移动。首先刀具以快速移动速度移动到减速点上，然后按 FL 速度移动到参考点。快速移动速度和 FL 速度由参数（1420、1421、1425）设定。如图 5.3.1 所示。

The return-to-zero mode is mainly used for setting the machine tool mechanical frame. Select the return-to-zero mode, and move the tool in the direction specified in the parameter (1006 ♯ 5) by using the switch for returning each axis to the reference point on the machine tool operator panel. First, the tool moves to the deceleration point at a quick moving speed, and then to the reference point at FL speed. The quick moving speed and the FL speed are set by parameters (1420, 1421, 1425). As shown in Figure 5.3.1.

2. 手动（JOG）方式

2. Manual (JOG) mode

在 JOG 方式，按机床操作面板上的进给轴和方向选择开关（一般为同一个键），机床沿选定轴的选定方向移动。手动连续进给速度由参数 1423 设定。按快速移动开关，以 1424 设定的速度移动机床。手动操作通常一次移动一个轴，但也可以用参数 1002♯0 选择同时 2 轴运动。如图 5.3.2。

In the JOG mode, the machine tool moves in the selected direction of the selected axis by pressing the feed axis and the direction selection switch (typically the same key) on the machine operator panel. The manual continuous feed speed is set by parameter 1423. Press the quick-moving switch to move the machine at the speed set by 1424. For manual operations, it typically moves one axis at a time, but also parameter1002 ♯ 0 can be used to select simultaneous 2-axis motion mode. As shown in Figure 5.3.2.

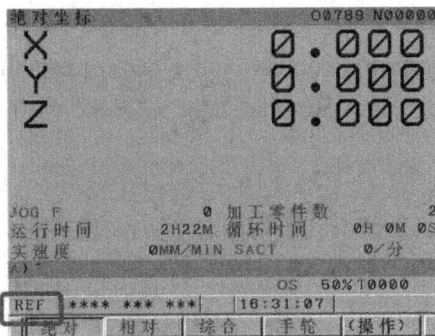

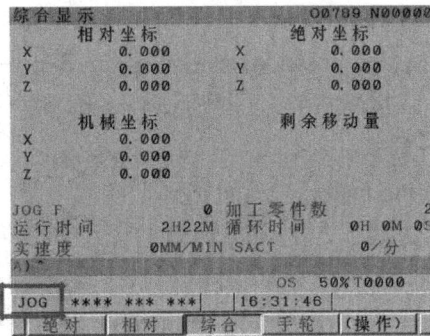

图 5.3.1 回零画面 图 5.3.2 JOG 方式画面
Figure 5.3.1 Return-to-zero screen Figure 5.3.2 JOG mode screen

3. 增量进给（INC）方式

3. INC mode

在增量进给方式，按机床操作面板上的进给轴和方向选择开关，机床在选择的轴选方向上移动一步。机床移动的最小距离是最小增量单位。每一步可以是最小输入增量单位的 1 倍、10 倍、100 倍或 1000 倍。当没有手摇时，此方式有效。如图 5.3.3 所示。

In the INC mode, the machine tool moves one step in selection direction of the selected axis according to the feed axis and direction selection switch on the machine tool operator panel. The minimum distance that a machine moves is the minimum incremental unit. Each step can be 1, 10, 100, or 1000 times the minimum input increment unit. This mode works when there is no hand-held device. As shown in Figure 5.3.3.

4. 手轮进给方式

4. Handwheel feed mode

在手轮方式，机床可用旋转机床操作面板升手摇脉冲发生器而连续不断地移动。用开关选择移动轴和倍率。如图 5.3.4 所示。

In the handwheel mode，the machine tool can continuously move by means of lifting the hand pulse generator by the operator panel of the rotating machine tool. Use the switch to select the moving axis and magnification. As shown in Figure 5.3.4.

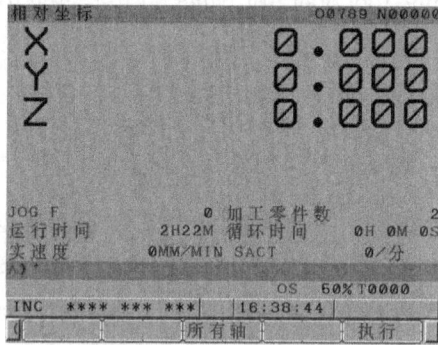

图 5.3.3　增量方式画面
Figure 5.3.3　Incremental mode screen

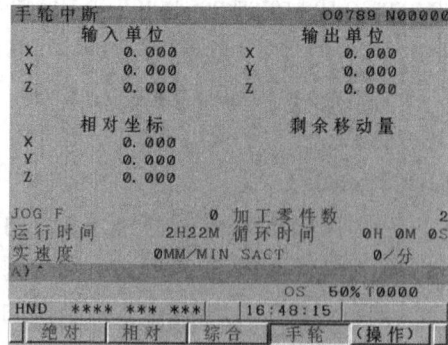

图 5.3.4　手轮方式画面
Figure 5.3.4　Handwheel mode screen

5. 存储器运行方式

5. Memory operation mode

在自动运行期间，程序预先存在存储器中，当选定一个程序并按了机床操作面板的循环启动按钮时，开始自动运行。如图 5.3.5 所示。

During automatic operation，the program is pre-stored in memory，and when a program is selected and cycle start button of operation panel is pressed，It starts to run automatically. As shown in Figure 5.3.5.

6. MDI 运行方式

6. MDI running mode

在 MDI 运行方式下，在 MDI 面板上输入 10 行程序段，可以自动执行，MDI 运行一般用于简单的测试操作。如图 5.3.6 所示。

In MDI mode，you can enter 10 lines of program segments on the MDI panel to execute automatically，and MDI run is typically used for simple test. As shown in Figure 5.3.6.

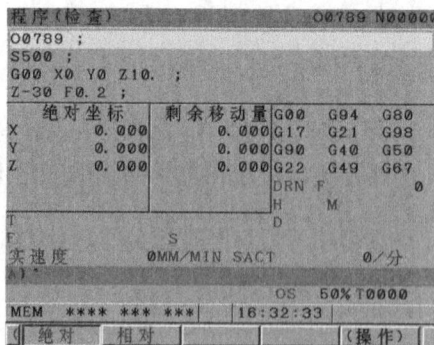

图 5.3.5　存储器方式画面
Figure 5.3.5　Memory mode screen

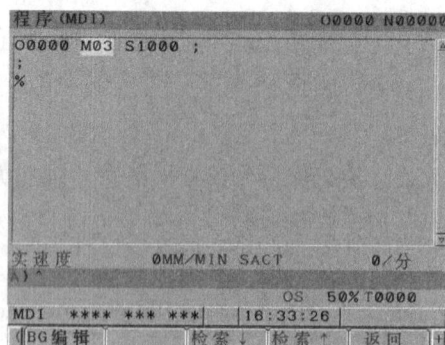

图 5.3.6　MDI 方式画面
Figure 5.3.6　MDI mode screen

7. 程序编辑（EDIT）方式

7. EDIT mode

在程序编辑方式下可以进行程序的编辑、修改、查找等功能。如图 5.3.7 所示。

In the EDIT mode，you can edit，modify，search the program etc. As shown in Figure 5.3.7.

（二）和机床维护操作有关的画面操作

（Ⅱ）Screen operations related to machine tool maintenance

1. 参数设定画面

1. Parameters setting screen

用于参数的设置、修改等操作，在操作时需要打开参数开关，按"OFSSET"键显示如图 5.3.8 所示画面就可以进行修改参数开关，参数开关为 1 时，可以进入参数进行修改。如图 5.3.9 所示。

It is used for parameter setting，modification and other operations. During operation，the parameter switch needs to be turned on. Press the "OFSSET" key to display the screen shown in Figure 5.3.8 to modify the parameter switch. When the parameter switch is 1，the parameter can be entered for modification. As shown in Figure 5.3.9.

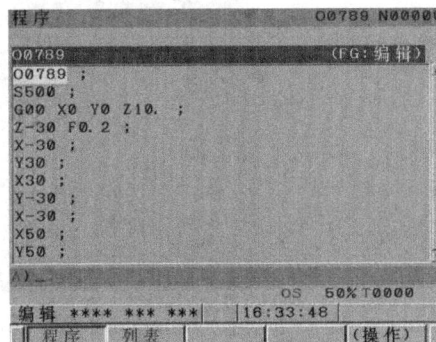

图 5.3.7 编辑方式画面

Figure 5.3.7 EDIT mode screen

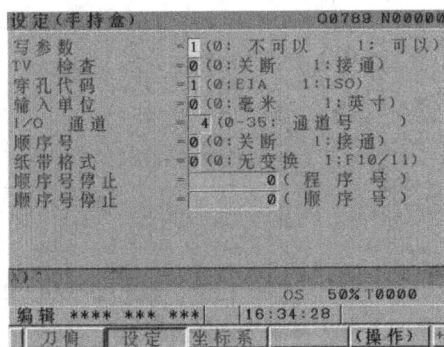

图 5.3.8 参数开关画面

Figure 5.3.8 Parameter switch screen

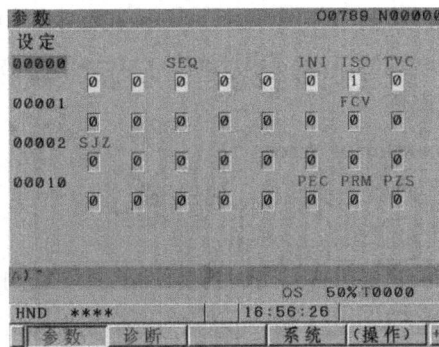

图 5.3.9 参数画面

Figure 5.3.9 Parameter screen

2. 诊断画面

2. Diagnostic screen

当出现报警时，可以通过诊断画面进行故障的诊断，按上图中的诊断键，如图 5.3.10 所示。

When an alarm occurs，the fault can be diagnosed through the diagnosis screen. Press the diagnosis key in the figure above，as shown in Figure 5.3.10.

3. PMC 画面

3. PMC screen

PMC 就是利用内置在 CNC 的 PC 执行机床的顺序控制的可编程机床控制器，PMC 画面是比较常用的一个画面，它可以进行状态查询、PMC 在线编辑、通信等功能。按"SYSTEM"键后按右扩展键出现 PMC，如图 5.3.11 所示。

PMC is a programmable machine tool controller which uses the PC built in CNC to execute the sequential control of the machine tool. PMC screen is a more commonly used screen. It can be used for status inquiry，PMC online editing，communication and other functions. PMC will appear when "SYSTEM" key is pressed and right extension key is pressed，as shown in Figure 5.3.11.

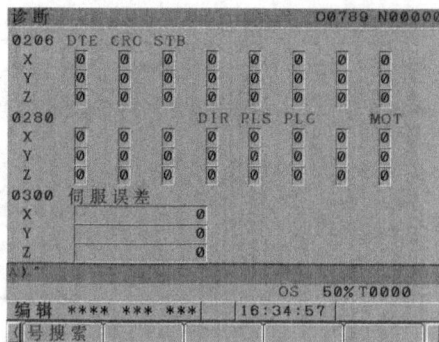

图 5.3.10　诊断画面

Figure 5.3.10　Diagnostic screen

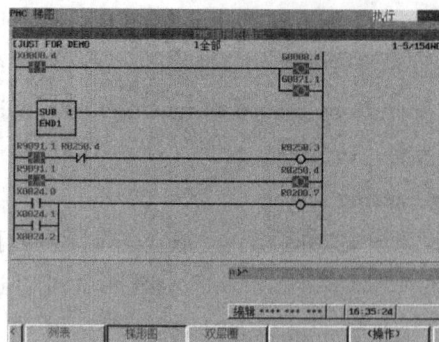

图 5.3.11　PMC 画面

Figure 5.3.11　PMC screen

4. 伺服监视画面

4. Servo monitoring screen

主要是进行伺服的监视，如位置环增益、位置误差、电流、速度等，按"SYSTEM"键后按右扩展键出现 SV 设定，如图 5.3.12 所示。

Mainly for servo monitoring，such as position loop gain，position error，current，speed，etc.，press the "SYSTEM" key and then press right extension key to display the SV setting，as shown in Figure 5.3.12.

5. 主轴监控画面

5. Spindle monitoring screen

主要是进行主轴状态的监控，如主轴报警、运行方式、速度、负载表等。按"SYSTEM"键后按右扩展键出现 SP 设定，如图 5.3.13 所示。

Mainly to monitor the status of the spindle，such as spindle alarm，operation mode，speed，load table and so on. The SP setting will appear when the "SYSTEM" key is pressed and the right extension key is pressed，as shown in Figure 5.3.13.

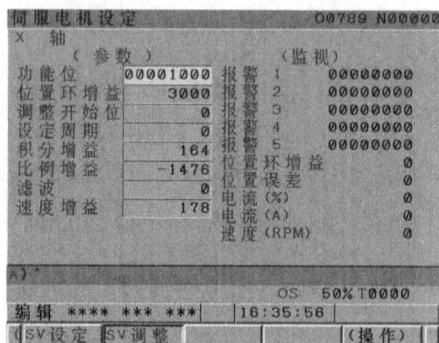

图 5.3.12　伺服监视画面

Figure 5.3.12　Servo monitoring screen

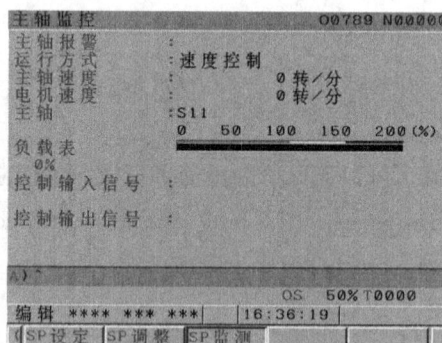

图 5.3.13　主轴监控画面

Figure 5.3.13　Spindle monitoring screen

（三）数控系统基本参数的含义

（Ⅲ）Meaning of basic parameters of CNC system

1. 数控机床与轴相关的参数

1. Axis-related parameters of CNC machine tools

1020：表示数控机床各轴的程序名称，如在系统显示画面显示的 X、Y、Z 等，一般设置是，车床为 88，90；铣床与加工中心为 88，89，90。见表 5.3.1。

1020: Represents the program name of each axis of a CNC machine tool, such as X, Y, Z displayed on the system display screen etc. Generally, the lathe: 88, 90; Milling machines and machining centers: 88, 89, 90. As shown in Table 5.3.1.

表5.3.1

Table 5.3.1

轴名称 Axis name	X	Y	Z	A	B	C	U	V	W
设定值 Setting value	88	89	90	65	66	67	85	86	87

1022: 表示数控机床设定各轴为基本坐标系中的哪个轴，一般设置为1，2，3。见表5.3.2。

1022: Represents the axis in the basic frame set for CNC machine tool, generally 1, 2 and 3. As shown in Table 5.3.2.

设定值及其含义

Setting values and meanings

表5.3.2

Table 5.3.2

设定值 setting value	含义 Meaning
0	旋转轴 Rotation axis
1	基本3轴的X轴 Axis X of basic axis 3
2	基本3轴的Y轴 Axis Y of basic axis 3
3	基本3轴的Z轴 Axis Z of basic axis 3
5	X轴的平行轴 Parallel axis of axis X
6	Y轴的平行轴 Parallel axis of axis Y
7	Z轴的平行轴 Parallel axis of axis Z

1023: 表示数控机床各轴的伺服轴号，也可以称为轴的连接顺序，一般设置为1，2，3，设定各控制轴为对应的第几号伺服轴。

1023: Represents the servo axis number of each axis of a CNC machine tool, which may also be referred to as the connecting sequence of the axis, and is generally set to 1, 2, and 3, and sets the corresponding number of servo axis for each control axis.

8130: 表示数控机床控制的最大轴数，由CNC控制的最大轴数。

8130: Represents the maximum number of axes controlled by a CNC machine tool; Maximum number of axes controlled by CNC.

2. 数控机床与存储行程检测相关的参数

2. Parameters related to memory stroke detection for CNC machine tools

1320: 各轴的存储行程限位1的正方向坐标值，一般指定的为软正限位的值。当机床回零后，该值生效，实际位移超出该值时出现超程报警。

1320：Positive coordinate value of memory stroke limit 1 for each axis. Generally, the value of soft positive limit is specified. When the machine tool returns to zero, the value takes effect, and the overrun alarm occurs when the actual displacement exceeds the value.

1321：各轴的存储行程限位1的负方向坐标值。同参数1320基本一样，所不同的是指定的是负限位。

1321：Negative coordinate value of memory stroke limit 1 for each axis. Basically the same as parameter 1320, except that a negative limit is specified.

3. 数控机床与 DI/DO 相关的参数

3. Parameters related to DI/DO of CNC machine tools

3003♯0：是否使用数控机床所有轴互锁信号。该参数需要根据 PMC 的设计进行设定。

3003♯0：Whether use all axes of CNC machine tools for interlocking signals. This parameter needs to be set according to the design of PMC.

3003♯2：是否使用数控机床各个轴互锁信号。

3003♯2：Whether use each axis of CNC machine tools for interlocking signals.

3003♯3：是否使用数控机床不同轴向的互锁信号。

3003♯3：Whether use different axes of CNC machine tools for interlocking signals.

3004♯5：是否进行数控机床超程信号的检查，当出现 506，507 报警时可以设定。

3004♯5：Whether to check overrun signal of the CNC machine tool, which can be set in case of 506 and 507 alarms.

3030：数控机床 M 代码的允许位数。该参数表示 M 代码后边数字的位数，超出该设定出现报警。

3030：Allowable digits of code M for CNC machine tool This parameter represents the number of digits after the code M, beyond which an alarm occurs.

3031：数控机床 S 代码的允许位数。该参数表示 S 代码后数字的位数，超出该设定出现报警。例如：当 3031＝3 时，在程序中出现 S1000，即会产生报警。

3031：Allowable digits of code S for CNC machine tool. This parameter represents the number of digits after the code S, beyond which an alarm occurs. For example：When 3031＝3, an alarm will be generated when i S1000 i appears in the program.

3032：数控机床 T 代码的允许位数。

3032：Allowable digits of code T for CNC machine tool.

4. 数控机床与显示和编辑相关的参数

4. Parameters related to display and editing of CNC machine tool

3105♯0：是否显示数控机床实际速度。

3105♯0：Whether to display the actual speed of CNC machine tool.

3105♯1：是否将数控机床 PMC 控制的移动加到实际速度中显示。

3105♯1：Whether the movement controlled by PMC of CNC machine tool is added to the actual speed display.

3105♯2：是否显示数控机床实际转速、T 代码。

3105♯2：Whether to display the actual speed of CNC machine tool, T code.

3106♯4：是否显示数控机床操作履历画面。

3106♯4：Whether to display CNC machine tool operation history screen.

3106♯5：是否显示数控机床主轴倍率值。

3106♯5：Whether to display the magnification value of spindle of CNC machine tool.

3108♯4：数控机床在工件坐标系画面上，计数器输入是否有效。

3108♯4：Whether the counter input is valid if the CNC machine tool is on workpiece frame screen.

3108♯6：是否显示数控机床主轴负载表。

3108♯6：Whether to display the spindle load table of CNC machine tool.

3108♯7：数控机床是否在当前画面和程序检查画面上显示 JOG 进给速度或者空运行速度。

3108♯7：Whether the CNC machine tool shows the JOG feed speed or no-load running speed on the current screen and program check screen.

3111♯0：是否显示数控机床用来显示伺服设定画面软件。

3111♯0：Whether to display CNC machine tool to display servo setting screen software.

3111♯1：是否显示数控机床用来显示主轴设定画面软件。

3111♯1：Whether to display CNC machine tool to display spindle setting screen software.

3111♯2：数控机床主轴调整画面的主轴同步误差。

3111♯2：Spindle synchronization error of CNC machine tool spindle adjustment screen.

3112♯2：是否显示数控机床外部操作履历画面。

3112♯2：Whether to display CNC machine tool external operation history screen.

3112♯3：数控机床是否在报警和操作履历中登录外部报警/宏程序报警。

3112♯3：Whether CNC machine tool logs in external alarm/macro program alarm in alarm and operation history.

3281：数控机床语言显示，15 为中文简体。

3281：CNC machine tool language display，15 is simplified Chinese.

3208♯0：MDI 面板的功能键"SYSTEM"无效。

3208♯0：Function key "SYSTEM" of MDI panel is invalid.

三、任务实施

Ⅲ. Task Implementation

（一）在实训设备上进行参数的查找，填入下表

（Ⅰ）**Search the parameters on the training equipment and fill in the following table**

（1）按下系统面板"SYSTEM"键，出现参数画面如图 5.3.14。

（1）Press the "SYSTEM" key of the system panel to display the parameter screen as shown in Figure 5.3.14.

（2）输入需要查找的参数号，按"号查询"软键。如图 5.3.15 所示。

（2）Enter the parameter number you want to find and press the "Number Query" softkey. As shown in Figure 5.3.15.

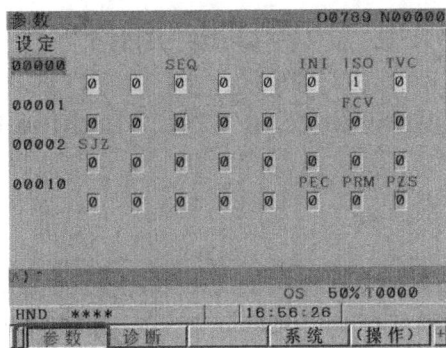

图 5.3.14　参数画面
Figure 5.3.14　parameter screen

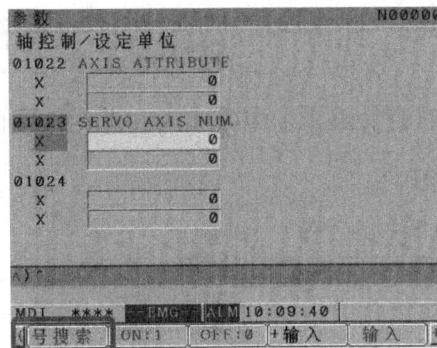

图 5.3.15　查找参数号
Figure 5.3.15　Find parameter number

3. 查询参数号，填入下表

3. Search the parameter number and fill in the following table

参数号 Parameter No.	参数值 Parameter value	含义 Meaning	备注 Remarks

（二）在实训设备上对以上介绍的画面进行操作，并写出操作步骤

（Ⅱ）**Operate the above-described screen on training equipment，and list out the operation steps**

（1）回零方式。

（1）Return-to-zero mode.

（2）手动方式。

（2）Manual mode.

（3）手轮方式。

（3）Handwheel mode.

（4）自动运行方式。

（4）Automatic operation mode.

（5）MDI 方式。

（5）MDI mode.

（6）诊断画面。

（6）Diagnostic screen.

（7）PMC 画面。

（7）PMC screen.

（8）伺服画面。

（8）Servo screen.

任务四　数控机床主轴模块电气故障诊断与维修

Task Ⅳ　Electrical Fault Diagnosis and Maintenance of CNC Machine tool Spindle

一、任务分析

Ⅰ. Task Analysis

（1）了解数控机床主轴基本工作原理及电气原理图。

（1）Understand the basic working principle and electrical schematic diagram of CNC machine tool spindle.

（2）掌握数控机床主轴简单电气故障的排查方法。

（2）Grasp the troubleshooting method of simple electrical faults of CNC machine tool spindle.

二、相关知识技能

Ⅱ. Relevant Knowledge and Skills

（一）数控机床主轴传动形式

（Ⅰ）Spindle drive form of CNC machine tool

1. 分段无级变速传动（图 5.4.1）

1. Stepless variable transmission（Figure 5.4.1）

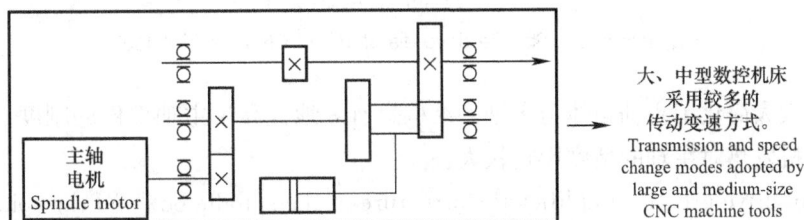

图 5.4.1　数控机床主轴传动图 1

Figure 5.4.1　Spindle drive diagram 1 of CNC machine tool

在带有齿轮变速的分段无级变速系统中：

主轴的正、反向启动与停止，制动是由伺服电机来实现；主轴变速由电机无级变速与齿轮有级变速相配合来实现；这种配置适合于大中型机床，确保主轴低速时输出大扭矩，高速时输出恒定功率特性的要求。对于这种配置形式，机械设计时都带有主轴换挡机构。

In a stepless variable speed system with gearshift：

Forward and reverse start and stop of spindle as well as the braking is realized by servo motor：Spindle speed change is realized by the combination of stepless speed change of motor and stepless speed change of gear. This configuration is suitable for large and medium-sized machine tools to ensure large torque output at low speed of spindle and output of constant power characteristics at high speed. For this configuration，the machine is designed with a spindle shift mechanism.

2. 带传动变速（图 5.4.2）

2. Belt drive speed change（Figure 5.4.2）

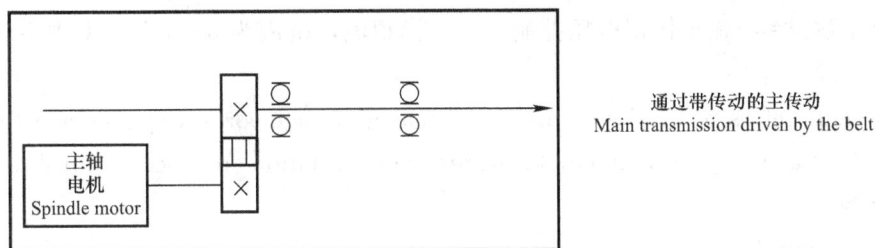

图 5.4.2　数控机床主轴传动图 2

Figure 5.4.2　Spindle drive diagram 2 of CNC machine tool

主要应用在小型数控机床上，可克服齿轮传动引起的振动与噪声，但只能适用于低扭矩特性要求的主轴。

Mainly used in small CNC machine tools，which can overcome the vibration and noise caused by gear drive，but only be suitable for the spindle with low torque characteristics.

3. 调速电机直接驱动（图 5.4.3）

3. Speed regulating motor direct drive (Figure 5. 4. 3)

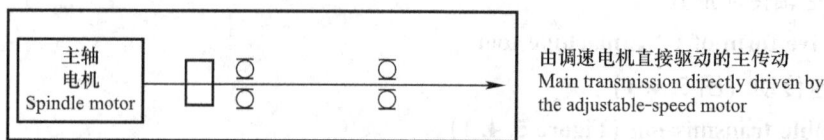

由调速电机直接驱动的主传动
Main transmission directly driven by the adjustable-speed motor

图 5.4.3　数控机床主轴传动图 3

Figure 5. 4. 3　Spindle drive diagram 3 of CNC machine tool

这种主传动方式大大简化了主轴箱体与主轴的结构，有效地提高了主轴部件的刚度，主轴转速高，但主轴输出扭矩较小，电机发热对主轴的精度影响较大。

This kind of main drive greatly simplifies the structure of the spindle box and the spindle, effectively improves the stiffness of the spindle components, the spindle speed is high, but the output torque of the spindle is small, and the motor heating has a great impact on the accuracy of the spindle.

4. 内装式电主轴

4. Built-in motorized spindle

机床主轴由内装式电机直接驱动，从而把机床主传动链的长度缩短为零，实现了机床的"零传动"。

The spindle of the machine tool is directly driven by the built-in motor, which reduces the length of the main drive chain to zero, and realizes the "zero drive" of the machine tool.

（二）数控系统主轴控制方式

（Ⅱ）CNC system spindle control mode

数控系统主轴控制主要有两大类：一类是系统输出模拟量控制，称为模拟主轴控制；另一类是系统输出串行数据控制，称为串行主轴。

There are two main types of spindle control in CNC system：One is the system output analog control, which is called analog spindle control；The other is the system output serial data control, shifted to the serial spindle.

1. 模拟主轴控制

1. Analog spindle control

模拟主轴控制指数控系统输出模拟电压控制主轴，模拟电压范围为 0～10V。实现主轴的启动、停止、正/反转以及调速等控制。

The analog spindle refers to CNC system output analog voltage control spindle and the analog voltage range is 0—10V. Start and stop the spindle，Forward/reverse rotation and speed control.

2. 串行主轴控制

2. Serial spindle control

串行主轴控制指数控系统与放大器之间数据控制和信息反馈采用串行通信。配套的主轴伺服电机也称为串行主轴电机。

Serial spindle refers to serial communication between CNC system and the amplifier for data control and information feedback. The supporting spindle servo motor is also called serial spindle motor.

（三）主轴电机类型及型号规格

（Ⅲ）Type and model specification of spindle motor

1. FANUC 数控系统配置主轴电机类型分为：ai 系列；Bi 系列；aCi 系列。

1. The type of spindle motor configured for FANUC CNC system includes: ai series; Bi series; aCi series.

2. 主轴电机代码、型号及主轴放大器匹配关系（表5.4.1）。

2. Spindle motor code，model and spindle amplifier matching relationship：(Table 5.4.1).

主轴电机规格类型 表5.4.1

Spindle motor specification & type Table 5.4.1

电机代码	电机型号	放大器型号
336	βil3/10000 (2000/10000min^{-1})	βiSVSP * −7.5
337	βil3/10000 (2000/10000min^{-1})	βiSVSP * −11
338	βil3/10000 (2000/10000min^{-1})	βiSVSP * −15
333	βil6/10000 (2000/10000min^{-1})	βiSVSP * −11
339	βil6/10000 (2000/10000min^{-1})	βiSVSP * −15
341	βil8/10000 (2000/10000min^{-1})	βiSVSP * −11
342	βil8/10000 (2000/10000min^{-1})	βiSVSP * −15
343	βil12/8000 (2000/8000min^{-1})	βiSVSP * −15
350	βil15/6000 (1200/6000min^{-1})	βiSVSP * −11
351	βil15/6000 (1200/6000min^{-1})	βiSVSP * −15
353	βil18/6000 (1000/6000min^{-1})	βiSVSP * −11
352	βil18/6000 (1000/6000min^{-1})	βiSVSP * −15

（四）主轴参数设定流程

（Ⅳ）Spindle parameter setting process

在进行主轴参数设定前，应确定与主轴硬件连接相关部件型号规格，具体有：CNC型号规格、主轴电机型号、电源模块型号、主轴放大器型号和主轴检测系统型号规格。在参数4133中输入电机代码，把3111＃1设定为1进行自动初始化，断电后再上电，系统自动加载部分主轴电机参数，然后再参考主轴参数说明书，对诸如主轴串行输出类型，主轴运行最高，最低钳制速度，主轴换挡方式，主轴反馈装置，主轴准停等参数进行手工设定。

Before setting the spindle parameters，the model and specifications of the components related to the spindle hardware connection shall be determined，specifically the CNC model and specifications，Spindle motor model，Model of power module，Spindle amplifier model，Model and specification of spindle detection system. Enter the motor code in parameter 4133，Set 3111 ＃ 1 as 1 for automatic initialization，Power off and then power on，the system automatically loads some parameters of spindle motor. Then the parameters such as the serial output type of spindle，the highest and lowest clamping speed of spindle，the shift mode of spindle，the spindle feedback device and the quasi-stop of spindle are set manually by referring to the spindle parameter specification.

（五）主轴设定画面操作

（Ⅴ）Spindle setting screen operation

（1）在急停状态下接通数控系统电源；

（1）Connect the power supply of CNC system in the state of emergency stop；

（2）按下功能键"SYSTEM"几次，进入参数设定支援画面；

（2）Press the function key "SYSTEM" several times to enter the parameter setting support screen；

（3）选择主轴设定，即进入主轴设定画面。

（3）Select Spindle setting to enter Spindle Setting screen.

（六）主轴设定画面参数说明（图5.4.4）

（Ⅵ）Spindle setting screen parameter descriptin (Figure 5.4.4)

1. 齿轮选择

1. Gear selection

显示机床一侧的齿轮选择状态。

Display the gear selection status on the machine tool side.

2. 主轴

2. Spindle

选择属于相对于哪个主轴的数据。

Select the data of the spindle to which it belongs.

3. 其他项目对应参数

3. Parameters corresponding to other items

齿轮比、主轴电高速度、C轴最高速度，对应参数号如下（表5.4.2）：

Gear ratio, electric spindle high speed, C-axis maximum speed, corresponding parameter numbers are as follows (Table 5.4.2):

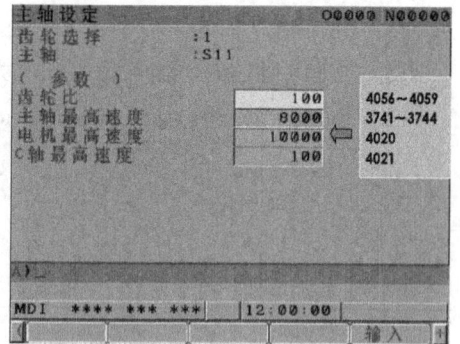

图5.4.4 主轴参数设定画面

Figure 5.4.4 Spindle parameter setting screen

主轴参数对照表   表5.4.2

Comparison table of spindle parameters   Table 5.4.2

	S11：第1主轴	S21：第2主轴	S31：第3主轴
齿轮比（HIGH）	4056	4056	4056
齿轮比（MEDIUM HIGH）	4057	4057	4057
齿轮比（MEDIUM LOW）	4058	4058	4058
齿轮比（LOW）	4059	4059	4059
主轴最高速度（齿轮1）	3741	3741	3741
主轴最高速度（齿轮2）	3742	3742	3742
主轴最高速度（齿轮3）	3743	3743	3743
主轴最高速度（齿轮4）	3744	3744	3744
电机最高速度	4020	4020	4020
C轴最高速度	4021	4021	4021

4. 其他主轴参数

4. Other spindle parameters

（1）主轴数设定参数：3701；

（1）Spindle number setting parameters：3701；

（2）齿轮选择参数：3705；

（2）Gear selection parameters：3705；

（3）主轴齿轮选择方式、主轴旋转方向参数：3706；

（3）Spindle gear selection mode, spindle rotation direction parameters：3706；

（4）与恒线速、主轴倍率相关参数：3708；

（4）Parameters related to constant linear speed and spindle magnification：3708；

（5）主轴倍率信号选择参数：3713；

（5）Spindle magnification signaling selection parameters：3713；

（6）主轴电机类型参数：3716；

（6）Spindle motor type parameters：3716；

（7）主轴配置参数：3717；

（7）Spindle configuration parameters：3717；

（8）主轴显示相关参数：3718；

（8）Parameters related to spindle display 3718；

（9）位置编码器相关参数：3720；

（9）Parameters related to position encoder 3720；

（10）主轴电机最低钳制速度相关参数：3735；

（10）Parameters related to spindle motor minimum clamping speed：3735；

（11）主轴电机最高钳制速度相关参数：3736；

（11）Parameters related to spindle motor maximum clamping speed：3736；

（12）分段无级变速主轴最大转速相关参数：3741～3744；

（12）Parameters related to maximum speed of stepless variable spindle：3741-3744；

（13）主轴换挡切换点速度相关参数：3751～3752；

（13）Parameters related to spindle shift point speed 3751～3752；

（14）周速恒定方式（G96）中主轴最低转速相关参数：3711；

（14）Parameters related to minimum spindle speed in constant speed mode（G96）：3711；

（15）主轴最高转速相关参数：3772。

（15）Parameters related to maximum spindle speed 3772.

三、任务实施

Ⅲ. Task Implementation

（一）变频调速系统的控制原理及连接、调试

（Ⅰ）Control principle，connection and commissioning of variable frequency speed control system

1. 数控车床变频调速系统的构成

1. Composition of variable frequency speed control system for CNC lathe

数控车床变频调速系统如图 5.4.5 所示：

The variable frequency speed control system of the CNC lathe is shown in Figure 5.4.5：

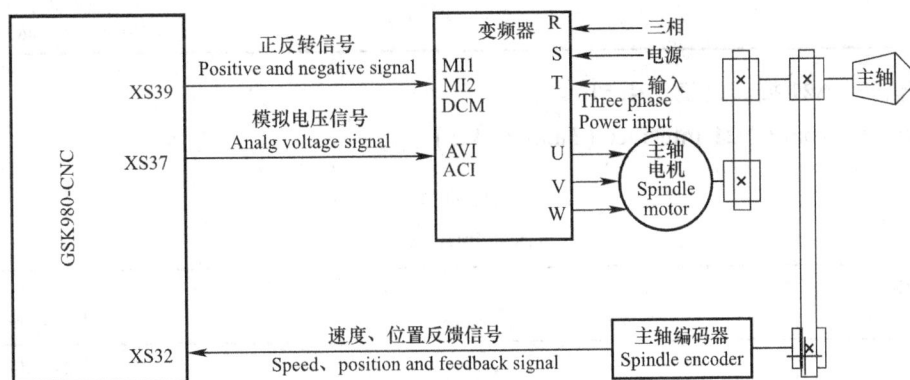

图 5.4.5　数控车床主轴变频器调速系统

Figure 5.4.5　Speed regulating system of spindle frequency converter in CNC lathe

变频器与数控装置的联系一般只需要两组线，一组是数控装置 CN62 接口到变频器的正反转信号，另一组为 CN15 接口的速度给定信号。数控装置根据 S 指令值，输出 0～10V 模拟电压给变频器，通过模拟电压的变化控制频率的变化，从而控制电机转速的变化，使主轴转速实现无级调速。主轴编码器一般通过同步齿形带与主轴相连，它测定主轴的旋转速度并通过接口 CN21 反馈给 CNC。

Generally, only two sets of wires are needed to connect the frequency converter with the CNC device. One is the positive and negative rotation signal from the CN62 interface to the frequency converter, and the other is the speed signal of the CN15 interface. CNC device outputs 0—10V analog voltage to frequency converter according to S instruction value, and controls frequency change through analog voltage change, thus controls motor speed change, and makes spindle speed realize stepless speed regulation. The spindle encoder is typically connected to the spindle via a synchronous toothed belt, which measures the speed of rotation of the spindle and feeds back to the CNC via interface CN21.

2. 转向指令信号 CN62 接口（表 5.4.3）

2. Turn instruction signal CN62 interface (Table 5. 4. 3)

CN62 接口

CN62 Interface

表 5.4.3

Table 5. 4. 3

插座针脚 Socket pin	地址 Address	功能 Function	说明 Description
4	Y0.3	M03	顺时针旋转 Clockwise rotation
5	Y0.4	M04	逆时针旋转 Anticlockwise rotation

3. 转速指令信号 CN15 接口（表 5.4.4）

3. Rotation speed instruction signal CN15 interface (Table 5. 4. 4)

CN15 接口

CN15 Interface

表 5.4.4

Table 5. 4. 4

插座针脚 Socket pin	功能 Function	说明 Description
12	GND	模拟电压 0V 端 Analog voltage 0V terminal
13	SVC—OUT1	模拟电压输出 1 Analog voltage output 1

4. 转速反馈信号 CN21 接口（表 5.4.5）

4. Speed feedback signal CN21 interface (Table 5. 4. 5)

CN21 接口

CN21 Interface

表 5.4.5

Table 5. 4. 5

插座针脚 Socket pin	名称 Name	说明 Description
7、8	*PAS/PAS	编码器 A 相脉冲 Phase A pulse of encoder
5、6	*PBS/PBS	编码器 B 相脉冲 Encoder phase B pulse

插座针脚 Socket pin	名称 Name	说明 Description
3、4	* PCS/PCS	编码器 C 相脉冲 Phase C pulse of encoder

（二）变频主轴控制原理（图 5.4.6）

（Ⅱ）Principle of variable frequency spindle control（Figure 5.4.6）

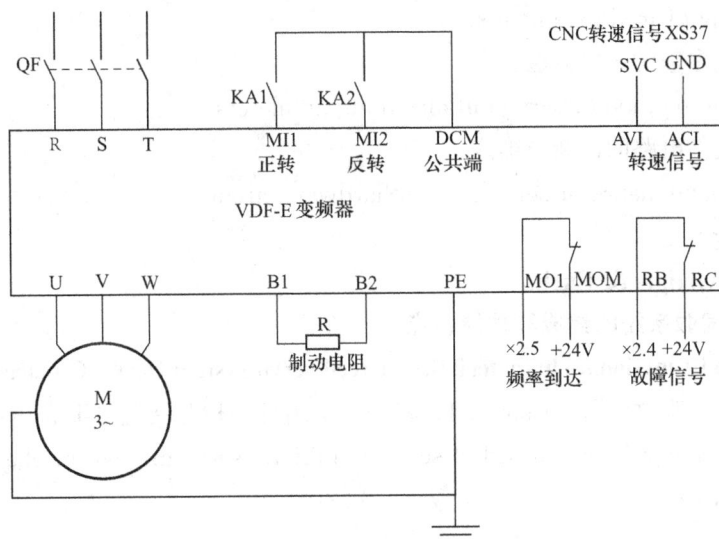

图 5.4.6　变频主轴控制原理图

Figure 5.4.6　Control principle diagram of variable frequency spindle

（1）合上 QF——三相电源引入变频器 R、S、T 端，变频器处于待机状态。

（1）Close the QF-introduce three-phase power supply into the frequency converter R，S，T terminals，the frequency converter is in standby state.

（2）正反转控制：数控系统发出正转信号 M03——XS39 的 7 号引脚输入低电平——KA1 得电——KA1 常开触点闭合——变频器端子 MO1 与 DCM 接通——主轴正转信号输入给变频器——变频器给主轴电机供电——主轴正转。反转时原理同上，但控制继电器为 KA2。

（2）Positive and negative rotation control：CNC system sends positive rotation signal M03-input low level for XS39 pin 7-KA1 energized-KA1 normally open contact closed-frequency converter terminal MO1 and DCM connected-spindle positive rotation signal input to frequency converter-frequency converter supplies power the spindle motor-spindle positive rotation. The principle is the same during the negative rotation，but the control relay is KA2.

（3）转速控制：变频器 AVI 端子接数控系统模拟量接口正信号，ACI 端子接负信号，信号为 0～10V 模拟电压信号。经数控系统调节的主轴速度输出值，通过模拟主轴接口电路，以模拟电压的方式输出至变频器，然后通过变频器变频，控制主轴速度。

（3）Rotating speed control：The AVI terminal of the frequency converter is connected with the positive signal of the analog interface of the CNC system，and the ACI terminal is connected with the negative signal，and the signal is 0—10V analog voltage signal. The output value of spindle speed regulated by CNC system is output to frequency converter by means of analog voltage through analog spindle interface circuit，and then the speed of spindle is controlled by frequency converter.

任务五　数控系统进给伺服系统故障诊断与维修

Task Ⅴ　Fault Diagnosis and Maintenance of Feed Servo System in CNC System

一、任务分析

Ⅰ. Task Analysis

（1）掌握进给参数的设置。

（1）Grasp the setting of feed parameters.

（2）掌握伺服初始化参数的设置方法。

（2）Grasp the setting method of servo initialization parameters.

（3）掌握伺服系统硬件电路的诊断与维修。

（3）Diagnosis and maintenance of servo system hardware circuit.

二、相关知识技能

Ⅱ. Relevant Knowledge & Skills

（一）数控机床进给伺服系统的组成和功能特点

（Ⅰ）Composition and functional characteristics of feed servo system for CNC machine tools

进伺服系统一般是由伺服放大器、伺服电机、机械传动组件和检测装置等组成。

Feed servo system is generally composed of servo amplifier, servo motor, mechanical transmission components and detection devices.

1. 伺服放大器

1. Servo amplifier

伺服放大器的作用是接收系统（伺服轴板）伺服信息传递信号，实施伺服电机控制，并采集检测装置的反馈信号，实现伺服电机闭环电流质量控制及进给执行部件的速度和位置控制。

The function of the servo amplifier is to receive the servo information transmission signal of the system（servo axis plate）, implement the servo motor control, and collect the feedback signal of the detection device, so as to realize the closed-loop current quality control of the servo motor and the speed and position control of the feed actuator.

2. 伺服电机

2. Servo motor

伺服电机是进给伺服系统电气执行部件，现代数控机床进给伺服电机普遍采用交流永磁同步电机，它是由定子部分、转子部分和内装编码器组成。

Servo motor is the electrical actuator of feed servo system. AC permanent magnet synchronous motor is widely used in feed servo motor of modern CNC machine tools, which is composed of stator part, rotor part and internal encoder.

3. 机械传动组件

3. Mechanical drive component

机械传动组件是将伺服电机的旋转运动转变为工作台或刀架直线运动，以实现进给运动的机械传动部件。主要包括电机与滚珠丝杠的连接装置、滚珠丝杠螺母副及其固定或支承部件、导向元件和润滑辅助装置。

The mechanical drive component is a mechanical drive part which converts the rotary motion of servo motor into linear motion of worktable or tool holder to realize the feed motion. It mainly comprises a coupling device between a motor and a ball screw, a ball screw nut pair and fixing or supporting parts thereof, a guide element and a auxiliary lubrication device.

4. 速度和位置检测装置

4. Speed and position detection device

速度和位置检测装置有伺服电机内装编码器和分离型检测装置（如分离型编码光栅尺）两种形式。

There are two types of speed and position detection devices: servo motor built-in encoder and separation type detection device (such as separation type encoding grating ruler).

（二）伺服单元的报警代码及故障原因分析

(Ⅱ) Alarm code and fault analysis of servo unit

当伺服单元出现故障时，系统将会出现"4♯♯"报警，如果伺服单元有状态窗口（LED），显示窗口将显示相应的报警代码。以βi系列伺服为例：

The "4 ♯ ♯" alarm will appear when the servo unit fails, and if the servo unit has a status window (LED), the corresponding alarm code will be displayed in the display window. Take βi series servo system as an example:

βi系列伺服放大器有STATUS1（主轴部分）、STATUS2（伺服部分）两个窗口，当伺服单元出现故障时这两个窗口也相应地显示出报警代码。例如βi系列伺服单元。

βi series servo amplifiers have two windows, STATUS1 (spindle part) and STATUS2 (servo part), and such windows will display corresponding alarm code when servo unit fails. Such as βi series servo units.

1. 伺服部分的报警

1. Alarm for servo part

（1）报警代码为1，故障原因：伺服单元输入的交流电源电压过高、制动电阻损坏或连接不良。

(1) Alarm code is 1, cause of failure: The AC supply voltage input to the servo unit is too high, the brake resistance is damaged, or the connection is poor.

（2）报警代码为2，故障原因：伺服DC24V电源电压过低。

(2) Alarm code is 2, and the fault is because of: The power supply voltage of the servo DC24V is too low.

（3）报警代码为3，故障原因：伺服单元主电路电压过低。

(3) Alarm code is 3, and the fault is because of: Servo unit's main circuit voltage is too low.

（4）报警代码为5，故障原因：伺服单元检测出过热输入信号。

(4) Alarm code is 5, and the fault is because of: The servo unit detects an overheating input signal.

（5）报警代码为8、9、b，故障原因：伺服单元检测出过电流。

(5) Alarm code is 8, 9, b, cause of fault: The servo unit detects an overcurrent.

（6）报警代码为8、9、b，故障原因：伺服单元过热、伺服单元的逆变块击穿短路或电机短路、伺服编码器内部短路。

(6) Alarm code are 8, 9 and b, and the fault are because of: Servo unit overheating, servo unit inverter block breakdown short circuit or motor short circuit, servo encoder internal short circuit.

2. 主轴部分的报警

2. Alarm for spindle part

报警发生时，STATUS1显示中报警LED（红）点亮，2位数的7段LED上显示报警代码。

When an alarm occurs, the LED (red) alarm light is on in the STATUS1 display, and the alarm code is displayed on the segment 7 of the 2-digit LED.

（1）报警代码为1，故障原因：主轴电动机内部温度大于标准温度多为主轴电机风扇故障或电机参数设定错误引起。

（1）Alarm code is 1，and the fault is because of：Spindle motor's internal temperature is higher than the standard temperature，which is mostly caused by spindle motor fan failure or motor parameter setting error.

（2）报警代码为2，主轴电机速度与指令速度差异过大多为加减速中时间参数设定不合理、速度检测器设定参数错误等引起。

（2）The alarm code is 2，the difference between the speed of the spindle motor and the instruction speed is mostly caused by the unreasonable setting of time parameters during acceleration and deceleration and the setting error of speed detector parameters.

（3）报警代码为6，电机温度传感器故障多为参数设置错误、电缆故障或控制电路板故障引起。

（3）Alarm code is 6，motor temperature sensor faults are mostly caused by parameter setting errors，cable faults or control circuit board faults.

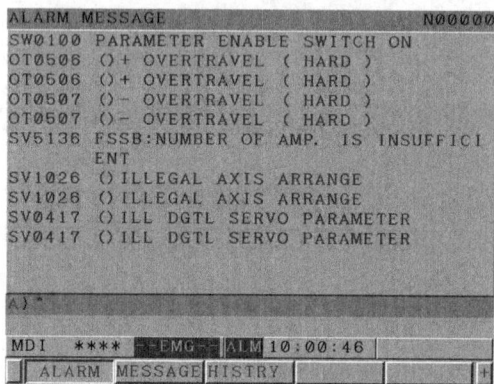

图5.5.1 开机报警画面

Figure 5.5.1 Startup alarm screen

（4）报警代码为12，主电路电流过大多为电机动力线故障、电机固有参数设定错误、功率元件损坏等引起。

（4）Alarm code is 12，the overcurrent of main circuit is mostly caused by motor power line fault，setting error of motor intrinsic parameter，power component damage and so on.

（5）报警代码为24，CNC与SVPM间串行通信数据异常，多为光缆或适配器故障引起。

（5）Alarm code is 24，abnormal serial communication data between CNC and SVPM is mostly caused by cable or adapter failure.

（6）报警代码为31，电机无法按指令速度旋转，而是停止或以极低速旋转，多由参数设定错误、电机相序错误、电机反馈线错误引起。

（6）Alarm code is 31，the motor cannot rotate according to the instruction speed，but stop or rotate at very low speed，which is mostly caused by parameter setting error，motor phase sequence error，motor feedback line error.

3. 新系统开机后出现的报警画面（图5.5.1）

3. Alarm screen after the new system is powered on（Figure 5.5.1）

4. 修改语言的操作（图5.5.2）

4. Operation of language modification（Figure 5.5.2）

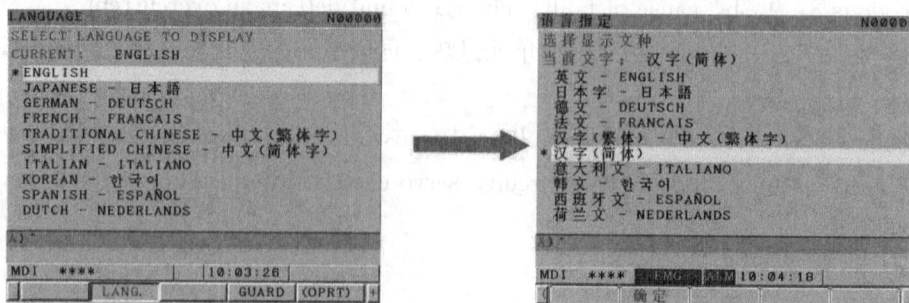

图5.5.2 语言修改画面三、任务实施

Figure 5.5.2 Language modification screen Ⅲ，task implementation

按下 "OFSSET" → "STTING" → "MENU" → "LANG" → "OPRT" → "APPLY"。

Press "OFSSET" → "STTING" → "MENU" → "LANG" → "OPRT" → "APPLY".

(三) 参数设置

(Ⅲ) Parameter settings

全清系统是同时按住 "RESET"+"DELETE"，全清前，请备份。

Press "RESET"+"DELETE" at the same time to clear system，before full-clear，please backup.

常用参数设置 (表 5.5.1)

Common parameter settings (Table 5.5.1).

<table>
<tr><td colspan="3" align="center">常用参数表
Common Parameter table</td><td align="right">表 5.5.1
Table 5.5.1</td></tr>
<tr><td>参数号
Parameter No.</td><td>一般设定值
General settings</td><td colspan="2">说明
Description</td></tr>
<tr><td>0000♯1</td><td>1</td><td colspan="2">输出数据位 ISO 代码
Output data bit ISO code</td></tr>
<tr><td>103，113</td><td>10</td><td colspan="2">波特率
Baud rate</td></tr>
<tr><td>20</td><td>4</td><td colspan="2">输入设备接口号，4 为存储卡
Enter the device interface number，4 is the memory card</td></tr>
<tr><td>1005♯0</td><td>1</td><td colspan="2">未回零执行自动运行，调试时为 1，否则有 (PS224) 报警
Automatic operation is performed if not returning to zero，it displays 1 at commissioning，otherwise there is (PS224) alarm</td></tr>
<tr><td>1006♯0</td><td>0</td><td colspan="2">直线轴，一般是直线运动的轴，千万不要想到是电机旋转，设为回转轴，回转工作台才是回转轴
Linear axis generally is the axis of linear motion. It is not motor rotation，set as the rotation axis，the rotation table is the rotation axis</td></tr>
<tr><td>1006♯3</td><td>1</td><td colspan="2">车床 X 轴，直径编程和半径编程
Lathe X-axis，diameter programming and radius programming</td></tr>
<tr><td>1020</td><td>88，90</td><td colspan="2">轴名称，设定值为轴名称的 ACSSII 码
Axis name，ACSSII code with setting value as axis name</td></tr>
<tr><td>1022</td><td>1，3</td><td colspan="2">设定各轴为基本坐标系中的那个轴，2 为 Y 轴，车床没有 Y 轴
Set each axis as the axis in the basic frame，2 as the Y axis，and the lathe has no Y axis</td></tr>
<tr><td>1023</td><td>1，2</td><td colspan="2">轴连接顺序；轴屏蔽设置为 -128，2009♯1=1
Axis connection sequence；Axis shield setting -128，2009♯1=1</td></tr>
<tr><td>3401♯0</td><td>1</td><td colspan="2">指令数值单位，mm，否则默认为 μm，后面所有数据要按 μm 设置，需要输入很多 0
Instruction value unit，mm，otherwise μm，in default；all subsequent data shall be in μm Setting，you need to enter a lot of zeros</td></tr>
<tr><td>1320</td><td>调试为 99999999
Adjust as99999999</td><td colspan="2">存储行程限位正极限，这个值调试为 99999999，在设置好参考点后，手摇方式移动轴接近机械极限位置，读取机械坐标值。超出 500 后报警
Positive limit of memory travel limit，which is adjusted to 99999999. After setting the reference point，move the axis by hand close to the mechanical limit position and read the mechanical coordinates. 500 alarm occurs in case of excess</td></tr>
<tr><td>1321</td><td>调试为 99999999
Adjust as99999999</td><td colspan="2">存储行程限位负极限，同上，超出 501 报警，如果设置 1320 小于 1321，500，501 报警自动忽略
Memory travel limit negative limit，same as above，501 alarm occurs in case of excess，if setting 1320 is less than 1321，500，501 alarm will be ignored automatically</td></tr>
</table>

续表

参数号 Parameter No.	一般设定值 General settings	说明 Description
1401♯0	调试为1 Adjust as1	未回零执行手动快速操作，未设置会发现快移键无效果，值为1420 Manual fast moving operation is performed if not returning to zero. In case of no setting, you will find the snapshot key ineffective，1420 value
1410	1000	空运行速度 No-load running speed
1420	3000	各轴快移速度 Fast moving speed of each axis
1421	1000	各轴快移倍率为 F0 的速度 Speed at which each axis is shifted at a magnification of F0
1423	3000	各轴手动速度 Manual speed of each axis
1424	同 1420 Same as 1420	各轴手动快移速度，也可以为 0，详细见参数手册 P76 Each axis can also be moved quickly manually at speed of 0. See Parameter Manual P76 for details
1425	300-400	各轴返回参考点 FL 的速度 Speed at which each axis returns to reference point FL
1430	1000	各轴最大切削进给速度 Maximum cutting feed rate for each axis
1620	50-200	快移时间常数，设置过大，会发现按键和轴移动反应有点慢 Fast moving time constant is set too large, and you will find that the key and axis movement reaction is a little slow
1622	50-200	切削进给时间常数 Cutting feed time constant
1624	50-200	JOG 时间常数 JOG time constant
1815♯4	1	机械位置和绝对位置编码器的对应关系未建立，出现 300 报警 The correspondence between the mechanical position and the absolute position encoder was not established and 300 alarms occurred
1815♯5	1	采用绝对值编码器，带电池 Absolute value encoder with battery is used
1820	2	CMR 值，指令倍乘比，详细见参数手册 P108 CMR value, instruction multiplication ratio, refer to Parameter Manual P108 for details
1821	5000	参考计数器容量，对绝对值编码器的意义不大，和回零有关 The counter capacity reference is of little significance to the absolute value encoder and is related to the return to zero
1825	3000	各轴位置环增益，这些参数不设会出现 417 报警 Gain of each axis position loop, 417 alarm may occur if these parameters are not set.
1826	20	各轴到位宽度 Width of each axis in place
1827	20	切削进给时的到位宽度 Width in place for cutting feed
1828	10000	各轴移动位置极限偏差 Limit deviation of moving position of each axis
1829	200	各轴停止位置极限偏差 Limit deviation of stopping position of each axis

参数号 Parameter No.	一般设定值 General settings	说明 Description
2003♯3	1	P-I 控制方式 P-I control mode
2003♯4	1	停止时微小振动设 1 The slight vibration is set as 1 during stop.
2020	256	电机代码 Motor code
2021	200	负载惯量比 Load inertia ratio
2022	111	电机旋转方向，反向为－111，见参数说明书 P135 Motor rotation direction, reverse direction－111, see Parameter Manual P135
2023	8192	速度反馈脉冲数 Number of feed feedback pulses
2024	12500	位置反馈脉冲数，半闭环设置 12500 Number of position feedback pulses, semi-closed loop setting 12500
2084，2085	1/200	柔性齿轮比 Flexible gear ratio
3003♯0	1	互锁信号无效，G8.0 Interlock signal is invalid, G8.0
3003♯2	1	各轴互锁信号无效，G130 Interlock signals in different axes are invalid, G130
3003♯3	1	不同轴向的互锁信号无效，G132，G134 Interlock signals in different axis directions are invalid, G132, G134
3004♯5	1	硬超程信号无效，出现 506、507 设置，G114，G116 Hard override signal is invalid with 506, 507 settings, G114, G116
3105♯0	1	实际进给速度显示 Actual feed speed display
3105♯2	1	主轴速度和 T 代码显示 Spindle speed and T-code display
3106♯4	1	操作履历画面显示 Operation history screen display
3106♯5	1	主轴倍率显示 Spindle magnification display
3108♯6	1	显示主轴负载表 Display spindle load table
3108♯7	1	实际手动速度显示 Actual manual speed display
3111♯0	1	伺服调整画面显示 Servo adjustment screen display
3111♯1	1	主轴设定画面显示 Spindle setting screen display
3111♯2	1	主轴调整画面显示 Spindle adjusting screen display
3111♯5	1	操作监控画面显示 Monitoring interface operation display

参数号 Parameter No.	一般设定值 General settings	说明 Description
3112♯2	1	外部操作信息履历画面显示 External operation information history screen display
8130	2	控制轴数 Number of control axis
8131♯0	1	手轮有效 Handwheel valid
7113	100	手轮进给倍率 m，不设置手轮的 X100 倍率无效 Handwheel feed magnification m, X100 magnification without setting handwheel is invalid
7114	0	手轮进给倍率 n Handwheel feed magnification n
3716	0	模拟主轴 Analog spindle
3717	1	主轴放大器号 Spindle amplifier number
3718	80	显示下标 Show subscript
3720	4096	主轴编码器脉冲数 Number of spindle encoder pulse
3730	1000	主轴速度模拟输出的增益调整 Gain adjustment of spindle speed analog output
3735	0	主轴最低钳制速度 Minimum clamping speed of spindle
3736	1400	主轴最高钳制速度 Maximum clamping speed of spindle
3741	1400	主轴最大速度 Maximum spindle speed
3772	0	主轴上限钳制。设为 0，不钳制 Upper limit clamping of spindle Set as 0, Not clamped
8133♯5	1	不使用串行主轴 Serial spindle not used

1. 伺服参数的设置

1. Setting of servo parameters

伺服初始化参数的设置，进入初始化界面操作方法如下：

Setting of the servo initialization parameters and method to enter the initialization interface：

首先连续按"SYSTEM"键 3 次进入参数设定支援画面如图 5.5.3。

First，press the "SYSTEM" key three times to enter the parameter setting support screen as shown in Figure 5.5.3.

将光标移动到伺服设定上，然后按操作键进入选择画面，如图 5.5.4 所示。

Move the cursor to the servo setting and press the Operation

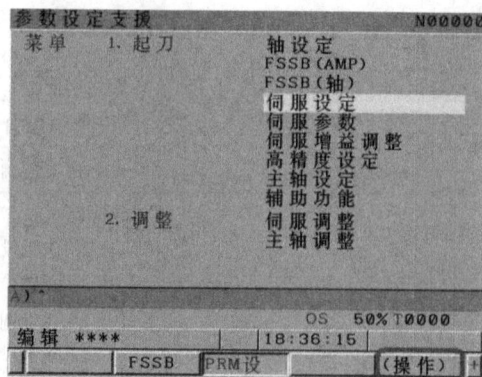

图 5.5.3　参数设定画面

Figure 5.5.3　Parameters setting screen

key to enter the selection interface, as shown in Figure 5.5.4.

在此界面按选择键进入伺服设定画面，如图 5.5.5 所示。

In this interface, press Select key to enter the servo setting screen, as shown in Figure 5.5.5.

图 5.5.4 选择画面

Figure 5.5.4 Selection interface

图 5.5.5 设定画面

Figure 5.5.5 Servo setting screen

在此界面按向右扩展键进入菜单与切换画面，如图 5.5.6 所示。

In this interface, press the Right Expansion key to enter the menu and switch screen, as shown in Figure 5.5.6.

图 5.5.6 切换画面

Figure 5.5.6 Mena and swich screen

在此界面按切换键进入伺服初始化画面，如图 5.5.7 所示。

Press the Switch key in this interface to enter the servo initialization interface, as shown in Figure 5.5.7.

在此界面便可以对伺服进行初始化操作，下面对每一项内容进行详细介绍。

The servo can be initialized in this interface and each item is described in detail below.

第一项为机床初始化位，初始化时设定为 0，也可以设定参数 1902♯0 位为 0。

The first item is the machine tool initialization bit, which is set to 0 at the time of initialization, or the parameter 1902 ♯ 0 bit may be set to 0.

设置完成后重启，此时应该无任何报警出现，表示设置完成。

Restart when setup is complete, no alarms should appear at this time, indicating that setup is complete.

2. 使用 Fanuc 参数一键设定功能

2. Use Fanuc parameters one-button setting function

当系统第一次通电时，需要进行全清处理（上电时，同时按 MDI 面板上"RESET"和"DEL"）。如图 5.5.8 所示。

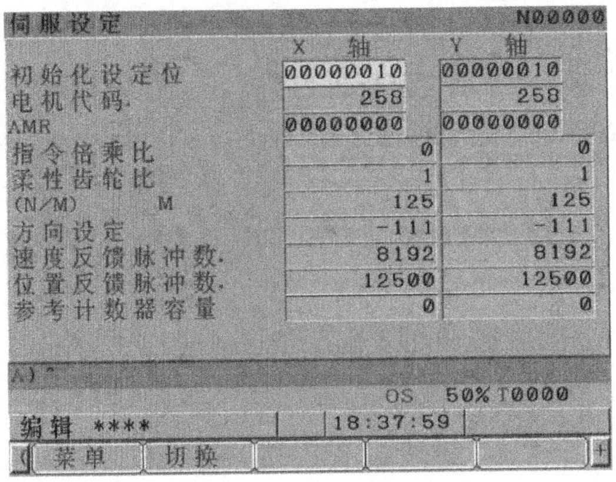

图 5.5.7 伺服初始画面

Figure 5.5.7 Servo initial screen

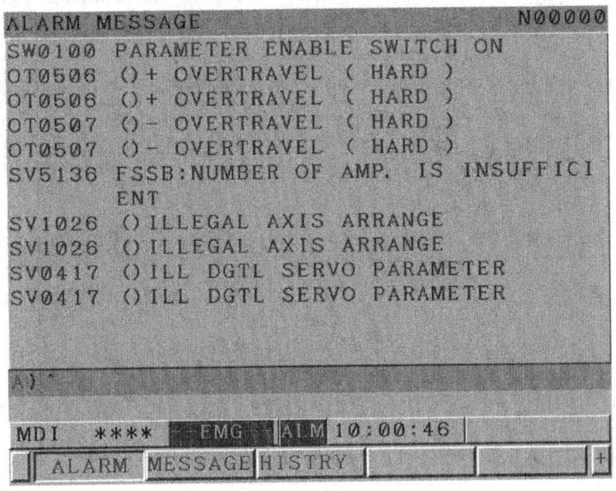

图 5.5.8 一键设定画面

Figure 5.5.8 One-batton setting screen

When the system is powered on for the first time，it needs to be fully cleared (During power-up, press "RESET and DEL" on the MDI panel at the same time). As shown in Figure 5.5.8.

系统基本参数设定可通过参数设定支援画面进行操作。参数设定支援画面是以下述目的进行参数设定和调整的画面。通过在机床启动时汇总需要进行最低限度设定的参数并予以显示，便于机床执行启动操作，通过简单显示伺服调整画面、主轴调整画面、加工参数调整画面，更便于进行机床的调整。

The basic parameter setting of the system can be operated through the parameter setting support screen. The parameter setting support screen is a screen for setting and adjusting parameters for the following purposes. By summarizing and displaying the parameters which need to be set at minimum when the machine tool is started，it is convenient for the machine tool to carry out the start-up operation. By simply displaying the servo adjustment screen, the spindle adjustment screen and the processing parameter adjustment screen，it is more convenient for the machine tool to adjust.

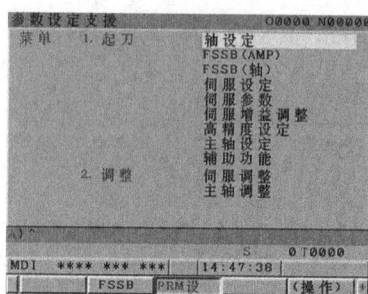

图 5.5.9　参数设定支持画面

Figure 5.5.9　Parameter setting support screen

3. 参数设定支援画面显示方法
3. Display method of parameter setting support screen

通过以下步骤可显示该画面。

You can display this screen by following these steps.

操作步骤：按下功能键"SYSTEM"后，按继续菜单键"＋"数次，显示软键"PRM 设定"。按下软键［PRM 设定］，出现参数设定支援画面。如图 5.5.9 所示。

Operation steps：Press the function key "SYSTEM" and press the Continue menu key "＋" several times to display the softkey "PRM Settings". Press the softkey "PRM Setting" to display the parameter setting support screen. As shown in Figure 5.5.9.

4. 标准值设定
4. Standard value setting

通过软键"初始化"，可以在对象项目内所有参数中设定标准值。

The softkey "Initialize" allows you to set standard values in all parameters within the object item.

（1）初始化只可以执行如下项目：轴设定、伺服参数、高精度设定、辅助功能。

（1）Initialization can only allow executing the following items：Axis Setting, Servo Parameters, High Precision Setting, Auxiliary Function.

（2）进行本操作时，为了确保安全，请在急停状态下进行。

（2）To ensure safety，only perform this operation in the emergency stop state.

（3）标准值是 FANUC 建议使用的值，无法按照用户需要个别设定标准值。

（3）Standard values are values recommended by FANUC and cannot be set individually as required by the user.

（4）本操作中，设定对象项目中所有的参数，但是也可以进行对象项目中各组的参数设定，或个别设定参数。

（4）In this operation，all parameters in the object item are set，but parameters of each group in the object item may be set，or parameters may be set individually.

（5）标准值设定操作步骤如下说明：

（5）The steps of standard value setting are as follows：

在参数设定支援画面上，将光标指向要进行初始化的项目。按下软键"操作"，显示如下软键"初始化"。如图 5.5.10 所示。

On the parameter setting support screen，point the cursor to the item to be initialized．Press the softkey "Operate"，Display the following softkey "Initialize"．As shown in Figure 5.5.10.

按下软键"初始化"。软键按如下方式切换，显示警告信息是否设定初始值？

Press the softkey "Initialize"．The softkey is switched as follows to display the warning message．Whether to set the initial value？

按下软键"执行"，设定所选项目的标准值。通过本操作，自动地将所选项目中所包含的参数设定为标准值。如图 5.5.11 所示

Press the softkey "Execute" to set the standard value for the selected item．Through this operation，it automatically sets the parameters contained in the selected item as standard values. As shown in Figure 5.5.11

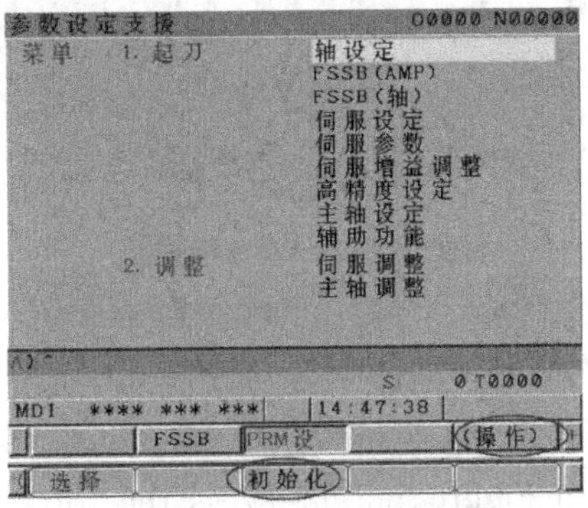

图 5.5.10　标准值参数设定 1　　　　图 5.5.11　标准值参数设定 2
Figure 5.5.10　Standard value setting 1　　Figure 5.5.11　Standard value setting 2

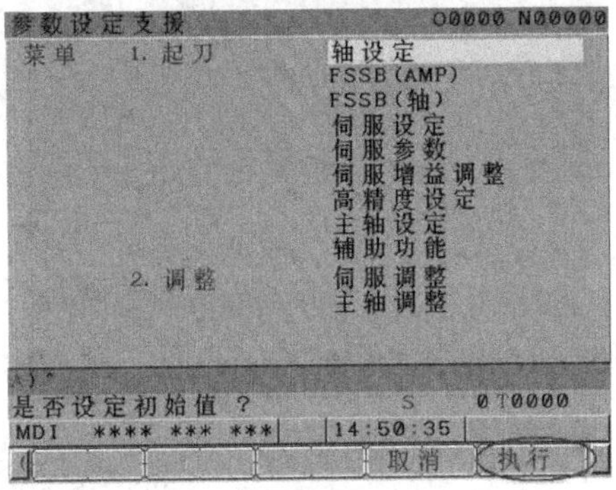

不希望设定标准值时，按下软键"取消"，即可中止设定。另外，没有提供标准值的参数，不会被变更。

When you do not want to set the standard value，press the softkey "Cancel" to exit the setting．In addition，parameters that do not provide a standard value are not changed.

（6）与轴设定相关的 NC 参数初始设定

（6）Initial setting of NC parameters related to axis setting

准备：进入参数设定支援画面，按下软键"操作"，将光标移动至轴设定处，按下软键"选择"，出现参数设定画面。此后的参数设定，就在该画面进行。如图 5.5.12 所示。

Preparation：Enter the parameter setting support screen，press the softkey "Operation"，move the cursor to the xis setting，press the softkey "Select" to display the parameter setting screen．Thereafter，the parameter setting is performed on the screen．As shown in Figure 5.5.12.

在参数设定画面上进行参数的初始设定。在参数设定画面上，参数被分为几个组，并被显示在每组的连续页面上。

Initial setting of parameters is performed on the parameter setting screen．On the parameter setting screen，the parameters are divided into groups and displayed on continuous pages of each group.

1）基本组

1）Basic group

参数设定，包含标准值和非标准值。

Parameter setting，including setting of standard and non-standard values.

① 标准值设定

① Standard value setting

进行基本组的参数标准设定。

Conduct parameter standard setting for the basic group.

按下 "PAGE UP/PAGE DOWN" 键数次，显示出基本组画面，如图 5.5.13 所示。而后按下软键 "GR 初期"。

Press the "PAGE UP/PAGE DOWN" key several times to display the basic group screen, as shown in Figure 5.5.13 and then press the softkey "GR Initial".

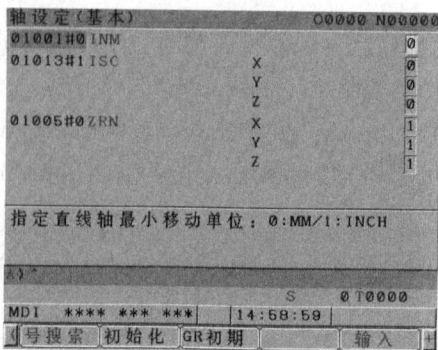

图 5.5.12　参数设定画面
Figure 5.5.12　Parameter setting screen

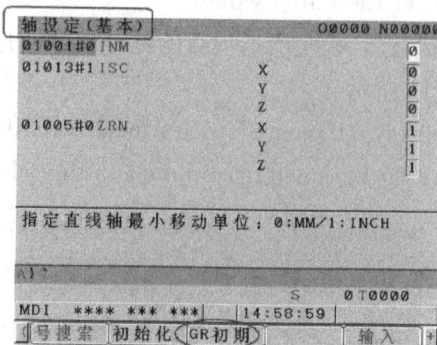

图 5.5.13　基本组画面
Figure 5.5.13　Basic group screen

画面上出现是否设定初始值? 提示信息。按下软键 "执行"。如图 5.5.14 所示。

图 5.5.14　提示信息
Figure 5.5.14　Prompt message

The screen displays: whether to set initial value Prompt message. As shown in Figure 5.5.14 Press the softkey "Execute" As shown in Figure 5.5.14.

至此，基本组参数的标准值设定完成。

At this point, the standard value setting of the basic group parameters is completed.

注释

Comment

(1) 无论从组内的哪个页面上选择 "GR 初期"，对于组内的所有页面上的参数，均进行标准值设定。

(1) No matter from which page in the group you select "GR Initial", standard values are set for parameters on all pages in the group.

(2) 有的参数没有标准值。即使进行了标准值的设定，这些参数的值也不会被改变。

(2) Some parameters do not have a standard value. Even if standard values are set, the values of these parameters are not changed.

(3) 根据标准值设定，有时会出现报警（PW0000）必需关断电源，并切换到报警画面，但是，此时不必立即切断电源。请按照（1）准备中的说明，重新显示出参数设定画面，进入下一步骤。

(3) According to the standard value setting, sometimes the alarm (PW0000) may occur, it must turn off the power supply and switch to the alarm screen, but it is not necessary to cut off the power supply immediately at this time. Follow the instructions in (1) Preparation to redisplay the parameter setting screen and proceed to the next step.

② 没有标准值的参数设定

② Parameter setting without standard value

有的参数是没有标准值的，还需要根据配置进行手工设定。见表 5.5.2。

Some parameters do not have standard values and need to be set manually according to the configuration. As shown in Table 5.5.2.

表 5.5.2

Table 5.5.2

无标准值参数设定

Parameters table 1 with standand value

参数号 Parameter No.	一般设定值 General settings	说明 Description
1001♯0	0	
1013♯1	0	
1005♯1	0	本设备中不用 Not used in this equipment
1006♯0	0	
1006♯3	0	
1006♯5	0	本设备中不用 Not used in this equipment
1815♯1	0	
1815♯4	1	
1815♯5	1	使用绝对值编码器 Use absolute value encoder
1825	3000	
1826	10	
1828	7000	
1829	500	

2）主轴组按下"PAGE"键进入主轴组①标准值设定

2）Press the "PAGE" key of the spindle group to enter the spindle group①Standard value setting

① 标准值设定

① Standard Value Setting

进行主轴组的参数标准值的设定。

Set the parameter standard value of the spindle group.

以与"基本组"的"标准值设定"相同的步骤进行设定。

Set it in the same steps as the "Standard Value Setting" of the "Basic Group".

② 没有标准值的参数设定（表 5.5.3）

② Parameter setting without standard value (Table 5.5.3)

表 5.5.3

Table 5.5.3

无标准值参数表 2

Parameters table 2 without standard value

参数号 Parameter No.	一般设定值 General settings	说明 Description
3716	0	
3717	1	
3718	80	
3720	4096	
3730	1000	
3735	0	
3736	1400	
3741	1400	
3772	0	
8133♯5	1	

3）坐标组

3）Coordinate group

① 标准值设定

① Standard value setting

进行坐标组的参数标准值的设定。

Set the parameter standard value of the coordinate group.

以与"基本组"的"标准值设定"相同的步骤进行设定。

Set it in the same steps as the "Standard Value Setting" of the "Basic Group".

② 没有标准值的参数设定（表5.5.4）

② Parameter setting without standard value （Table 5.5.4）

无标准值参数表3   表5.5.4

Parameters table 3 without standard value   **Table 5.5.4**

参数号 Parameter No.	一般设定值 General settings	说明 Description
1240	0	
1241	0	
1320	99999999	调试时设置 Settings during commissioning
1321	99999999	调试时设置 Settings during commissioning

4）进给速度组

4）Feed speed group

① 标准值设定

① Standard value setting

进行进给速度的参数标准值的设定。

Set the parameter standard value of the feed speed.

与"基本组"的"标准值设定"相同的步骤进行设定。

Set it in the same steps as the Standard Value Setting of the Basic Group.

② 没有标准值的参数设定（表5.5.5）

② Parameter setting without standard value （Table 5.5.5）

无标准值参数表4   表5.5.5

Parameters table 4 without standard value   **Table 5.5.5**

参数号 Parameter No.	一般设定值 General settings	说明 Description
1410	1000	
1420	5000	
1421	1000	
1423	1000	
1424	5000	
1425	150	

参数号 Parameter No.	一般设定值 General settings	说明 Description
1428	5000	
1430	3000	

5）进给控制组

5) Feed control group

该组无标准参数，需要手工设定（表5.5.6）。

This group has no standard parameters and needs to be set manually (Table 5.5.6).

无标准值参数表 5

Parameters table 5 without standard value

表 5.5.6

Table 5.5.6

参数号 Parameter No.	一般设定值 General settings	说明 Description
1610＃0	0	
1610＃4	0	
1620	100	
1622	32	
1623	0	
1624	100	
1625	0	

断开 NC 的电源，而后再接通。通过上述操作，与轴设定相关的 NC 参数的初始设定到此结束。

Disconnect the power to the NC and connect it again. Through the above-described operation, the initial setting of NC parameters related to the axis setting is finished here.

此时，轴还是不能移动，还需要设置（PMC 正确的前提下），还需要设置如下参数（表5.5.7）：

At this time, the axis still cannot be moved and needs to be set (under the premise that the PMC is correct). The following parameters need to be set as well (Table 5.5.7)：

无标准值参数表 6

Parameters table 6 without standard value

表 5.5.7

Table 5.5.7

参数号 Parameter No.	一般设定值 General settings	说明 Description
3003＃0	1	
3003＃2	1	
3004＃5	1	
3003＃3	1	

（7）FSSB 的设定：在伺服设定画面、伺服调整画面上进行下列设定。如图 5.5.15 所示。

(7) Setting of FSSB: The following settings are made on the servo setting screen and the servo adjustment screen. As shown in Figure 5.5.15.

（8）手摇的设定（表 5.5.8）。

(8) Hand-held device settings (Table 5.5.8).

在急停状态下接通NC 的电源
Turn on the power supply of NC in the emergency stop state

在急停状态下接通NC 的电源
Turn on the power supply of NC in the emergency stop state

初始化设定位
Initialization setting
bit　　　　　　　　00000000
电机代码
Motor code　　　　256
AMR　　　　　　　00000000
指令倍乘比
Instruction
multiplication ratio　2
方向设定　　　　　111(从检测器端看沿顺时针方向)
Direction setting　　111(Clockwise from detector end)
　　　　　　　　　-111(从检测器端看沿逆时针方向)
　　　　　　　　　-111(Anticlockwise from detector end)

参考计数器容量
Reference　　counter
capacity　　　　　5000

全闭环　　　　　　全闭环/半闭环　　　　　半闭环
Full closed loop　　Full/semi-closed loop　　Semi-closed loop

使用分离型检测器
Use detached detectors
(No.1815#1=)

设定柔性齿轮比　　　　　　　　　　设定柔性齿轮比
Set flexible gear ratio　　　　　　　　Set flexible gear ratio

速度反馈脉冲　　　　　　　　　　　速度反馈脉冲
Speed feedback　8192　　　　　　　Speed feedback　8192
pulses　　　　　　　　　　　　　　pulses
位置反馈脉冲　　　　　　　　　　　位置反馈脉冲
Position　　　　　　　　　　　　　Position
feedback pulse　Ns　　　　　　　　feedback pulse　12500

电源OFF/ON
Power supply OFF/ON

参数设定结束
End of parameter setting

电源OFF/ON
Power supply OFF/ON

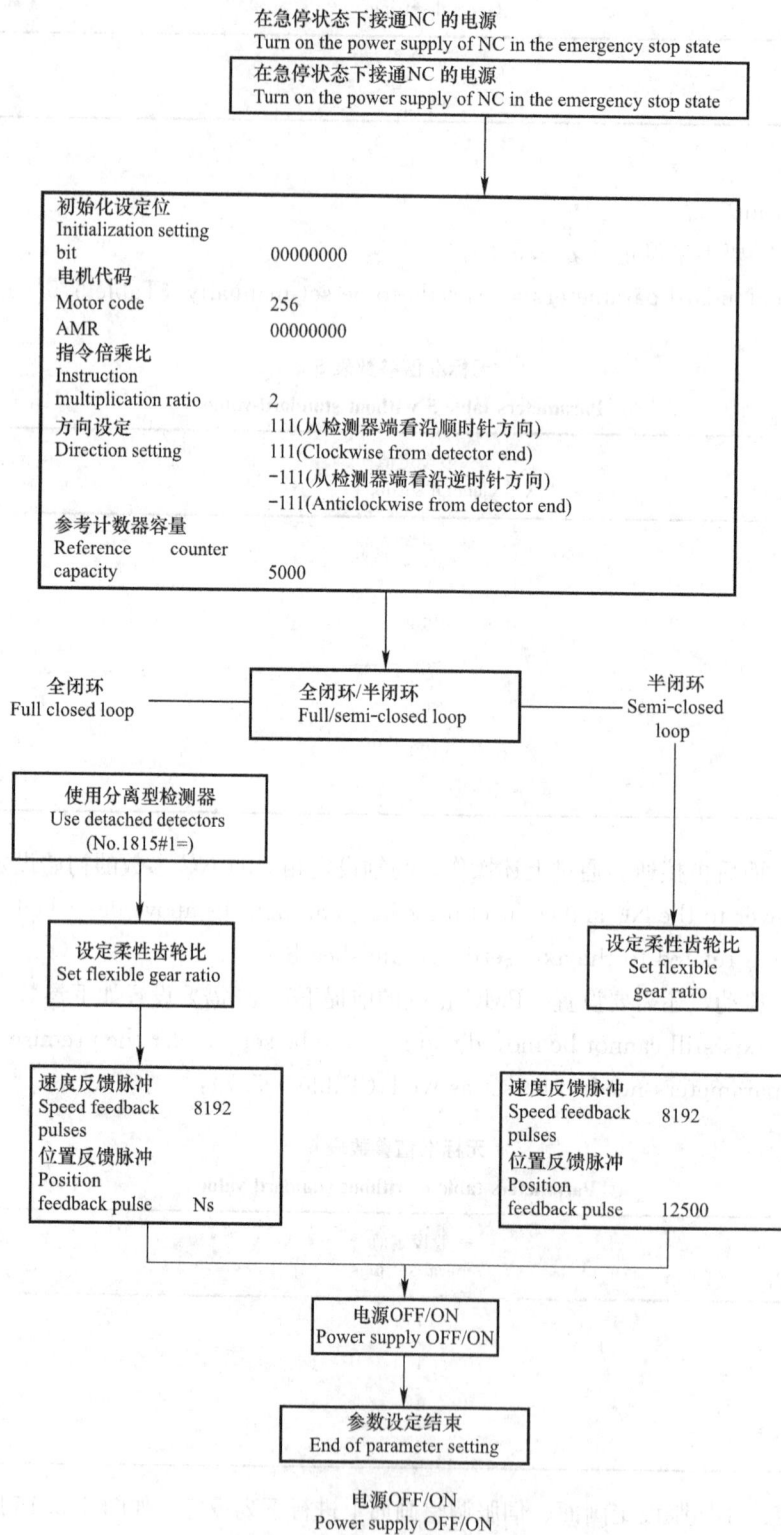

图5.5.15　参数设定结束
Figure 5.5.15　End of parameter setting

表 5.5.8
Table 5.5.8

手轮参数
Handwheel parameters

参数号 Parameter No.	一般设定值 General settings	说明 Description
8131#0	1	
7113	100	
7114	0	

任务六　数控机床换刀装置故障维修

Task Ⅵ　Troubleshooting of CNC Machine Tool Changer

一、任务分析

Ⅰ. Task Analysis

1. 了解四工位电动刀架的工作原理

1. Understand the working principle of four-station electric tool holder

2. 掌握刀架电气故障诊断与维修

2. Master electrical fault diagnosis and maintenance of tool holder

二、相关知识技能

Ⅱ. Relevant Knowledge & Skills

（一）四工位电动刀架的机械结构及原理

（Ⅰ）Mechanical structure and principle of four-station electric tool holder

1. 四位电动刀架的机械结构

1. Mechanical structure of four-station electric tool holder

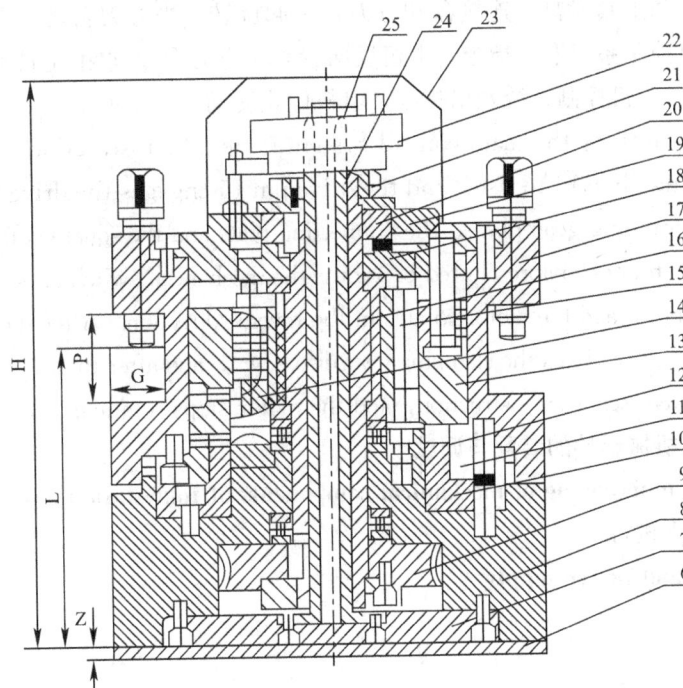

图 5.6.1　四工位电动刀架机械结构原理图（一）

Figure 5.6.1　Mechanical structure schematic diagram of four-station electric tool holder（Ⅰ）

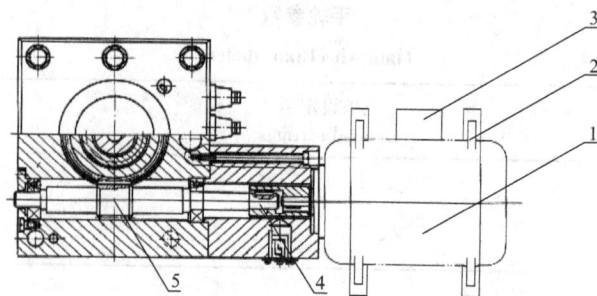

图 5.6.1 四工位电动刀架机械结构原理图（二）

Figure 5.6.1 Mechanical structure schematic diagram of four-station electric tool holder（Ⅱ）

如图 5.6.1 所示，四工位电动刀架机械部件一般由以下部件组成：1—电机；2—电机座；3—接线盒；4—传动套；5—蜗杆轴；6—调整垫；7—底盘；8—底座；9—蜗轮；10—定齿盘；11—动齿盘；12—定位盘；13—夹紧齿盘；14—定位销；15—传动销；16—螺杆；17—方刀台；18—传动盘；19—连接盘；20—固定环；21—螺母；22—机械开关；23—罩；24—立轴；25—螺母

As shown in Figure 5.6. 1 above, the mechanical component of the four-stvation electric tool holder generally consists of the following parts：1—Electric motor；2—Motor base；3—Junction box；4—Drive sleeve；5—Worm axis；6—Adjustable pad；7—Chassis；8—Base；9—Worm gear；10—Fixed gear disc；11—Movable gear disc；12—Positioning plate；13—Clamping gear disc；14—Positioning pin；15—Driving pin；16—Screw；17—Square knife stand；18—Drive disc；19—Connecting plate；20—Fixed ring；21—Screw nut；22—Mechanical switch；23—Cover；24—Vertical axis；25—Screw nut；

2. 四工位电动刀架的工作过程及原理

2. Working process and principle of four-station electric tool holder

基本动作是主机控制系统发出指令，电机 1 转动，螺杆 16 带动夹紧盘 13 升起，传动销 15 与螺杆 16 上的传动盘 18 接合，从而带动平紧齿盘，连接盘和方刀台一起旋转。当方刀台转动到预先选好的位置时，机械开关 22 发出信号，主机控制系统发出指令，电机反向旋转，方刀台反向粗 定位后，螺杆带动夹紧齿盘下降，三个齿盘啮合锁紧后，主机控制系统发出指令，电机停止转动。

The basic operations are that：the main control system issues the instruction，the motor 1 rotates，the screw 16 drives the clamping plate 13 to rise，and the drive pin 15 engages the drive plate 18 on the screw 16，thereby driving the flat tightening gear plate，the connecting plate and the square cutter table to rotate together. When the square cutter table rotates to apreselected position，the mechanical switch 22 sends a signal，the host control system sends an instruction，and the motor rotates in the reverse direction. After the square cutter table rotates in the reverse direction，the screw drives the clamping gear disc down，and after the three gear discs are engaged and locked，the host control system sends an instruction，and the motor stops rotating.

（二）四工位电动刀架机械部件的装配与调试

（Ⅱ）Assembly and commissioning of mechanical components of four-station electric tool holder

1. 必须在断电的情况下进行

1. It must be done without power supply

2. 电机的拆卸

2. Motor removal

打开接线盒 3，拨开线头，松开电机 1 螺钉，取下电机。

Open junction box 3，pull off wire head，loosen motor 1 screw，remove motor.

3. 方刀台的拆卸

3. Square tool table removal

取下上端防护罩 23，松开螺母 25 拆下机械开关 22、螺母 21、再松开连接盘 19 螺钉，拆下连接盘、固定环 20 和传动盘 18，旋转平紧齿盘使其脱离螺杆，即可向上取出方刀台，安装与之方法相反。

Remove upper end shield 23, loosen nut 25, remove mechanical switch 22, nut 21, loosen connecting plate 19 screws, remove connecting plate, retaining ring 20 and drive plate 18, rotate flat tightening gear disc to remove it from screw rod and remove square tool table upwards, or you may install in opposite steps.

4. 蜗杆轴的拆卸

4. Worm axis removal

卸下电机座 2，拆下堵 26、27 将蜗杆轴向右端推出。

Remove motor housing 2, remove plugs 26, 27 and push worm axis to the right end.

5. 底部的拆卸

5. Bottom removal

取下上端罩 23，松开螺母 25 拆下机械开关 22、抽出导线，松开底盘 7 螺钉即可将立轴和底盘卸下。

Remove upper end shield 23, loosen nut 25, remove mechanical switch 22, pull out lead, loosen chassis 7 screws to remove vertical axis and chassis.

6. 润滑

6. Lubrication

刀架装配时，各相对运动部件已注入润滑脂，用户在完成一次拆装后，必须重新进行全面的润滑。刀架使用一段时间后，必须打开注油孔注入润滑脂，以保证刀架润滑状态良好。每班拨下刀架上罩 23，身螺母 21 注入少许机油。

Each relatively moving part has been filled with grease during assembly of the tool holder, and the user must re-lubricate completely after each disassembly and assembly. After the tool holder has been in use for a period of time, the oil injection hole must be opened to inject grease to ensure that the tool holder is in good lubrication condition. Each shift pulls off the upper cover 23 of the tool holder and the body nut 21 is filled with a little engine oil.

7. 刀架走线

7. Tool holder routing

刀架内部的电线安排，要认真细心，以防止电线被损坏或刮伤。

Carefully lay out the wires inside the tool holder to prevent damage or scratches to the wires.

8. 注意事项

8. Points for attention

（1）刀架方向连续运转时间不允许超过 3 分钟。

（1）Continuous running time in tool holder direction shall not exceed 3 minutes.

（2）刀架运转次数每分钟不得超过五次，否则容易造成电机烧损。

（2）The running times of the tool holder shall not exceed five times per minute, otherwise the motor will be easily burned.

（三）刀架位置检测装置的种类和原理

（Ⅲ）Classification and principle of tool holder position detection device

1. 刀架检测装置的种类

1. Classification of tool holder position detection device

四工位电动刀架在旋转到指定的刀位时，需要发出信号让系统知道已经到了需要的刀位。这就需要使用到检测装置。数控机床的刀位检测装置常见的有以下几种：编码器、发讯开关、霍尔发讯盘。其中最常用的是霍尔发讯盘。

When the four-station electric tool holder is rotated to the specified tool position, it needs to send a signal to let the system know that the required tool position has been reached. So the detection device is used. There are several kinds of tool position detection devices commonly used in CNC machine tools, such as: Encoder, messaging switch, hall messaging disk. One of the most commonly used is the Hall messaging disc.

2. 霍尔发讯盘的工作原理

2. Working principle of Hall messaging disc

霍尔发讯盘主要由霍尔元件组成，每个刀位对应的装有一个霍尔元件，当刀架旋转到某个刀位时，该刀位的霍尔元件向数控装置输出信号，数控装置将此信号与指令信号进行比较，当两信号相同时，刀架转动到位。

Hall messaging disc is mainly composed of Hall elements. Each tool position is installed with a Hall element. When the tool holder rotates to a certain tool position, the Hall element of the tool position outputs a signal to the CNC device. The CNC device compares the signal with the instruction signal, and when the two signals are the same, the tool holder rotates into place.

（四）刀架电气线路的组成及控制原理分析（图 5.6.2）

（Ⅳ）Analysis of the composition and control principle of the electrical circuit of the tool holder (Figure 5.6.2)

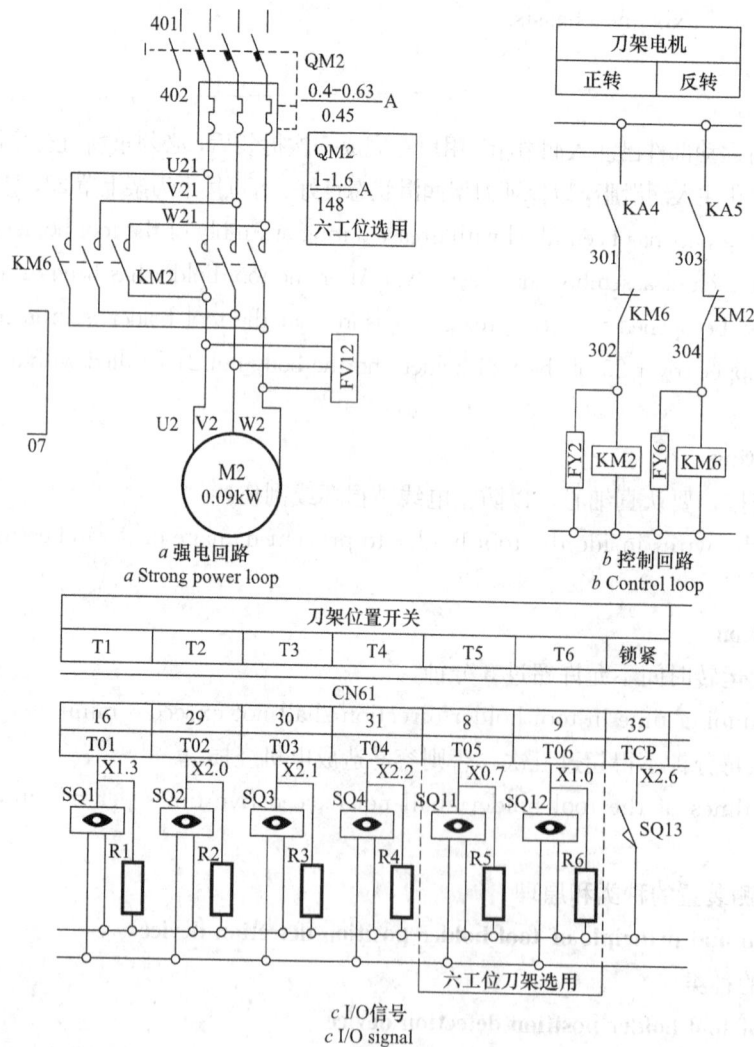

图 5.6.2 刀架电气原理图

Figure 5.6.2 Electrical schematic diagram of tool holder

2. 刀架电气控制原理
2. Electric control principle of tool holder

（1）执行换刀操作后，系统输出刀架正转信号 TL＋并开始检测刀具到位信号，检测到刀具到位信号后，关闭 TL＋输出，延时后输出刀架反转信号 TL－。然后检查锁紧信号 TCP，当接到到此信号后，延时后关闭 TL－信号，换刀结束。

（1）After tool change operation，the system outputs the tool holder positive rotation signal TL ＋ and starts to detect the tool position signal. After detecting the tool position signal，the system closes the TL ＋ output and outputs the tool holder negative signal TL－after the delay. Then the lock signal TCP is checked. When the lock signal is received，the TL－signal is closed after the delay，and the tool change is finished.

（2）在规定的时间内，未完成换刀，系统将产生报警并关闭刀架反转信号。

（2）If the tool change is not completed within the specified time，the system will generate an alarm and close the tool holder negative rotation signal.

（3）若无刀架锁紧信号时，可通过参数屏蔽锁紧信号，此时刀架锁紧信号一直有效。

（3）If there is no tool holder locking signal，the parameter can be used to shield the locking signal，at which time the tool holder locking signal is always valid.

三、任务实施
Ⅲ. Task Implementation
（一）根据故障现象分析故障原因及范围（表 5.6.1）
（Ⅰ）Analyze the cause and scope of the fault according to the fault phenomena（Table 5.6.1）

故障现象和原因　　　　　　　　　　　　　　　　表 5.6.1
Faults phenomena and the cause　　　　　　　　　　Table 5.6.1

故障现象 Faults	故障原因 Cause
电动刀架每个刀位都转不停 Electric tool holder rotates continuously at each tool position	刀位信号＋24V 电源回路故障 Tool position signal＋24V power supply circuit fault
	霍尔元件故障 Hall element fault
	反转控制回路故障 Negative control loop fault
电动刀架转不转 Electric tool holder can't rotate	刀架电机或刀架电机三相电源故障 Fault of tool holder motor or three-phase power of tool holder motor
	刀架正转控制信号回路故障 Tool holder positive control signal circuit fault
	刀位信号回路故障 Tool position signal circuit fault
电动刀架到某一刀位转不停 Electric tool holder rotates continuously at a certain knife position	此刀位的霍尔元件故障 Hall element fault at this tool location
	此刀位信号回路故障，系统无法接收信号 The tool position signal circuit is faulty and the system cannot receive the signal
	系统刀位接收电路故障 System tool position receiver circuit failure
刀架偶尔转不动 The tool holder can't rotate occasionally.	刀架的控制信号受干扰 The control signal of the tool holder is disturbed

（二）通过诊断信息缩小故障排查范围

（Ⅱ）Narrow down the scope of troubleshooting by identifying information

1. CNC 诊断界面（图 5.6.3）

1. CNC diagnostic interface（Figure 5.6.3）

CNC 和机床的输入/输出信号的状态，CNC 和 PLC 之间传送的信号状态，PLC 内部数据及 CNC 内部状态等都可以通过诊断显示出来。按 键进入 CNC 诊断页面显示，CNC 诊断页面显示有键盘诊断；状态诊断及辅助机能断诊参数等内容。可通过 键、 键查看。

The status of input/output signals of CNC and machine tools，the status of signals transmitted between CNC and PLC，the internal data of PLC and the internal status of CNC can be displayed by diagnosis. Press key to enter the CNC diagnosis page display，there is keyboard diagnosis on CNC diagnosis page Condition diagnosis and auxiliary function diagnosis parameters etc. View through the key， key.

在 CNC 诊断显示页面，页面的下部有两行诊断号详细内容显示，第一行显示当前光标所在的诊断号的某一位的中文含义，可以按 键或 键来改变显示的诊断位；第二行显示当前光标所在诊断号所有位的英文缩写。

On the CNC diagnosis display page，there are two lines of diagnosis number details at the bottom of the page. The first line displays the Chinese meaning of a certain digit of the diagnosis number where the current cursor is located，and the displayed diagnosis digit can be changed by pressing key or key；The second line displays the abbreviations for all digits of the diagnosis number where the current cursor is located.

2. PLC 状态界面（图 5.6.4）

2. PLC status interface（Figure 5.6.4）

在 PLC 状态界面的页面依次共显示 X0000～X0029、Y0000～Y0019、F0000～F0255、G0000～G0255、A0000～A0024、K0000～K0039、R0000～R0999 等地址状态。反复按 键、 键即可查看到 PLC 各地址的信号状态。

The address states such as X0000-X0029，Y0000-Y0019，F0000-F0255，G0000-G0255，A0000-A0024，K0000-K0039，R0000-R0999 are displayed on the page of PLC status interface. Press the ， key repeatedly to view the signal status of each address of the PLC.

在 PLC 状态页面，页的下部有二个详细内容显示行，第一行显示当前光标所在的地址号的某一位的中文含义，可以按 键或 键来改变显示的地址位；第二行显示当前光标所在地址号所有位的英文缩写。

On the PLC status page，there are two lines of details at the bottom of the page. The first line displays the Chinese meaning of a certain digit of the address number where the current cursor is located，and the displayed address digit can be changed by pressing key or key；The second line displays the abbreviations for all digits of the address number where the current cursor is located.

3. PLC 数据界面（图 5.6.5）

3. PLC data interface（Figure 5.6.5）

在 PLC 数据界面的页面依次显示 T0000～T0099、D0000～D0999、C0000～C0099、DT000～DT099、DC000～DC099 等寄存器的数值。反复按 键进入 PLC 数据界面。按 键、 键即可查看到 PLC 各数据值。

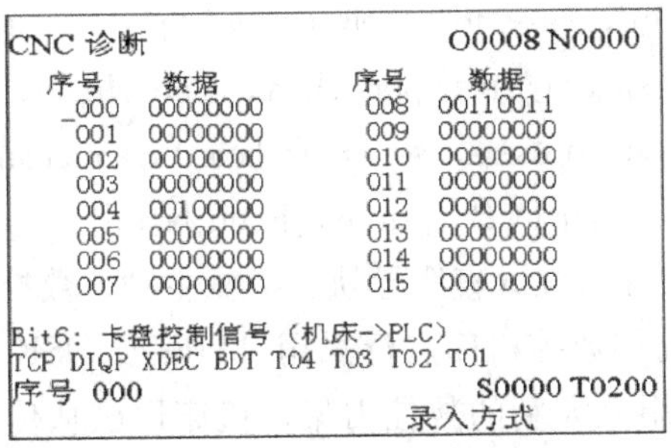

图 5.6.3　CNC 诊断画面
Figure 5.6.3　CNC diagnostic interface

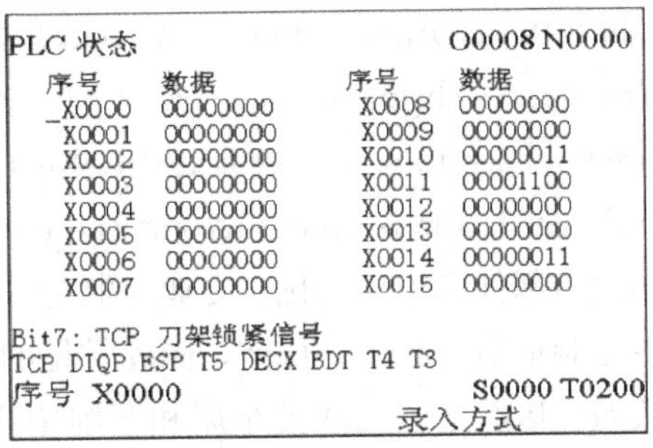

图 5.6.4　PLC 状态画面
Figure 5.6.4　PLC status interface

The values of registers such as T0000—T0099，D0000—D0999，C0000—C0099，DT000—DT099，DC000—DC099 are displayed on the page of PLC data interface. Press ⌗ key repeatedly to enter the PLC data interface. Press the ⌗, ⌗ key to view the PLC data values.

在 PLC 数据页面中，页面的下部有一行中文提示行，显示当前光标所指参数的含义。

On the PLC data page，there is a Chinese prompt line at the bottom of the page that shows the meaning of the parameters indicated by the current cursor.

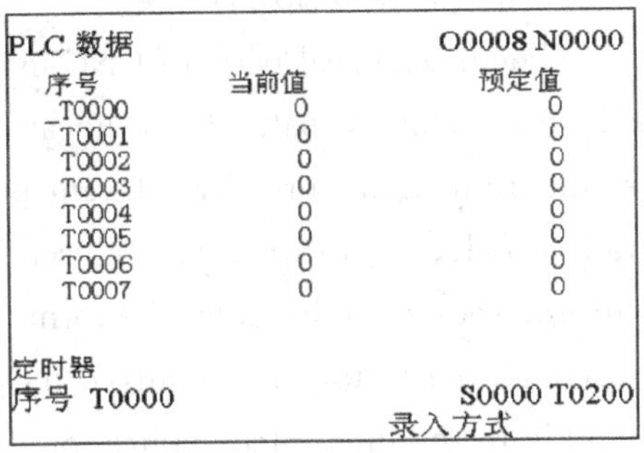

图 5.6.5　PLC 数据界面
Figure 5.6.5　PLC data interface

任务七　数控机床整机电气故障维修

Task Ⅶ　Electrical Trouble Maintenance of CNC Machine Tool

一、任务分析
Ⅰ. Task Analysis

1. 解数控机床的常见故障

1. Understand the common faults of CNC machine tools

2. 掌握通过系统诊断的方法进行故障维修

2. Master the troubleshooting method through system diagnosis

二、相关知识技能
Ⅱ. Relevant Knowledge & Skills

通过上面的学习，我们可以总结出 FANUC 数控系统的某些重要特征，从而重新认识数控机床维修这一概念。

Through the above study，we can summarize some important characteristics of the CNC system，so as to re-understand the concept of CNC machine tool maintenance.

首先数控系统采用专用的总线结构、专用的伺服驱动，所以它不像通用 PC 的总线那样，具有标准通用、备件易采购、有互换性、代码开放、参考文献渠道多等特点。数控系统备件供应渠道单一，图纸、程序协议不对用户开放，专用 LSI 不对用户开放，所以线路板维修非常困难，一般数控制造商不建议用户维修 PCB（印刷线路板）。

First of all，the CNC system adopts special bus structure and special servo drive，so unlike the bus of general PC, it has the characteristics of universal standard, easy to purchase spare parts, interchangeability, open code, many reference channels and so on. CNC system spare parts supply channels are single, drawings, program protocols are not open to users, dedicated LSI is not open to users, so circuit board maintenance is very difficult, general CNC manufacturers do not recommend users to repair PCB (printed circuit board).

数控机床的机械部件采用模块化、专业化制造，如滚珠丝杠、直线导轨、机械主轴、数控刀塔、数控转台等均是由各专业制造商来制造。目前，国内常见的中高档数控机床广泛采用 THK 或 NSK 的滚珠丝杠和直线导轨，原长城机床厂生产的数控车床和车削中心采用意大利的数控刀架，机床厂已从传统的零部件设计、生产、组装、面面俱到的生产方式，转变为机电一体化集成应用商。所以作为数控机床的维修人员，修复上述这些专业化生产的机械部件非常困难，例如直线导轨磨损后，我们最终用户没有手段修磨直线导轨的滑道，也无法修复损坏的滑块。

The mechanical parts of CNC machine tools are manufactured by modularization and specialization method, for example, ball screw, linear guide rail, mechanical spindle, CNC knife tower, CNC rotary table and so on are manufactured by relevant professional manufacturers. At present, the ball screw and linear guide rail of THK or NSK are widely used in middle-high-grade CNC machine tools in China. The CNC lathe and turning center produced by the former Great Wall Machine Tool Factory adopt Italian CNC tool holder. The machine tool factory has changed its production mode from the traditional parts design, production and assembly to the integrated application of mechatronics and electronics. Therefore, as the maintenance personnel of CNC machine tools, it is very difficult to repair the above-mentioned professionally produced mechanical parts, for example, in case of linear guide wear, our end-users have no means to repair the straight guide slideway, nor repair the damaged sliding block.

对于我们上面谈到的新技术应用部件直线电机、扭矩电机、电主轴等，由于现场的工艺条件和现有的技术手段的限制，现场设备维修人员修复这些部件也是非常困难的，例如 FANUC 的高速电主轴对装配调试工艺要求非常高，必须经过专门的培训后才可拆装，否则主轴速度达不到出厂指标。

For the new technology application component mentioned above, such as linear motor, torque motor, motorized spindle and so on, due to the limitations of the field process conditions and the existing technical means, it is very difficult for the field equipment maintenance personnel to repair these parts. For example, the high-speed motorized spindle of FANUC requires very high assembly and commissioning process, and can only be disassembled and assembled after special training, otherwise, the speed of the spindle cannot satisfy the factory index.

线路板不能修，很多机械件也不能修，机电一体化部件更碰不得，那么我们现场维修人员修什么呢？这就需要我们从传统的维修概念中摆脱出来。二十世纪七八十年代的数控维修人员需要对模拟电路、数字电路有比较深刻的了解，由于那个时代的制造技术还是基于模拟电路和中规模数字电路搭建的硬件环境，器件大都采用标准器件，他们通过电烙铁、万用表、示波器修理损坏的线路板。而现今的数控技术紧随着 IT 业的进步而改变，目前 FANUC 数控系统除了 CPU 和存储器采用标准制造商的产品外，CPU 周边以及大量的外围芯片均由自己设计开发，例如数字伺服处理、RS232 通信、字符及图形显示等。另外，系统各环节之间的数据传送也由 20 年前的并行传送为主，改变为目前的串行传送为主。在串行传送的环境下，用示波器已无法诊断信号的来龙去脉，万用表更是无能为力。目前示波器和万用表仅作为一些并行信号或静态信号的检测工具，对伺服放大器或电源模块的维修还有些帮助，但是对于 CNC 系统本身和数字伺服部分的维修帮助非常有限。

Circuit board cannot be repaired, many mechanical parts cannot be repaired, electromechanical parts also cannot be touched, so what our on-site maintenance personnel can repair? This needs us to have a new understanding of the traditional maintenance concept. CNC maintenance personnel in 1970s and 1980s need to have a deep understanding of analog circuits and digital circuits. Because the manufacturing technology of that era was still based on the hardware environment of analog circuits and medium-scale digital circuits, most of the devices use the standard devices. These personnel use electric soldering iron, multimeter and oscilloscope to repair damaged circuit boards. Nowadays, the CNC technology changes with the development of IT industry. At present, FANUC CNC system CPU and memory adopt the products of standard manufacturers, and CPU surrounding and a large number of peripheral chips are designed and developed by ourselves, such as digital servo processing, RS232 communication, character and graphic display, and so on. In addition, the data transmission between each link of the system has changed from parallel transmission 20 years ago to serial transmission at present. Under the circumstance of serial transmission, it is impossible to diagnose the signal with oscilloscope, and the multimeter also does not work. At present, oscilloscopes and multimeters are only used as detection tools for some parallel signals or static signals, which is helpful for maintenance of servo amplifiers or power modules, but has very limited effect on maintenance of CNC system itself and digital servo parts.

现今最有效的维修诊断手段是由数控系统制造商来提供的，如 FANUC 0i 系列的 PMC TRACER（接口信号跟踪诊断），数字伺服波形画面等功能（图 5.7.1）。

Nowadays, the most effective maintenance and diagnosis methods are provided by CNC system manufacturers, such as PMC TRACER (Interface Signal Tracking Diagnosis) of FANUC 0i series, digital servo waveform screen and other functions. (Figure 5.7.1).

归纳前面所描述的数控机床的结构和特点，我们不难发现现场维修人员的主要工作不是修复线路板，而是利用现有手段（数控制造商提供的各种监控或诊断方法），特别是借助计算机或人机界面，及时准确地判断出故障类型，确定维修方向（机械—电气—液压—工艺）。如果是电气故障应及时判断出是 CNC、伺服部分还是 PMC 接口电路出现了故障，并找出故障点。接下来就要能够利用最直接有效的渠道迅速买到备件，正确更换备件。

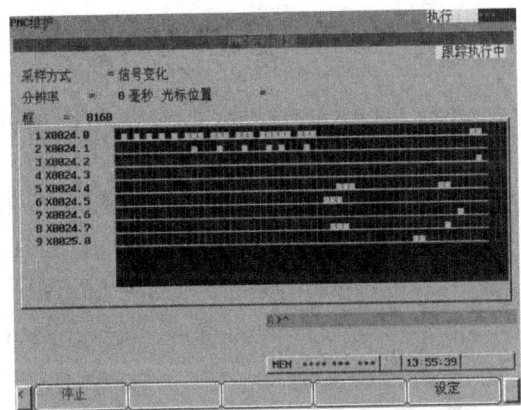

图 5.7.1　Fanuc 信号跟踪画面

Figure 5.7.1　Fanuc signal tracking screen

Summing up the structure and characteristics of the CNC machine tool described above, it is not difficult to find that the main job of the field maintenance personnel is not to repair the circuit board, but to use the existing means (various monitoring or diagnosis methods provided by the CNC manufacturer), especially take advantage of computer or man-machine interface, to timely and accurately judge the fault type and determine the maintenance direction (Mechanical-electrical-hydraulic-process). For electrical fault, the personnel should timely judge whether the CNC, servo part or PMC interface circuit is faulty, and find out the fault point. The next step is to use the most direct and effective channels to buy spare parts quickly and replace them correctly.

而正确更换备件也是一件需要重视的工作，在前面我们曾经提到数控系统的某些重要数据是存放在 SRAM 中的，数据有易失性，更换 CNC 主板或存储器板会造成数据丢失，那么修好硬件后恢复数据，就成为我们正确更换备件的工作之一。作为一个维修工程师如果仅会更换硬件，而不会恢复数据，等于不会修理数控机床，因为你无法使机床进入正常工作状态。

The correct replacement of spare parts is also a work to be paid attention to, and we have mentioned it before that some important data of the CNC system is stored in SRAM. Due to data volatility, replacement of the CNC main board or memory board will cause data loss, so repairing the hardware and recovering data has become a part of our correct replacement work of spare parts. A maintenance engineer who only replaces hardware and does not recover data is deemed unable to repair CNC machine tools because you cannot put the machine tool into normal working state.

从维修实践中体会到，随着数控机床的发展，机械和控制系统的结构越来越简单，能够处理的硬件越来越少，而对各类软件的使用要求越来越高。如现场维修人员需要掌握 FANUC 梯形图编程软件、FLADDER Ⅲ以及各种随机诊断软件和网络通信软件。

Based on the experience from the maintenance practice and with the development of CNC machine tools, the structure of the mechanical and control system is becoming simpler, less hardware can be processed, and the requirements for the use of various types of software are becoming higher and higher. For example, field maintenance personnel need to master FANUC ladder diagram programming software, FLADDER Ⅲ and various random diagnosis software and network communication software.

过去维修人员更多地使用改锥、钳子，而现今我们维修人员离不开计算机。过去的维修人员很少介入备件管理，但是今后对于数控机床的维修无论是电气还是机械、液压，备件选型和正确更换将是维修工程师重要的工作内容。数控机床维修将融入更多的非技术因素，因为我们维修数控机床的目的并不是为了单纯地显现我们技术有多么的出色，我们的最终目的是最有效地减少故障停机时间，提高设备的无故障运转时间。

In the past, maintenance personnel used screwdrivers and pliers more often, but nowadays we can't complete maintenance works without computers In the past, maintenance personnel seldom participated in spare parts management, but in the future, the maintenance of CNC machine tools, whether electrical or mechanical, hydraulic, spare parts selection and correct replacement will be the important work of maintenance engineers. Maintenance of CNC machine tools will incorporate more non-technical factors, because the purpose of our maintenance of CNC machine tools is not simply to show how advanced our technology is, and our ultimate goal is to minimize downtime and improve equipment failure-free operation time.

三、任务实施
Ⅲ. Task Implementation

维修情景一——刀架类故障诊断与维修

Maintenance scenario 1——Tool holder fault diagnosis and maintenance

（一）情景描述
（Ⅰ）Scenario description

一台数控车床在进行换刀时，找不到 1 号刀，经过 1 号刀位时，刀架不停，刀架一直转，过一段时间后，出现寻不到刀报警。

One CNC lathe cannot find the No. 1 knife when changing the knife. After passing No. 1 knife position, the tool holder does not stop and rotates continuously. After a period of time, there is an alarm that the knife cannot be found.

（二）故障流程分析（图 5.7.2）
（Ⅱ）Fault analysis flow (Figure 5.7.2)

（三）资料收集与维修计划制定
（Ⅲ）Data collection and maintenance planning

1. 电动刀架工作原理

1. Working principle of electric tool holder

数控车床使用的回转刀架是最简单的自动换刀装置，有四工位和六工位刀架，回转刀架按其工作原理可分为机械螺母升降转位、十字槽转位等方式，其换刀过程一般为刀架抬起、刀架转位、刀架压紧并定位等几个步骤。回转刀架必须具有良好的强度和刚性，以承受粗加工的切削力。同时还要保证回转刀架在每次转位的重复定位精度。在 JOG 方式下，进行换刀，主要是通过机床控制面板上的手动换刀键来完成的，一般是在手动方式下，按下换刀键，刀位转入下一把刀。刀架在电气控制上，主要包含刀架电机正反转和霍尔传感器两部分，实现刀架正反转的是三相异步电机，通过电机的正反转来完成刀架的转位与锁紧；而刀位传感器一般是由霍尔传感器构成，四工位刀架就有四个霍尔传感器安装在一块圆盘上，但触发霍尔传感器的磁铁只有一个，也就是说，四个刀位信号始终有个为 1 或为 0。

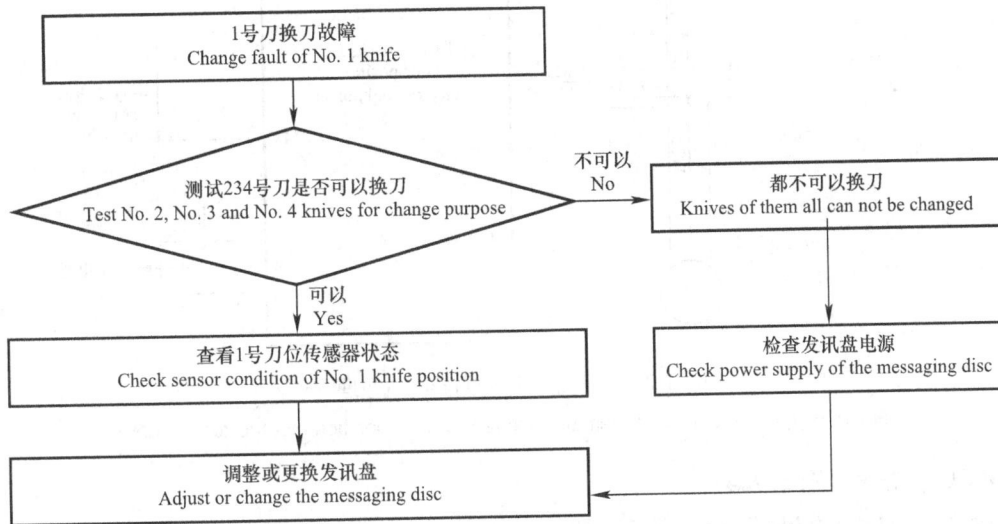

图 5.7.2　故障分析流程

Figure 5.7.2　Fault analysis flow

The rotary tool holder used in CNC lathe is the simplest automatic tool changer, equipped with four-station and six-station tool holders. According to its working principle, the rotary tool holder can be divided into mechanical nut lifting and lowering transposition, cross-groove transposition and so on. The process of tool change is generally divided into several steps, such as tool holder lifting, tool holder transposition, tool holder pressing and positioning. Rotary tool holder must have good strength and rigidity to withstand rough cutting force. At the same time, the repeated positioning accuracy of the rotary tool holder in each transposition should be ensured. In JOG mode, tool change is mainly accomplished by pressing the manual tool change key on the control panel of the machine tool. Generally, the tool change key is pressed in manual mode, and the next tool goes to the tool position. In the electric control, the tool holder mainly includes two parts: the positive and negative rotation of the tool holder motor and the Hall sensor. The positive and negative rotation of the tool holder is realized by the three-phase asynchronous motor, which completes the transposition and locking of the tool holder through the positive and negative rotation of the motor. The tool position sensor is generally composed of Hall sensors. Four Hall sensors are mounted on a disk for four-station tool holder, but there is only one magnet to trigger the Hall sensor, that is, one of the four tool position signals is always 1or 0.

2. 电动刀架的 PMC 连接

2. PMC connection of electric tool holder

　　下图是电动刀架与 PMC 的连接图，包含输入与输出两部分，输入主要是刀位信号，输出是刀架电机的正反转，对应的控制逻辑由 PMC 设计完成。（图 5.7.3）。

The following is the connection diagram between the electric tool holder and PMC, including input and output. The input part is mainly the tool position signal，the output part is the positive and negative rotation of the tool holder motor，and the corresponding control logic is designed by PMC. (Figure 5. 7. 3).

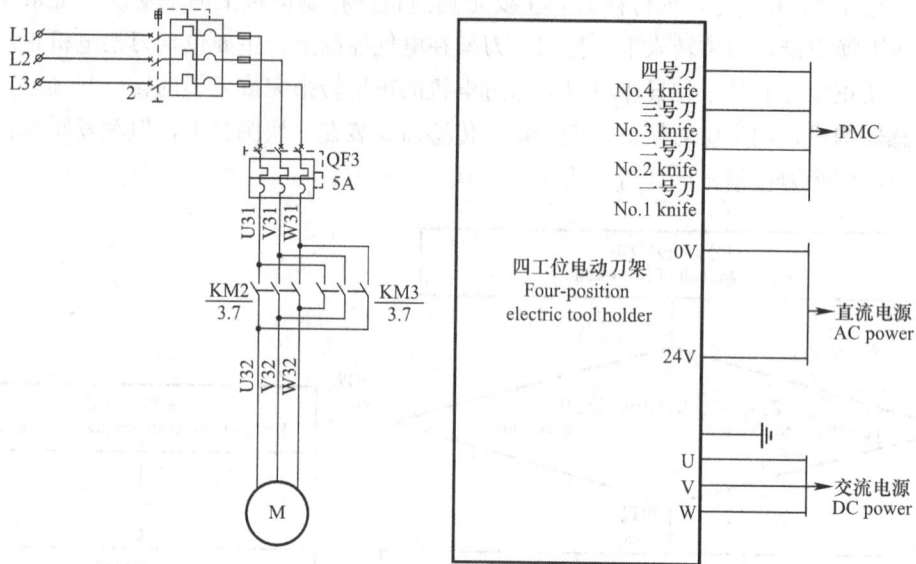

图 5. 7. 4　四工位电动刀架电气原理图

Figure 5. 7. 4　Electrical schematic diagram of four-position electric tool holder

3. 查看 PMC 状态表 （图 5. 7. 4）

3. View the PMC status table （Figure 5. 7. 4）

发那科系统提供 PMC 状态查询，我们可以依次按系统面板上的 "SYSTEM"—"PMC"—"信号"，搜索 X3 查询现有地址的状态。正常状态下的刀架是有一位是低电平，三个为高电平，如果四位相同，那么就表示刀架信号异常，就会产生不能换刀的故障，这时候，就需要用检查发讯盘与线路了。Fanuc 提供的信号状态查询功能，可以很好地进行信号状态的查询，对判断故障原因提供很大的方便。这个功能是需要我们牢固掌握的。

TheFanuc system provides PMC status query function，and we can search X3 for the status of existing addresses by pressing "SYSTEM"—"PMC"—"Signals" on the system panel. Normally，one bit of the tool holder is at low level and three bits are at high level. If the four bits are the same，it means that the tool holder signal is abnormal and the tool holder cannot be changed. At this time，it is necessary to check the messaging panel and circuit. The signal state inquiry function provided by Fanuc can allow the signal state inquiry，which provides great convenience for judging the cause of the fault. This function needs to be grasped firmly.

4. 计划实施

4. Plan implementation

拆卸电动刀架发讯盘

Remove electrical tool holder messaging panel

① 拆卸与更换发讯盘（图 5. 7. 5）。

① Remove and replace messaging disc (Figure 5. 7. 5).

② 拆卸发讯盘盖（图 5. 7. 6）。

② Remove messaging disc cover (Figure 5. 7. 6).

图 5.7.4　PMC 信号状态表

Figure 5.7.4　PMC signal status table

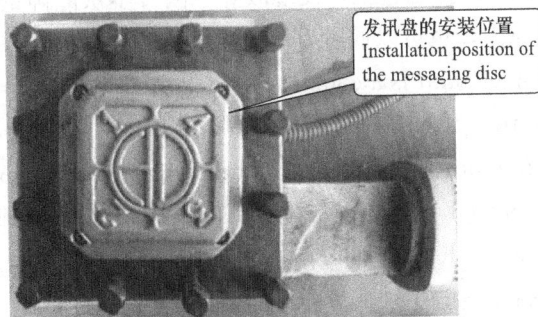

发讯盘的安装位置
Installation position of
the messaging disc

图 5.7.5　发讯盘安装位置

Figure 5.7.5　Installation location of messaging disc

用十字螺丝刀拧下发询盘上的4颗螺钉
Unscrew four screws on the messaging disc with
a cross screwdriver

图 5.7.6　发讯盘盖拆卸

Figure 5.7.6　Remove messaging disc cover

③ 调整发讯盘位置（图 5.7.7）。

③ Adjust the position of the messaging disc（as shown in Figure 5.7.7）.

调整发询盘位置，通过PMC状态表来判断发询盘的好坏，同时也
需要测量发询盘的电源电压，本项目中是发询盘位置不准，导致
1号刀位丢失，重新调整发询盘位置，排除故障。
Adjust the position of messaging disc and judge whether the messaging
disc is good via PMC status table. Measure power supply voltage of the
messaging disc as well. The fault of this project is missing of No. 1 knife
position caused by inaccurate messaging disc position,which can be
eliminated by adjusting position of the messaging disc again.

图 5.7.7　发讯盘

Figure 5.7.7　Messaging disc

5. 小结

5. Brief summary

刀架故障是常见的数控车床故障，原因很多，也有是因为刀架电机正反转不良造成的，所以需要仔细掌握刀架与 PMC 的控制过程，发现故障原因。本次故障的维修主要是通过 PMC 状态表，查看刀位信号，从而判断故障原因，也可以通过万用表测试相关信号的电平来进行判断。

Tool holder fault is a common CNC lathe fault, caused due to many reasons, or due to poor positive and negative rotation of holder motor. So it is necessary to carefully grasp the control process of tool holder and PMC, and find out the reasons for the failure. The maintenance process of this fault is mainly to look at the tool position signal through the PMC status table, to judge the cause of the fault, or use the multimeter to test related signal levels to judge the fault.

维修情景二——主轴类故障诊断与维修

Maintenance scenario 2——main axis fault diagnosis and maintenance

（一）情景描述

（Ⅰ）Scenario description

一台数控车床的主轴采用变频器控制，在进行主轴控制的过程中出现可以反转，不可以正转。

The spindle of a CNC lathe is controlled by a frequency converter, and negative rotation (but not positive rotation) is allowed in the process of spindle control.

1. 故障流程分析（图 5.7.8）

1. Fault flow analysis chart（Figure 5.7.8）

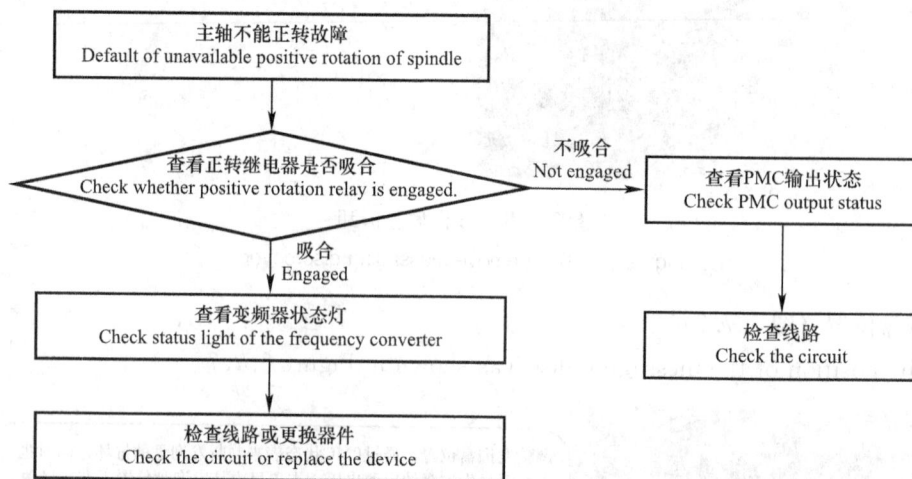

图 5.7.8　主轴故障分析流程图

Figure 5.7.8　Flow chart of spindle fault analysis

2. 资料收集与维修计划制定

2. Data collection and maintenance planning

（1）数控机床对变频器控制原理

（1）Control principle of CNC machine tool to frequency converter

FANUC 数控系统对模拟主轴的控制，主要包含速度与方向控制，速度控制的来源是由系统根据速度指令转化为 0～10V 的电压给变频器进行控制，如在 MDI 方式下输入 M03 S1000，该程序段中的 S1000 会通过系统转换为 0～10V 的模拟电压，输出给变频器的模拟量控制接口；而主轴旋转的方向，是由 PMC 根据指令进行输出正反转继电器吸合来完成的。上例中的 M03 就是由 PMC 进行译码，输出一个信号给继电器，继电器吸合后，闭合变频器上的正转端子，完成主轴正转的控制。

The control ofFanuc CNC system to the analog spindle mainly includes speed and direction control. As for the source of speed control, the system transforms the voltage of 0—10V to the frequency converter according to the speed instruction. For example，when M03 S1000 is input in MDI mode，S1000 in this program section will be transformed to the analog voltage of 0—10V by the system which will be output to the analog control interface of the frequency converter；The rotation direction of the spindle is accomplished by the PMC pulling in the output positive and negative rotation relays according to the instructions. The M03 in the above example is decoded by the PMC，and outputs a signal to the relay. After the relay is pulled in，the positive rotation terminal of the frequency converter is closed，and the positive rotating control of the spindle is completed.

（2）变频器连接原理（图5.7.9）

（2）Connection principle of frequency converter（Figure 5.7.9）

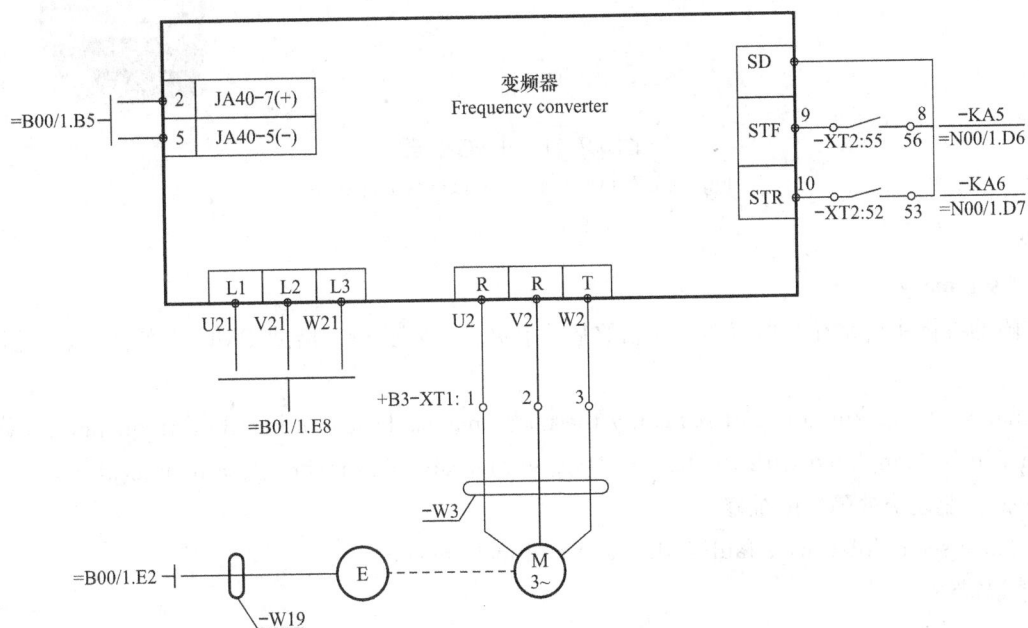

图5.7.9　主轴电气原理图

Figure 5.7.9　Electrical schematic diagram of spindle

3. 计划实施

3. Plan implementation

（1）查看正转继电器（图5.7.10）

（1）View positive rotation relay（Figure 5.7.10）

图5.7.10　主轴正转继电器状态

Figure 5.7.10　Positive rotation relay status of spindle

（2）查看变频器状态（图5.7.11）

（2）View the status of the frequency converter（Figure 5.7.11）

> 变频器接收到正反转信号后，iRUNi 灯会亮起。本例中，正反转吸合，iRUNi 不亮，说明是继电器触点线路故障，通过检查发现时继电器触点故障，更换排除故障。
> The iRUNi light will be lightened after the frequency converter receiving positive and negative rotation signals. In this example, the fault of relay contact circuit causing engaged positive and negative rotation and unlighted iRUNi,can be eliminated by replacing it.

图 5.7.11　主轴变频器

Figure 5.7.11　Spindle frequency converter

4. 小结

4. Brief summary

本类故障的维修主要是在掌握变频器控制的基础上进行故障分析，借助 PMC 状态表与变频器状态指示来完成。

Maintenance of this kind of fault is mainly the fault analysis based on based on frequency converter control，which will be completed with the help of PMC status table and inverter status indication.

维修情景三 驱动类故障判断维修

Maintenance scenarioIII-drive fault judgement and maintenance

（一）情景描述

（Ⅰ）Scenario description

一台数控车床使用 0i-mate TD 系统，系统开机停止或运行时偶尔出现 401 报警。

A CNC lathe uses the 0i-mate TD system，and 401 alarms occasionally appear when the system is turned on，turned off or running.

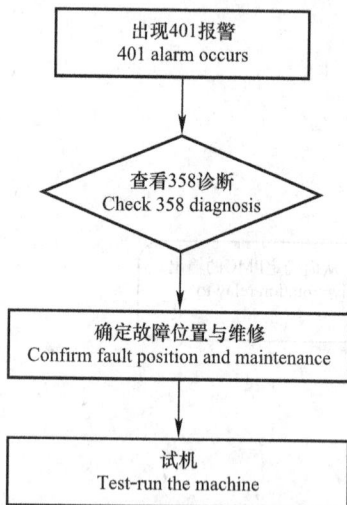

1. 故障流程分析（图5.7.12）

1. Fault flow analysis (Figure 5.7.12)

2. 资料收集与维修计划制定

2. Data collection and maintenance planning

（1）报警说明

（1）Alarm description

401 报警原理分析：如图 5.7.13 所示，其中红色箭头和信号名，表示指令，蓝色箭头和信号名表示反馈信号，当 CNC 发出 MCON 指令后，一定时间内没有接收到 DRDY 信号，将发生 401 号报警（DRDY OFF）。

401 alarm principle analysis：As shown in the figure 5.7.13，the red arrow and the signal name indicate the instruction，and the blue arrow and the signal name indicate the feedback signal，401 alarm (DRDY OFF) occurs when the CNC fails to receive the DRDY signal for a certain period of time after issuing the MCON instruction.

```
出现401报警
401 alarm occurs
   ↓
查看358诊断
Check 358 diagnosis
   ↓
确定故障位置与维修
Confirm fault position and maintenance
   ↓
试机
Test-run the machine
```

图 5.7.12　驱动故障分析流程图

Figure 5.7.12　Fault analysis flow chart of the driver

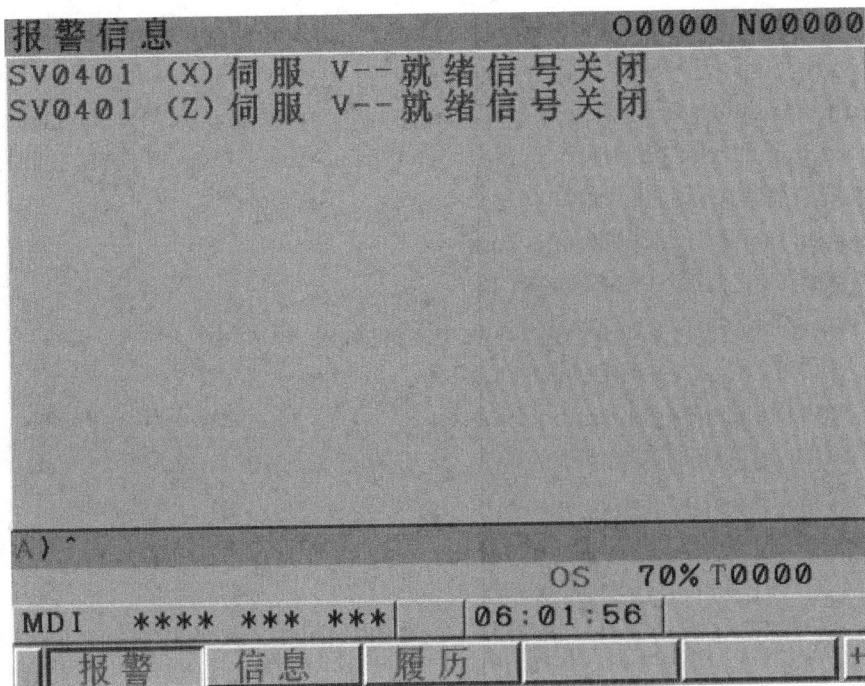

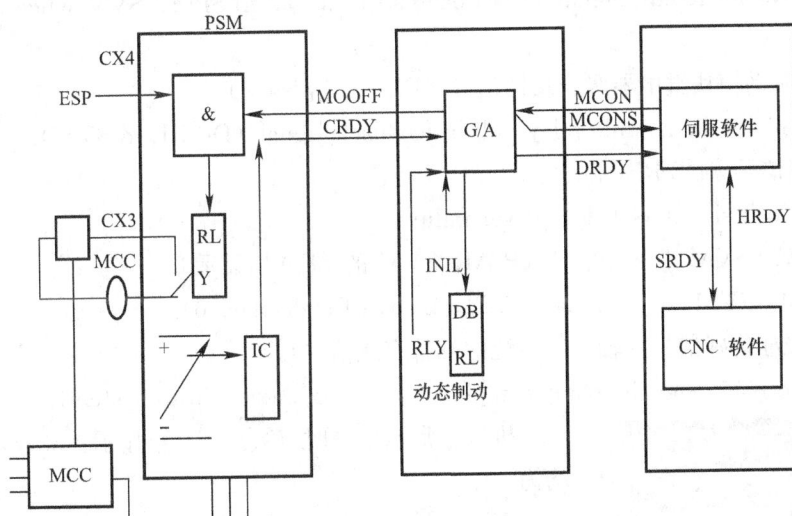

图 5.7.13 系统报警画面及伺服电气原理
Figure 5.7.13 System alarm screen and servo electrical principle

上图中信号的状态在诊断 358 可以查看，如果使用的伺服软件是 90B0/D 以后版本，可根据诊断内容判断具体哪个信号断开（红色或蓝色）。诊断号 358 是用一个 10 进制数表示一个 16 位的二进制数，所以在实际应用中需要换算成二进制。具体信号名称见表 5.7.1。

<table>
<tr><td colspan="8" align="center">信号表</td><td align="right">表 5.7.1</td></tr>
<tr><td colspan="8" align="center">Signal table</td><td align="right">Table 5.7.1</td></tr>
</table>

#15	#14	#13	#12	#11	#10	#9	#8
	SRDY	DRDY	INTL	RLY	CRDY	MCOFF	MCONS
#7	#6	#5	#4	#3	#2	#1	#0
MCON	*ESP	HRDY					1

The status of the signal shown in the above figure can be viewed at diagnosis 358, and if the servo software used is the version above 90B0/D, the specific off signal (red or blue) can be judged based on the diagnosis content. Diagnosis number 358 is a 16-bit binary number represented by a decimal number and therefore needs to be converted to binary in practice. Specific signal names are shown in Table 5.7.1.

#5HRDY：系统监控程序启动。

#5HRDY：System monitoring program is started.

#6*ESP：外部急停信号（从 PSM 的 CX4 输入）。

#6*ESP：External emergency stop signal (input from CX4 of PSM).

#7*MCON：MCON 信号（系统给伺服的）。

#7*MCON：MCON signal (system to servo).

#8MCONS：MCON 信号（伺服给系统的）。

#8MCONS：MCON signal (servo to system).

#9MCOFF：MCC 断开信号（PSM 给 SVM）。

#9MCOFF：MCC OFF signal (PSM to SVM).

#10*CRDY：逆变器准备就绪信号（当 PSM 的 DCLINK 电压约 300V 启动，PSM 把该信号传递给 SPM、SVM）。

#10*CRDY：Inverter Ready Signal (PSM passes this signal to SPM, SVM when the PSM's DCLINK voltage is approx. 300V).

#11RLY：动态制动模块继电器吸合反馈信号（DB RL 给 SVM）。

#11RLY：Dynamic brake module relay pull-in feedback signal (DB RL to SVM).

#12INTL：连锁信号（DB RL 掉电）。

#12INTL：Interlock signal (DB RL power-failure).

#13DRDY：PSM、SVM 准备完信号（PSM、SVM 的 LED 均显示 0）。

#13DRDY：PSM, SVM ready signal (PSM, SVM LED displays 0).

#14SRDY：伺服准备好信号（轴卡给系统的准备完成信号）。

#14SRDY：Servo ready signal (ready complete signal from axis clip to system).

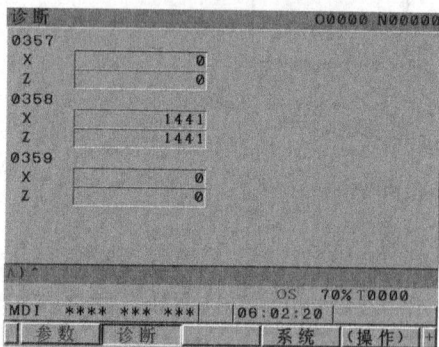

图 5.7.14　诊断状态界面

Figure 5.7.14　Diagnosis status interface

机床正常准备好时,诊断 358 号显示：32737（即#5——#14 均为 1）。

When the machine tool is normally ready, diagnosis No. 358 displays：32737 (ie. 1 for #5-#14).

3. 计划实施

3. Plan implementation

（1）查看诊断状态，按 "SYSTEM"—"诊断"，输入 358 搜索。如图 5.7.14 所示。

(1) View the diagnosis status, press "SYSTEM"-"Diagnosis" and enter 358 for search. As shown in Figure 5.7.14.

（2）确定位置

(2) Position confirmation

正常情况下，358 应该显示的是 32737，我们可以把 32737 转换为二进制为 0111111111100001，而本例中显示为 1441，把 1441 转换为二进制是 0000010110100001，第 6 位 ESP 为 0，表示无急停信号。经检查是伺服放大器的 CX30 接口的转接端子，接触不良，修复后，排除故障。

Normally，358 should display 32737 that can be converted to binary 01111111100001；but in this case，it displays 1441，after converting 1441 to binary 0000010110100001，bit #6 ESP is 0，indicating no emergency stop signal. By means of inspection，the adapter terminal of the servo amplifier CX30 interface is poorly connected，and such fault has been repaired and solved.

4. 小结

4. Brief summary

通过系统的自诊断，可以很快地发现故障的位置，所以我们需要掌握识读与分析类似诊断的方法。

Through the system self-diagnosis，we can quickly find the location of the fault，so we need to grasp the method of identification and analysis similar diagnosis methods.

参 考 文 献
Reference

［1］ 廖常初. S7-1200 PLC 编程及应用（第 3 版）［M］. 北京：机械工业出版社，2017.

［2］ S7-1200 可编程控制器系统手册 ［M/CD］. 德国：Siemens AG 数字化工厂集团，2016-09.

［3］ SINAMICS G120C 参数手册 ［M/CD］. 德国：Siemens AG 数字化工厂集团，2017-09.

［4］ SINAMICS G120C 操作说明 ［M/CD］. 德国：Siemens AG 数字化工厂集团，2017-09.

［5］ ASDA-B2 系列标准泛用型伺服驱动器应用技术手册 ［M/CD］. 中国：中达电通股份有限公司，2014-11.

［6］ Kinco 步进产品型录 ［M/CD］. 中国：上海步科自动化股份有限公司，2019-10.

［7］ MCGS 嵌入版用户手册 ［M/CD］. 中国：深圳昆仑通态科技有限责任公司，2016-10.